T0332621

SUN ABOVE THE HORIZON

Pan Stanford Series on Renewable Energy

Series Editor
Wolfgang Palz

Titles in the Series

Published

Vol. 1
Power for the World: The Emergence of Electricity from the Sun
Wolfgang Palz, ed.
2010
978-981-4303-37-8 (Hardcover)
978-981-4303-38-5 (eBook)

Vol. 2
Wind Power for the World: The Rise of Modern Wind Energy
Preben Maegaard, Anna Krenz, and Wolfgang Palz, eds.
2013
978-981-4364-93-5 (Hardcover)
978-981-4364-94-2 (eBook)

Vol. 3
Wind Power for the World: International Reviews and Developments
Preben Maegaard, Anna Krenz, and Wolfgang Palz, eds.
2013
978-981-4411-89-9 (Hardcover)
978-981-4411-90-5 (eBook)

Vol. 4
Solar Power for the World: What You Wanted to Know about Photovoltaics
Wolfgang Palz, ed.
2013
978-981-4411-87-5 (Hardcover)
978-981-4411-88-2 (eBook)

Vol. 5
Sun above the Horizon: Meteoric Rise of the Solar Industry
Peter F. Varadi
2014
978-981-4463-80-5 (Hardcover)
978-981-4613-29-3 (Paperback)
978-981-4463-81-2 (eBook)

Forthcoming
Vol. 6
Biomass Power for the World
Wim P. M. van Swaaij, Sascha R. A. Kersten, Wolfgang Palz, eds.
2015

Pan Stanford Series on Renewable Energy
Volume 5

Meteoric Rise of the Solar Industry

SUN ABOVE
THE HORIZON

Peter F. Varadi

PAN STANFORD PUBLISHING

Published by

Pan Stanford Publishing Pte. Ltd.
Penthouse Level, Suntec Tower 3
8 Temasek Boulevard
Singapore 038988

Email: editorial@panstanford.com
Web: www.panstanford.com

British Library Cataloguing-in-Publication Data
A catalogue record for this book is available from the British Library.

Sun above the Horizon: Meteoric Rise of the Solar Industry

Disclaimer
The regulatory information contained in this book is intended for academic learning and market planning and is subject to change frequently. Translations of laws and regulations are unofficial.

ISBN 978-981-4463-80-5 (Hardback)
ISBN 978-981-4613-29-3 (Paperback)
ISBN 978-981-4463-81-2 (eBook)

Printed in the USA

Contents

Act 2: Sunrise—1985–1999

ACT 3: TOWARDS HIGH NOON—2000–2013

Acknowledgments

I am writing this acknowledgment exactly 40 years after June 1973, when Joseph Lindmayer and I walked into our boss's office at COMSAT Laboratories to hand him our resignation and informed him that we did this as we were starting a new company. I vividly remember how that happened:

Joseph summarized our decision: "You see, we escaped from a communist country and came here to the center of capitalism. Therefore we decided to start a company and become capitalists."

"What business are you planning to start?"

"We are going to start a business to produce solar cells and systems for terrestrial utilization," said Joseph.

"But there is no market for that and no technology to make inexpensive solar cells."

"You are right. But that makes the venture more interesting. You see, to become a capitalist is easy, but to become a capitalist in a field and market which does not exist will be a challenge," I chimed in.

Forty years later, I now know that I was absolutely right. It was a challenge and I would repeat what I said when in 2004 I received the Bonda Prize from EPIA: "We realized, and by now you know also that this (PV) is one of the most interesting business that anybody can be in. I am feeling very good that I got into it and I am still in it. This is a business where there are limitless applications, where one can use imagination, and on top of that we are also working on the future salvation of mankind. I am glad to be in it and it was an incredible thing that we got into it."

It was a pleasure to write the history of the past 40 years. It was a pleasure to remember many details of the events as a result of which PV became a real industry, changed the life of mankind, and became a part of our electricity supply. I felt I must write this book. Many of us old timers are still around to tell how in spite of tremendous headwinds, an industry that started with perhaps 20 people in 1973 grew up in 40 years to employ probably more than one million people.

It was also a pleasure to write this book because I had the opportunity to get in touch with people I worked with during the past 40 years and with many new ones in this business. I would like to thank all of those who were kind enough to help me assemble this book and provided information or helped me with their memories and those who reviewed parts and even the entire book.

First, I would like to thank Richard Kay and Helene Ramo for their invaluable help in editing many chapters and especially for their encouragement which I needed many times because of the complexity of the work. I thank Ramon Dominguez, with whom I had the pleasure to work for 50 years and who has an extraordinary collection of files and pictures, which he researched and provided for me to be used in my work.

I would like to thank Anil Cabraal, Dave Carlson, Maurice Juillerat, Wolfgang Palz, Jean Posbic, Loretta Schaeffer, and John Wohlgemuth, to whom I turned many times for help and who edited chapters and gave me valuable information also describing their memories of the "old" days.

I would like to thank Denis Curtin and Joseph Pelton for providing information and help so I could write about the space-oriented use of PV.

There are many others who helped me assemble the material for this book or gave me permission to use some of the material they had. Among them, I would like to mention Walter Ames, Joachim Benemann, Bob Edgerton, Robert Edmund, Allan Hoffman, Winfried Hoffmann, Sarah Kurtz, Bernard McNelis, Paula Mints, Jodie Roussell, Robert Strauss, Qin Haiyan, Philip Wolfe, Chuck Wrigley, and Bill Yerkes.

I would like to thank Arvind Kanswal for his very professional help in preparing this manuscript for publication.

Introduction

Do you know that if solar cells, which convert light directly to electricity, had not existed, the GPS navigation system used worldwide would not have existed either?

Most people do not know and only a few people remember how the terrestrial utilization of solar electricity advanced in 40 years from a scientific curiosity to replace nuclear electricity. This book is dedicated to those people who achieved this and to tell the story how difficult it was to achieve it.

The age of electricity started during the middle of the nineteenth century and by now in the twenty-first century our life and the quality of our life is unimaginable without the use of electricity. In order to satisfy our appetite for electricity, we are drilling holes miles deep under the surface of the oceans, fracturing stones in the crust of the Earth, and transporting fuel in huge ships and pipelines from one side of the Earth to the other to operate machinery to produce electricity.

But 60 years ago it was found that a scientific curiosity discovered in 1839 to convert light, the energy from the sun, into electricity can be used to obtain meaningful quantity of electricity without any fuel or any moving parts utilizing Silicon, the second most abundant element on Earth. The scientists called it photovoltaic (PV) cells, but in simple English it can be called solar cells.

Today most people know that in contrast to oil or nuclear, solar energy is a "renewable" clean energy source. But even today only few people realize that without the direct conversion of solar energy to electricity by solar cells, many important things we are using in our life such as the global phone service, cell phones, TV, Internet, global weather service, the GPS system, manned space station, and machinery exploring the surface of

Mars would not be possible. Solar cells opened up for mankind the utilization and exploration of the universe.

Soon it was realized that solar cells could be also used anywhere on the Earth where electricity is needed in small quantities to operate a calculator, larger quantities to operate a street light, in houses to provide people with electricity and also in very large size to be used in factories to manufacture everything we need, including automobiles, airplanes, etc., and to replace nuclear and some of the fossil fuel we use. It was true that many great possibilities existed for the terrestrial utilization of solar cells but the production technology developed for space use resulted in prohibitively expensive prices for those applications. The idea and the basics of the technology were known, the problem was that for the widespread terrestrial utilization of solar cells they had to be produced inexpensively and in large quantities.

Several people believed that a lot of research will be needed to achieve this, but a few had the idea that simply with changed technology very inexpensive solar cells could be made for terrestrial applications and what was needed was to start a manufacturing industry the purpose of which should be to utilize a changed technology to reduce the manufacturing cost and find market and with increased volume, prices will come down and that will open new markets. The result was that these few people started in 1973 the terrestrial "PV industry" namely two small companies to achieve these lofty goals.

These two companies constituted at that time the entire "terrestrial PV industry" of the entire World—employed probably about 20 people and the total production was not more than 500 W for that year. Today in 2013—40 years later—the PV industry of the World is consisting of several thousand companies employing more than a million people and 100 GW of solar electric systems are already deployed which produces about the same amount of electricity as 20 nuclear electric power plants and the *yearly* production of PV systems is at least 30 GW, equivalent to five or more nuclear power plants. It is expected that the production will increase at least 30% yearly.

This book is different from the books written about solar (PV) electricity. Those books usually deal with the research leading to its development, its technology, its application, the politics, the benefits environmental, or others. This book describes what is missing from those books, namely the 40-year history of the terrestrial PV industry bringing an expensive technical curiosity and without any market at that time, into a mass manufactured product, which is able to replace or minimize the use of electricity-generating materials, which are dangerous to humanity or threatening the climate of our planet.

The terrestrial PV industry started 40 years ago, but practically nothing is written about the first 20 extremely interesting years. This book describes how the "terrestrial PV industry" was started, how it developed the needed technology, and market to be able to compete with established electricity-generating systems fueled by oil, gas, or nuclear energy. Survived the various changes in the global government's policies and ultimately winded up as a global industrial force surprising all of those in the government, in electric utilities and the public who still thinks PV is a scientific curiosity. The book is based on solid facts and research and describes all of the ups and downs, the rocks, the successes and failures, the people who made it happen, the influence of the governments, the oil giants, the utilities, and finally the result that the terrestrial utilization of PV became unstoppable.

This book is written by one of the people who started the terrestrial PV industry in 1973—40 years ago—and participated in its history to this day.

The history of these 40 years can be separated in three parts:

- Act 1: Dawn—1972–1984
- Act 2: Sunrise—1985–1999
- Act 3: Toward High Noon—2000–2013

This book's narrative is as non-technical as possible and the Annexes provide material for those who would like to learn more about the technical part.

Act 1

Dawn

1972–1984

Chapter 1

Damn the Torpedoes and Full Steam Ahead[1]

After the 1953 discovery at Bell Laboratories of how to produce useful solar cells, converting light directly to electricity, and their first utilization was envisioned for satellites and other space projects and the research was directed to improve their efficiency to convert light to more electricity and to improve their durability in the space environment. For this application, it was important to produce more electricity from an area, long life in space and low weight. The cost of the solar cell system was not important, because whatever it was, it was negligible compared to the cost of the satellite that they powered.

Communication Satellite Corporation (COMSAT), incorporated in 1963 as a publicly traded company, was created by the Congress of the USA Government in 1962 with the responsibility for the development of a global satellite communication

---

[1]Source material for this chapter:

P. Varadi, *Diary and Notes*.

P. Varadi (June 7–11, 2004), *Lecture 19th European Photovoltaic Solar Energy Conference*, Paris.

---

system. COMSAT established a research laboratory in 1968[2] and Dr. Joseph Lindmayer became the head of the physics department and I became the head of the chemistry department. Joseph was an excellent physicist and he made a great breakthrough in the efficiency of solar cells for space use. My department provided help in some of that work.

The Bell Telephone system's 1956 advertisement[3] looked into the future and predicted: "The Bell Telephone Laboratories invention has great possibilities for telephone service and for all mankind." This was true. Many great possibilities existed for the terrestrial utilization of solar cells, but the production technology developed for the solar cell's space use resulted in prohibitively expensive solar cells for these applications. Therefore, only a small quantity of solar cells, which did not qualify for space, could be used for a few terrestrial applications. The problem was that for the wide spread of the terrestrial utilization of solar cells, a new production technology was needed to produce them inexpensively and in large quantities.

The use of solar cells in space would doubtlessly increase with the increasing use of satellites and space objects, but Joseph Lindmayer and I thought that the utilization of terrestrial solar cells had an unlimited future, since humanity would need it to reduce the use of non-renewable energy sources because they would be exhausted and were environmentally harmful, and also to reduce or eliminate the utilization of nuclear energy, which was considered dangerous, and was proved that those fears were right, for example in Japan.[4] The use of electricity converted from solar energy would also be the means to provide electricity for 1.6 million people, who had no electricity and for whom it was and will be almost impossible to be connected by traditional power lines.

---

[2]COMSAT Research Laboratory, Clarksburg, MD outside of Washington, DC. COMSAT headquarters was in Washington, DC. COMSAT was merged into Lockheed Martin Global Telecommunications in 2000.

[3]*National Geographic Magazine*, September 1956.

[4]Fukushima, following the Tōhoku earthquake and tsunami on 11 March 2011. It is the largest nuclear disaster since the Chernobyl of 1986.

In 1972, we spoke many times about solar cells which could be used for terrestrial purposes. As a result of our discussions, Joseph submitted a proposal to the management of COMSAT, to give permission and money to develop the technology for solar cells to be used for terrestrial applications, which required solar cells to be manufactured inexpensively and in large quantities. Therefore, it required a different technology from the one used for solar cells for satellites. COMSAT's management understandably and rightfully wanted to spend no resources to develop the terrestrial use of solar cells, since they only wanted to spend manpower and money for what was in the interest of the company, which was, the use of solar cells for satellites, to increase their quality, durability, and performance.

After this became evident, during lunch breaks, and when our families met, Joseph and I spoke a lot about the terrestrial potential of PV. There were two companies in the USA, Spectrolab and Centralab, which manufactured solar cells for satellites, and they sold for terrestrial use only the small amount of rejects from their production. We discussed that perhaps we should start a new company the purpose of which should be the manufacturing of solar cells to be used exclusively for terrestrial purposes.

This was also the subject of our discussion when our families celebrated together New Year's Eve in 1972. I do not know after how many glasses of excellent French champagne during the first hours of 1973 we agreed that if we left the crystallized bureaucracy of Hungary marching toward communism and we came here to the USA, the center of capitalism, we should leave our jobs to form a company and become capitalists. To this, we clinked our glasses saying that, "the time is now or never." So, we agreed that it was an excellent idea. With this we stood up, shook hands, and filled our glasses with more champagne, clinked them, and announced our decision to our spouses who dismissed the idea, considering that it was the result of a good party.

We got paper and pencil, filled our glasses again with champagne, and sat down. Firstly, we wanted to discuss the company's name. Joseph stated that it did not matter to him what the company's name was but insisted that it should end with an x. We wrote down on a paper several possibilities, all of them ending with an x. At the end, we settled that the name should be Solarex.

The next issue which we had to decide was who should do what in Solarex. I stated that both of us had been in science, but I thought that, since he was a great solid-state physicist, as is evident from his book on *Solid State Physics*[5] and recently the development of the "violet solar cell," he should do the technology, research, and development, and since as of that time I already had lots of publications and patents but in spite of that I had not yet received the Nobel prize in chemistry, I should stop working in science and go into another field, maybe I should run the business. It sounded an excellent idea to run operations such as the production, sales, marketing and accounting, administration, and so on. I never did this sort of thing before, but on the other hand, why should I not to be able to do it? So, we agreed that this distribution of duties was acceptable to both of us.

I proposed we should start Solarex in 6 months from that date. It was agreed, and we started to draw up a plan. The next decision was the timetable, and how to meet it. We proceeded to write down point by point what we should do to complete it, and who should be responsible for what. Our goal was that we should quit our jobs at the end of June 1973 and soon thereafter open the doors of Solarex as one of the very first terrestrial solar cell development and manufacturing companies.[6] We decided that the date ought to be August 1, 1973.

---

[5]Lindmayer J, Wrigley C (1965) *Fundamentals of Semiconductor Devices.* Van Nostrand, New York.

[6]Also in the same year (1973), at practically the same time and also with the same purpose as Solarex, Elliot Berman founded in the USA "Solar Power Corporation," which was financed by the oil company Exxon. These two companies (Solar Power Corporation and Solarex) launched the terrestrial solar energy business. (*Switching to Solar*, Prometheus Books, 2011, 24). Berman left Solar Power Corporation in 1975. Exxon decided to close Solar Power, and Solarex purchased all of Solar Power Corporation's assets on June 15, 1984.

Next day after this New Year's night, on Monday, January 1, 1973 we met again in Joseph's home, looked at the two pages of the pencil-written program both of us had decided on the previous night to start Solarex, and we pondered and simply did not understand by the light of day, why well-paid scientists should get into such a harebrained idea for terrestrial utilization of solar cells, when there was no technology for new low-cost solar cells. Solar module assembly manufacture equipment was also necessary to be developed for these new products. We had to create a market for these new miracles for which no market existed and on top of that, we had no money. After considering the pros and cons of the situation, we decided "damn the torpedoes and full steam ahead." So the first step was to develop a business plan and we agreed that, based on our experience of writing fiction, such as successful proposals to get grants, we could easily write another fiction, an excellent business plan. In spite of what business schools teach, the similarity between the fiction that Jules Verne wrote—*Around the World in 80 Days*[7]—and a business plan was that both described what may happen in the future, the only difference being that Verne was writing to sell books, and the business plan was written to sell shares in the company. As we had no money, the question was where would we get it? It sounded very simple, we would send the excellent business plan to venture capital companies and they would love the idea, and surely stand in line to provide the necessary capital.

After we discussed and re-discussed all the issues from every angle again and again, we decided to try it. If it did not work out we could always go back to scientific work. Eventually we decided that for two people like us, who survived in Hungary during World War II, the bloody years of communism and a revolution, the development, manufacture, and market to a non-existent market of terrestrial solar cells and systems, could not create a problem. We decided that we would devote several hours a day to develop the details.

That is what we did. We had lunch together practically every day and after work at 5 p.m., Joseph came to my office and

---

[7]*Around the World in 80 Days* by the French writer Jules Verne, was first published in 1873.

we spent several hours discussing the details. He always came to my office, because I had an American secretary, Janice Campanaro, whose husband was an Italian and because of that she made the world's finest espresso coffee and brought it to my office, before she went home.

The first question was how much capital we would need and from where were we going to get it. Given that neither of us had any money, we thought that first we should develop a budget, how much money we would need for research, manufacturing, offices, sales, and salaries for as long as enough money came in from the sale of our products. After the budget was completed we incorporated Solarex. The incorporation of a business in the USA is a very simple procedure and does not require any capital. We were also going to complete a business plan and necessary papers for the offering of the company's stocks.

As regards raising the capital, we were confident that the business plan was very good. We planned to send to about 20 venture capital firms, and we were sure that these investors would be delighted to offer money, because this was a new industrial area, in which Solarex would be the first privately funded company in the world to start and that our position at COMSAT and our academic achievements were the guarantees of success. So the first task was the compilation of the budget and writing the business plan.

The first question was how much it would cost to purchase the necessary equipment and installation. The cost of the research equipment and installation was easy to estimate because establishing our departments at COMSAT laboratories a few years earlier gave us experience. The amount required for the manufacture of solar cells and panels was more difficult to determine since even though we knew the cost of the equipment needed for solar cells and modules for the satellites, we did not know the price of the equipment for the terrestrial use since we had to develop a new technology to bring down the manufacturing cost of solar cells, modules, and systems drastically.

The first decision we had to make was what should be the material to be used for the terrestrial solar cells?

Starting with Chapin's discovery in 1953, the material of the solar cells for satellites was silicon (Si) wafers used for producing transistors. When Solarex was started in 1973, both Si and cadmium sulfide (CdS/Cu_2S) were utilized as materials for solar cells. For space solar cell applications mostly Si, but in a few instances CdS material was also used.

Only a few terrestrial applications were already attempted at that time. The two American companies, Centralab and Spectrolab, were producing Si solar cells for space projects, and were selling a few solar modules made of solar cells rejected from space use for terrestrial applications. In France, the SAT company made a few terrestrial solar modules utilizing the CdS system.[8] In the USA, Karl Böer promoted the use of CdS/Cu_2S solar modules for terrestrial use,[9] and in 1973 a few of those were mounted on the roof of a house, which was completed on June 12, 1973. About the same time, Solar Energy Systems, Inc., (SES) was formed in Delaware, USA, with the purpose of producing CdS/Cu_2S solar modules for terrestrial purposes.[10]

When it was decided to start Solarex, dedicated to manufacture solar modules for terrestrial application, we had to select which material, Si or CdS, should we use. Both of us, Joseph from the semiconductor physics and I from the chemistry point of view, agreed that the CdS/Cu_2S would be a very bad choice as it would not be able to survive under terrestrial conditions.[11] Therefore, the decision was that Solarex should use only Si material for our terrestrial venture, because we believed that Si solar cells when properly fabricated could have a life expectancy exceeding 20 years.

We decided that in spite of the different requirements between the space use and at that time the non-existent

---

[8]Palz W (2011) *Power for the World*, Pan Stanford Publishing Pte. Ltd., Singapore.

[9]Böer KW (1972) *Future Large Scale Terrestrial Use of Solar Energy*, Proceedings of the 25th Power Sources Symposium, page 145.

[10]Böer KW *The Life of the Solar Pioneer*, iUniverse, Bloomington, NY.

[11]Shell Oil Company invested $80 million in the development and in a production line of SES, but CdS/Cu_2S modules were never able to be produced to be sold. SES was closed down by 1980.

technology for solar cells to be used for terrestrial applications, Si would be the basic material for solar cells for satellites and we assumed that solar cells for terrestrial use for many years to come would also remain the same, despite their manufacturing technology obviously becoming very different.

We made this decision in the beginning of 1973, and actually we were right. Until the beginning of the 1980s, Si material was used for solar cells for space and for terrestrial applications, but the development and the production technology ran on two different tracks. The efficiency of terrestrial solar cells was important, but more important was the cost. Even as of 2010, the large majority (80%)[12] of the solar cells manufactured for terrestrial use was still based on Si. On the other hand, solar cells manufactured for space use, where the cost was not important, but the efficiency, that is the production of electrical energy, and also that under very severe space conditions the solar cell should have a relatively long life, were extremely important. Si to be used for solar cells for satellites from 1980s onwards became increasingly less likely, and instead in 1990s increasingly complex systems, elements, compounds, and multi-layered structures were used as the material for space solar cells and Si was no longer used.[13]

Thus, it was clear to us that Si would be used for terrestrial solar cells, and on this basis we would assemble Solarex's business plan and estimate the necessary seed capital. Following our plans, we strictly adhered to the schedule we established on our January 1 meeting. In the last week of February, we incorporated Solarex (I received Solarex stock certificate #1 dated March 3, 1973). On March 26, we finished and printed Solarex's business plan, which, including two appendices, became 30 pages long.

Looking back on our business plan, it is interesting to see that a total of only three companies were mentioned, which were engaged in manufacturing, or started to manufacture

---

[12]Kurtz S (2010) NREL/PR-520-49176 (*Source*: Photon International).

[13]Bailey SG and Viterna LA (2011) Role of NASA in photovoltaic and wind energy, in *Energy and Power Generation Handbook* (KR Rao, ed) Chap. 6, ASME Press, New York, NY.

solar cells. One of them was Spectrolab, which was a part of Textron company at that time. Now it is a part of the Boeing aircraft manufacturing company. The other was Centralab, a division of Globe Union, Inc., which does not exist anymore. These two companies manufactured solar cells for satellites and spacecraft. For terrestrial use as mentioned, they only sold the rejects from that production. That was naturally a very small quantity. A third company, previously referred to as Solar Power Corporation, was founded about the same time as Solarex. In Germany at that time also existed, but was not mentioned in the business plan, AEG/Telefunken's small operation producing solar cells for space use. The company does not exist anymore.[14]

As we planned, I assembled a list of 20 venture capital firms and established by telephone who would be the person who would deal with this sort of investment. I mailed to each person a copy of our business plan. Subsequently I called them and asked if they had any questions. When they informed me that they would be interested, we discussed the date when we could meet and try to persuade them to invest in our venture.

The first venture capital company which we visited was a division of Textron company. This meeting was set up by Bill Yerkes. Bill, who was the President of the already mentioned Spectrolab, the manufacturer of space-oriented solar cells, was a good friend of Joseph Lindmayer. When we decided that we were planning to start a company for the manufacturing of solar modules to be used for terrestrial applications, Joseph called Bill and asked him for advice. Bill said that this is a great idea, and he already wanted Spectrolab, which at that time was a division of Textron, to also start to do it, but he did not receive permission from Textron's management. He offered that if Solarex, which was by that time incorporated, was planning to raise money, he would be very happy to introduce us to Textron's venture capital division to fund us.

---

[14]AEG is listed as the predecessor of Schott Solar (History of Schott Solar. http://www.schottsolar.com/us/about-us/history/. Schott Solar announced in 2012 that it is withdrawing from the crystalline photovoltaic business).

He envisioned that if we started Solarex, and Textron would have a relationship with both companies, sooner or later they would merge, which would be a solar cell and panel manufacturer, which would also produce space as well as terrestrial products.

Bill suggested that before we visit any venture capital company he would arrange a meeting with people from the Textron venture capital division. So when we sent out our business plan, Bill arranged a meeting with the people from Textron. He of course wanted to be there to give more weight to the matter. We were very happy with this, since this division of Textron started as the first publicly owned venture capital company, founded by the legendary George Doriot[15] in 1946, and only recently was bought by Textron.

The meeting took place in Textron's Investment office in Providence, RI. We had thought that it was a very successful meeting, since Bill strongly supported our program and also gave a very good description of Joseph and myself. A few days later, Bill called on the phone and told Joseph that Textron decided they would not invest money in Solarex. According to Bill, the reason they did not want to support it was because they did not see a great future in solar cells, neither in space nor for terrestrial use. It depressed us, but still we had 19 other venture capital companies which received our business plan. Joseph and I visited many of them, but to several I went by myself.[16]

---

[15] Ante SE (2008) *Creative Capital: George Doriot and the Birth of Venture Capital*, Harvard Business Press, Boston, MA.

[16] Not much later, in 1975, Textron sold Spectrolab to the Hughes Aircraft company and Bill Yerkes was immediately let go. Given that Bill by that time knew our experiences with venture capital companies, he did not even attempt to get seed capital from them, but received money from his family and invested the little money he saved, formed in 1975 Solar Technology International (STI) and began to produce terrestrial solar cells. In 1978, ARCO (the big American oil company) bought STI and named it Arco Solar. In 1989, Arco Solar was sold to Siemens, Shell took over from Siemens, and then finally in 2006 SolarWorld, a German company, became the owner.

It was already the first week of May when I visited the offices of our list's last venture capital firm J. H. Whitney & Company, in New York City. The person whom I talked to on telephone expressed his interest in our venture so we agreed that I would visit him in his office on the afternoon of May 3 at 2 p.m. This was very good for me, because that same evening at 7:30 p.m. I planned to fly from New York—JF Kennedy airport to Lisbon for a week-long trip.

I flew with a small bag and my briefcase from Washington, DC., where I lived, to New York and went by taxi from the airport to Manhattan to 630 Fifth Avenue to the Rockefeller Center where in one of its skyscrapers was the office of J. H. Whitney & Co., which was established in 1946 by the industrialist and philanthropist, John Hay "Jock" Whitney. It was one of the oldest in the venture capital business. I was familiar with the Whitney name from the Whitney Museum in New York.

Exactly at 2 p.m. I opened the door of the Whitney venture capital firm and entered a large room where almost the entire surface of the parquet floor was covered by a Persian rug. The room had no windows, the walls were carved wood. A large chandelier hung from the middle of the ceiling. Opposite the entrance door was also a large wall with two large closed carved doors. Between the two doors far enough away from the wall on the Persian carpet was a huge desk made probably of rosewood and behind it a respectable white-haired lady was sitting.

I went there, said hello and told her the person's name with which I had an appointment. The lady politely greeted me, apparently she was aware that I had an appointment, picked up the phone, dialed, and told whoever was on the other end that I had arrived.

Having hung up she told me that a few minutes later someone would come to escort me back to the office, until then I could sit down and she pointed toward the armchairs near the wall.

As I went there, I looked around and saw that on the walls of the room hung four or five French Impressionist paintings. Instead of sitting down I went to the pictures. One of them I realized was a Monet, the other one was a Renoir. When I turned around to go to the opposite wall to look at those paintings too, I had to pass in front of the lady and I told her that these were beautiful reproductions. The lady looked at me contemptuously and said that they were genuine.

I realized that the Whitney's founded the New York Whitney Museum on Madison Avenue, and these were maybe paintings here because they had run out of wall space there. Then a young girl came and told me to follow her and escorted me to a beautiful office also with a Persian rug on the floor. Don, with whom I already talked on the phone, was elegantly dressed. He stood up and approached me, shook hands, and led me to a table, which had four thick wooden legs and a thick glass top. Four leather armchairs were around it.

By the time we sat down, the young girl who escorted me to this office came in with a tray, on which were a porcelain coffee set, two cups, a coffee pot, milk jug, and a sugar bowl. She deposited the tray on the table. On the tray were also small spoons for the coffee. She asked if I would rather have tea. I said no, thank you.

When the girl walked out, my host put on the table the business plan I had sent him, opened it, and apparently already had read the whole thing, because on the side of the pages I saw that he had some notes and it seemed that he used his notes to ask questions.

The discussion lasted about 2 hours. He asked a few times if I would like to have more coffee. I said no. He once stood up and asked for coffee by phone, which the girl brought in a few minutes. When he reached the end of the business plan, even then he had more questions. I felt he was well prepared and really very interested in our venture.

After he ran out of questions, he sat back in the leather armchair and said that the thing was very interesting and, in his opinion, we would be successful, but he did not invest money in a company at such an early stage. However, he was sure that he would invest 2 years later. With this he gave me his business card and told me that at that time I should not forget about him. We stood up and Don escorted me through the room with the genuine French Impressionist paintings to the exit door, where we shook hands.

When I reached the elevator I looked at my watch which showed 4:10 p.m. So I thought that when I reached the airport I would still have lots of time until the plane left. When I arrived to the building's exit door, I saw with horror the pouring rain. I knew New York quite well, and I knew it was a weekday, as it was 4 p.m. and at that time it is pretty hard to get a taxi on a beautiful sunny day, but on a rainy day it was practically impossible. I was there with a small suitcase, a briefcase, and tried to keep the umbrella above my head.

The time passed slowly, and of course I could not get a taxi. The problem now became not only that there would not be a lot of time for me at the airport, but I would be late and miss my flight. Since I was already soaked through, it made no difference if the umbrella was open or not. Eventually I managed to fight to get a taxi. When I told the taxi driver to go to Kennedy Airport, he just shook his head and said that it is almost hopeless on such a rainy day because of the traffic. I asked him for how many years he was driving a taxi cab in New York. He said for 20 years. I said to him, "Then I am not afraid. I am going to get there on time."

Fortunately, in those days we did not have the airport security system that we have now and I got to the gate in time and even had time to call Lindmayer. I told him that after all, we had been very successful with the venture-capital companies. All 20 managed to learn how to spell the word P-H-O-T-O-V-O-L-T-A-I-C-S, but unfortunately we had not received a penny from any of them.

I told Joseph that I would be back in a week. In the meantime both of us should think how to reduce the necessary capital to the absolute minimum so we could survive one or one and a half years. I told him that I was soaked through because of the rain, but I had made a decision that when I got home I would get the amount that was needed within 2 weeks and I would never in my life visit the office of a venture capitalist again to try to obtain capital.

I can now say that all this was done as I said. We reduced the capital requirement to a minimum of $210,000 but it was plenty, because after 9 months of operation Solarex became profitable. We collected the required capital in 2 weeks from friends and acquaintances. One was Al Nerkin, who invested 25% of the needed money. Al was an engineer and he and his partner started Veeco Instruments, a vacuum equipment company, in 1946. By 1973, it had become a very large enterprise. When investing in Solarex, Al said that he was giving a lot of money to charity, and he thought that to invest in 1973 in solar energy was the same as giving it to a charity.

Chapter 2

Cook Book: How to Make Solar Cells and Modules

Readers who are familiar with "photovoltaic (PV) solar cells" please move to the next chapter.

This chapter is for those people who heard or read nothing or maybe just something about "photovoltaic cells" (PV cells) which are also called as "solar cells" and they think it is a very complicated subject. This chapter is to dispel the mystery about the subject and to show that it is only as complicated as sharpening a pencil.

To mention the word "photovoltaic" could be understandable for scientists or experts, but it is probably meaningless and may frighten away people.

The word "photo" is the Greek word for light and everybody knows the word "electricity." So if one would use *photoelectric* cells" the understanding would be easy. We would talk about a device that produces electrical energy when it is exposed to light. But to complicate matters the word "*photovoltaic* cells" or the expression "solar cells" is used for the devices whose function is exactly that. Why not call them simply "photoelectric" or "solarelectric" cells? The reason is very simple. Scientists

Sun above the Horizon: Meteoric Rise of the Solar Industry
Peter F. Varadi
Copyright © 2014 Peter F. Varadi
ISBN 978-981-4463-80-5 (Hardback), 978-981-4613-29-3 (Paperback), 978-981-4463-81-2 (eBook)
www.panstanford.com

in the nineteenth century were inconsiderate to the people of twenty-first century.

Edmond Becquerel, a French physicist in 1839, while studying the effect of light on "*electrolytic* (battery) *cells*" discovered a relationship between light and electronic properties of materials that certain material converts light directly into electrical energy. The culprit was actually Alfred Smee, who 10 years later coined the term "photovoltaic"[17] from "photo" and adding the word "Volta"[18] who was an Italian physicist (died more than 10 years before Becquerel discovered this effect), because the battery invented by Volta is credited as the first electrochemical cell. By working with electrolytic cells, which have a positive and a negative electrode, Smee named this device also having a positive and negative electrode, "photovoltaic cell," being a derivative of Volta's "electrochemical cell," operated not by chemicals, but by light.

Heinrich Hertz in 1887 discovered another effect, namely, that when light hits a material, the electrons escape from the material. They first named this as "Hertz effect," but it was soon renamed as "photoelectric" effect.

Figure 2.1 helps us to understand that the similarity in both the inventions of Becquerel's and Hertz's is that light produces electrons. Hertz's invention called "photoelectric" refers to the *emission*, or *ejection*, of electrons from the surface of a metal in response to incident light which means the electrons escape from the metal. Becquerel's invention is in which the electrons in response to incident light escape from the material, but like in a battery cell in an existing electrical field move toward the positive electrode. The chemical battery cell was invented by Volta, so Becquerel's invention was given the name by Alfred Smee a "photovoltaic" cell.

As an interesting note, that nobody understood how the Hertz and the Becquerel effect works, until Einstein in a paper in 1905 described it. This was in importance not comparable to the *special relativity* theory he published in the same year,

---

[17]Palz W (2011) *Power for the World*, Pan Stanford Publishing, Singapore.

[18]Count Alessandro Giuseppe Antonio Anastasio Gerolamo Umberto Volta (18 February, 1745, 5 March, 1827) was an Italian physicist known especially for the invention of the battery in the 1800s.

and in the final form of *general relativity* published in 1916. By 1920, it was a general agreement that he deserves the Nobel Prize but for political reasons in science they could not give it to him for the relativity theory, therefore, he received the Nobel Prize in 1921 for the explanation of the PV/photo-electric effect, the work he published 16 years before.

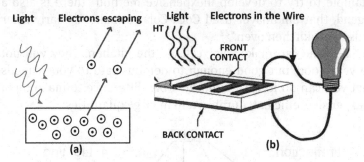

Figure 2.1 (a) Hertz's Photo-'electric'[19] effect and (b) Becquerel's Photo-'voltaic' effect

The issue was dormant for almost a half of a century, when Daryl Chapin of the Bell Telephone Laboratory discovered, that Silicon (Si)[20] material used for transistors is converting light to electricity and was able to produce a useful device with about 6% conversion efficiency of light into electricity. This invention and the need for electricity for space use focused the attention on this device the "PV cell."

This entire book deals with this device, the "PV cell" which is like a battery in which there is a positive and a negative electrode and the electrons are knocked out by the energy of the light from an atom and start to move in the direction of the positive electrode establishing an electric current. As mentioned above, it is a simple device as evidenced from the following recipe how one can make it.

How to Make Solar Cells?

One would think that the "PV" which is simply called "solar" or "PV" cells are complicated devices. But in reality they are

[19]Image from NASA Science
 (http://science.nasa.gov/science-news/science-at-nasa/2002/solarcells/)
[20]Silicon (Si) is the second most abundant element in the Earth's crust.

simple to make, as simple that even one could make it in the kitchen. As we decided to go into business to produce solar cells and modules for terrestrial purposes utilizing Si as the base material, we knew that among the semiconductor[21] devices, for example, transistors, integrated circuits, etc., solar cells are the simplest and the easiest devices to make. As an example to try to develop inexpensive method "there is also a legend, that Lindmayer used Coca-Cola to make his early solar cells in his kitchen oven."[22]

Yes, one can make solar cells in the kitchen. They will not be very good, but good enough to demonstrate to your friends that you can do it. It is naturally a big difference to make them inexpensive, efficient, durable, and in large quantities.

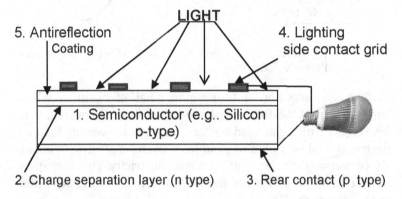

Figure 2.2 Schematic of a silicon solar cell.

In order to easily understand how a solar cell works and how to make it, a recipe is provided here to create it even in your kitchen. I will refer to Fig. 2.2, which shows its basic structure. Please do not be afraid that it looks complicated, if I would draw a picture showing the cross section of an "apple turnover" it would be more complicated, but as it is advertised

---

[21]Semiconductors are materials of electrical resistance which falls between conductors and insulators. There are elemental semiconductor materials which in pure state, have the above characteristics: germanium (Ge), silicon (Si), and selenium (Se) and there are solid solution-type semiconductors: gallium arsenide (GaAs), gallium aluminum arsenide (GaAlAs), silicon carbide (SiC), lead or cadmium telluride (CdTe or PbTe), and so on (Wikipedia).

[22]Yergin D (2011) *The Quest*, The Penguin Press, New York, NY.

"apple turnovers" are "Delicious, yet so easy to make. Anyone can do these classic apple turnovers!"[23] Si-based solar cells are not delicious, but they are easy to make. Anyone can make them.

One should remember for a battery one needs a positive and a negative electrode. As PV battery cells operated by light, they also need a positive and a negative electrode.

Take a pure Si wafer which is doped, that means the extremely pure Si (this will be discussed in Chapter 8) containing an extremely small amount of the element for example, Boron (chemical symbol B), and because of that it will be slightly positively charged (p type). I am not proposing that you should attempt to make the Si wafers. For cooking you do not make everything, for example you buy the yeast but do not attempt to make it. To make these wafers, it requires complicated and expensive equipment as will be described later. But these wafers can be purchased, because large amount of these are also made for the semiconductor industry.

The direct conversion of sunlight into electrical energy happens in the Si wafer (1), in such a way that the absorbed sun light directly creates electrical charges in the material. The next step (2) is that on the side of the B doped Si, the one which will be exposed to light a thin negative charge separation layer has to be formed. This is very simple.

The thin "charge separation layer" is created by diffusing phosphorus (chemical symbol P) into the Si wafer. This can be achieved by coating the wafer with a compound containing phosphorus and bringing the coated wafer to elevated temperature to about 800–900°C (1470–1600°F) to let the P diffuse into the wafer and form a slightly negatively charged area. For this the best is to use phosphoric acid (chemical symbol HPO_3), which is a clear, colorless, and odorless liquid with a syrupy consistency, it is easy to coat the wafer's surface. Where can one find phosphoric acid in the kitchen?

The legend is quoted above, that Lindmayer even tried Coca-Cola to make solar cells in his kitchen. One would think that he was a mad scientist, who tried everything. But he was not a mad scientist. As a matter of fact, he was one of the most

---

[23]http://allrecipes.com/recipe/apple-turnovers/.

ingenious physicists I ever met who invented the most efficient (at time) solar cell for space use. The truth is that many people, including Lindmayer, knew that Coca-Cola does have phosphoric acid as one of its ingredients. Phosphoric acid is used as an acidifying agent to give colas their tangy flavor. There is nothing wrong in Coca-Cola to have phosphoric acid as one of its ingredient, because phosphorus is needed very much by the human body, all multivitamins contain phosphorus. Not only in Coca-Cola but in many other materials you have at home containing P (e.g., fertilizers, etc.) could also be used.

In order to let the P diffuse into the Si wafer, as described, elevated temperature is required. Many people have a kiln at home which they use to make pottery, if not, one can be borrowed. Kilns can be set to operate at the required temperature. Kilns are also made so that gas can be flown into it to establish a protective atmosphere. In this case one can use nitrogen. This can be easily obtained in a bottle or borrowed from the dermatologist, who has it in every of his examination rooms and uses to freeze small skin cancers such as superficial basal cells.

When the diffusion of P is completed the interface between the area which contains P and the original wafer which contains B is called in scientific language the "p/n junction."

After this is completed, we can turn to make the positive rear contact (3). One can buy a paint spray, in which aluminum powder is sprayed to the painted object. Just spray a little aluminum paint on the back of the cell and heat it in the kitchen oven.

This basically completes the solar cell. The electrons released by the sunlight can drift in the internal field created by the top negative field and the bottom positive field. The only thing left is, to add electrical contacts, so the produced electrons should be able to leave the Si wafer.

In step 4, on the side where the light reaches the solar cell, a grid of metal contact should be created, which should not block all the light from the wafer. The back of the solar cell can be fully covered with metal. To add the front contact can be easily done at home. A conductive, for example silver paint, can be purchased and the grid can be painted with a little nail polishing brush on the front. This paint has also to be heated according to the paint manufacturer's instruction, but the home cooking oven is sufficiently good for this purpose.

There is one more step needed, which is not shown in the figure. It is advisory to take sandpaper and use it to sand the edge of the wafer. The purpose of this is to eliminate any material used to make the p/n junction, front and back contact, which may have inadvertently winded up on the edge.

When light strikes the solar cell, it develops a voltage difference between the light side and the back side. This voltage difference depends on the solar cell material. The Si base solar cell, for example, when illuminated produces approximately 0.5 V. How much power will the solar cell supply depends on the size of the device, its efficiency, and obviously the intensity of the illumination, that is, the percentage of light converted into electrical energy.

There is only one step (5) in (Fig. 2.2) which has not been explained yet. It is the anti-reflection coating. This has no functional effect on the cell operation. Its purpose is, like for photographic camera lenses or for eyeglasses, to increase the amount of light able to enter the Si material, thereby improving its efficiency.

This was a simplistic description of how solar cells are made, to demonstrate that they are simple and easy to make, but the reality is not far from it.

Jumping from the kitchen-produced solar cells to today's fully automated turnkey Si solar cell production facilities which can be purchased from several manufacturers (a list of these manufacturers can be found on the Internet), they actually use the same technology which was initiated over 30 years ago, but many changes are introduced to achieve high efficiency, be able to handle very large and very thin Si wafers [300 mm ("11.8 inch," usually referred to as "12 inch") and to save material their thickness is very thin, for example 775 µm].

Today, PV cells are mass produced in factories which are fully automated and require no manual handling of the wafers at least 98% mechanical and electrical yield and their wafer throughput can be 2–4 thousand wafers per hour.

How to Make Solar Modules

Today, direct current equipment operates at 6, 12 (as used in car batteries) or 24 or 48 V. As it was mentioned a single solar cell produces about 0.5 V, therefore the solar cells have to be connected in series and/or parallel to provide the necessary voltage and power. These units are called solar modules. The assembly of these PV modules is also very simple.[24] One can do it at home in the kitchen. "Do it yourself kits" to make at home solar modules were sold since 1980. The base can be window glass. One can easily cut it in the required size. The solar cells have to be connected with small pieces of flexible metal wires. One end of this flexible wire should be soldered, with a small soldering iron and standard solder material, which can be obtained in a store to the bottom contact of the solar cell. The other end should be soldered to the electric connection on top of another solar cell. (see Fig. 2.3). As the 1980, "Build your own solar module" kit describes: "it is recommended that you have basic soldering skills."

Figure 2.3   Solar cell interconnection.

In order to make a PV "module" providing for example 6 V, one has to solder at least 12 solar cells in series, one to the other. In order to charge a 12 V battery usually 18 cells are being used as shown on Fig. 2.4.

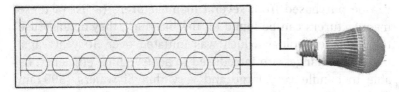

Figure 2.4   Photovoltaic module (shown nominal 6 V).

After the solar cells are connected—this we can call a solar array—it should be placed on a glass plate so that the side of

---

[24]They are also referred to as "solar panels," or "PV (photovoltaic) modules."

the solar cell which has the connecting grid and the surface area should see the light and should face the glass plate. But in order to keep the solar cell assembly in place and protect it from the atmosphere, for example rain, hail, etc., the solar cells are placed either between two glass sheets where they are surrounded with a plastic material, or only the top, which is facing the sun, is glass and on the back of the module a plastic sheet is laminated on.

Today several companies exist offering turnkey installation of automatic manufacturing plants for crystalline and thin film solar cells and modules. The basic of all these systems is similar to the one described in a very crude form above, but in order to achieve high efficiency, and long life, materials used in the production are obviously quite different.

Solar Cell and Module Types

Silicon Solar Cells and Modules

The types described above is the type predominantly used for terrestrial PV systems. More than 80% of all PV systems use solar cells made from Si wafers.

Thin-Film Solar Modules

The rest of the terrestrial PV systems utilize the so called "*thin-film*" solar modules. The operating structure of these thin-film modules is similar to the Si-based ones, described above but they are made not in a solid structure such as Si wafers, the differences are that they are deposited as thin films and no module assembly is needed. Several thin-film systems were and are in existence, but there are only three types which are used in larger quantity. They are described in Chapters 22 and 34.

Space Solar Cells and Modules

As mentioned, solar cells used for terrestrial applications require a different technology from those used for space. A different technology was needed, because, for space, the solar cells need to have as high efficiency as possible and high resistance to radiations the cells are exposed to. The price of the solar cells used for space applications was and is immaterial compared to the price of the payload they power. For terrestrial purposes, however, the pre-dominant criterion is the price: It had to be inexpensive and the terrestrial environment is not as harsh as in space and the weight of the material to protect it is not critical.

Chapter 3

Betting on Horses

After we realized that we did have the money to start Solarex, we reviewed again what we should do. Naturally, writing about it 40 years later has the advantage of providing a much more complete picture which can include other peoples' opinions.

At the time, Joseph Lindmayer and I were considering and were in the process of starting a company devoted to producing photovoltaic solar cells and systems for terrestrial application, the idea of doing something about the terrestrial utilization of solar energy also occurred to other people. We just did not know that our thinking was totally contrary to what the "prevailing view" was.

In 1972/73, in the USA and in Europe the solar cell research conducted in the past 20 years in government and non-government research laboratories was focused exclusively on space applications, and small companies[25] (in Europe and in the USA) existed producing PV cells and modules for space use.

About this industry Wolfgang Palz writes[26]: "The photovoltaic industry at that time was involved in space activities and in

---

[25]Curtin DJ (1972) *25th Power Sources Symposium*, 134.

[26]Palz W (11–15 May 1981) *IEEE Photovoltaic Specialist Conference*, Houston, TX.

Sun above the Horizon: Meteoric Rise of the Solar Industry
Peter F. Varadi
Copyright © 2014 Peter F. Varadi
ISBN 978-981-4463-80-5 (Hardback), 978-981-4613-29-3 (Paperback), 978-981-4463-81-2 (eBook)
www.panstanford.com

electronics but had no interest at all in energy matters. It was obvious that this industry had great difficulties moving into the terrestrial business: After all, a silicon chip achieves more than a hundred times the price when it is sold for a satellite generator or integrated circuits than it does for a terrestrial generator."

One of the first considerations of photovoltaics for terrestrial application was a paper presented in 1971,[27] and three papers presented at the IEEE Photovoltaic Specialist Conference in 1972. There was one paper presented at the 25th Power Sources Symposium in 1972.[28]

In January 1972,[29] a Solar Energy Panel was organized jointly by the National Science Foundation (NSF) and NASA and comprised nearly 40 scientists and engineers. The panel was charged with assessing the potential of solar energy as a national energy resource and the state of the technology in the various solar energy application areas, and recommending necessary research and development programs to develop the potential in those areas considered important.

The panel's report, *Solar Energy as a National Energy Resource*, was published in December 1972. The discussion about the production of terrestrial electricity utilizing photovoltaics was described only on 13 pages of the 85-page report.

The recommendations of the utilization of PV for terrestrial use did not consider the initiation of a market survey to find out what applications would be available for the commercial production of PV cells to build up the business, but was apparently based only on the experience and background of the members. As a result of this, the recommendations suggested conducting research to achieve price reduction of the solar cells used for space applications and, after that was achieved, the utilization of PV for terrestrial applications for "buildings, central ground stations, and central space stations."

---

[27]Ralph EL (May 1971) *Solar Energy Society Conference.*

[28]Böer KW (1972) *Future Large Scale Terrestrial Use of Solar Energy*, 25th Power Sources Symposium, 145.

[29]*NSF/NASA Solar Energy Panel*, PB-221-659: NSF/RA/N-73-001 (http://ntrs.nasa.gov/archive/nasa/casi.ntrs.nasa.gov/19730018091_1973018091.pdf).

The distributed nature of PV was realized, making it useful for buildings especially if it was combined with a solar thermal system. The limitation was mentioned that an electricity storage system would be needed.

The space orientation of the panel is evident in that the report used 5 of the 13 pages of the "Photovoltaic Solar Energy Conversion Systems" to describe a space central station (Fig. 3.1), because "a satellite in synchronous orbit around the earth's equator receives solar energy for 24 hours a day (which is) six to ten times the amount of solar energy available in suitable terrestrial locations in the U.S." [Quotation from the Report].

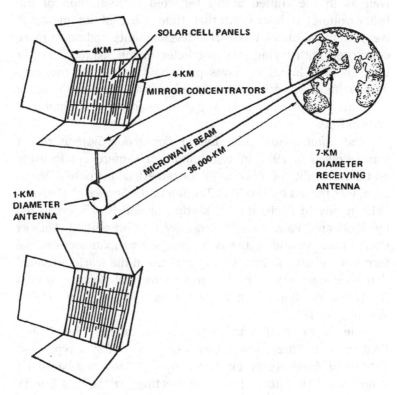

Figure 3.1   Space station concepts to produce 10,000 MW (Reproduced from the report).

The electric power generated by a space PV central station would be converted to microwave and beamed to the earth, where it would be reconverted to electricity. The length of this central space electric power station was estimated to be 8–10 miles (13–16 km), and because of its size it was nicknamed "pie in the sky." A timetable for "low cost solar array development" was estimated to be 7 years and "low cost solar array pilot plant" 14.5 years.

This report is a prime example of what Wolfgang Palz in his paper—mentioned above—in which he gave an excellent retrospective of what was the situation of the terrestrial utilization of photovoltaics in 1972/73/74: "It is interesting to note that in 1972, the "photovoltaic community" in Europe as well as in the United States reflected the situation of the (photovoltaic) industry: At that time most of its members were skeptical about terrestrial developments and were more concerned with trying to shoot holes in the "balloon" rather than giving a hand. This was probably one reason, too, why photovoltaics was critically assessed over and over again, much more than nuclear power or any other energy sources had ever been before."

The "*Photovoltaic Solar Energy Conversion Systems* report was followed in 1973 by two important conferences focusing on the possibility of utilizing PV for terrestrial purposes. One of them was hosted by UNESCO, *The Sun in the Service of Mankind*, held in July in Paris, the PV section of which was organized by Wolfgang Palz. Joseph Lindmayer representing Solarex Corporation as one of the two "prospective manufacturer" of terrestrial photovoltaic cells and systems in the world attended that meeting, while I tried to make sure that Solarex opened its doors as planned on August 1 in a location outside of Washington, DC.

The other conference was the so-called "Cherry Hill Conference": *Photovoltaic Conversion of Solar Energy for Terrestrial Applications*, organized by the California-based Jet Propulsion Laboratory (JPL). The meeting was held at Cherry Hill, NJ, on October 23–25, 1973, under the sponsorship of the "Research Applied to National Needs" (RANN) program of the NSF. It was also not planned, but this conference, by coincidence, was held at the beginning of the oil embargo and brought the

US Government's attention to the possibility of the utilization of PV to avert future energy crises. While this conference was an extremely important event directing the spotlight to the utilization of PV as a viable alternative to nuclear and fossil fuel energies, it did not consider the initiation of a market survey to find out what applications would be available for the commercial production of PV cells to build up the business. It mentioned several small applications, but the conclusion was[30]:

(a) "Estimates of commercially acceptable prices for (solar) arrays have varied from <$0.50/peak W to $0.10/peak W because the basis for comparison, the extrapolation of prices of electricity generated by other means, are uncertain."

(b) "Expected users, such as the power utilities and building construction industry."

(c) The time frame was estimated to be about 10 years.

(d) The conference conclusion was that the *government should fund and manage* the overall program.

I believe that the unfortunate effect of the Solar Energy Panel's December 1972 Report was that as it was written by very prestigious people, this conference wound up with similar recommendations, including a proposal for a "satellite solar power station."

Based on these recommendations, the US Government's activity was focused on this, the result of which is well described by T. Surek[31]: "The first decade (1973–1983) of focused PV research was indeed one of great expectations, and, as it turned out, false promises. Spurred by the oil crises in the 1970s, funding increased rapidly, mostly aimed at cost reductions in crystalline silicon technology and applications development. The goal was for PV-generated electricity to be cost competitive for utility central power applications by the end of the 1980s."

The US program considered the utilization of photovoltaics for terrestrial generation of electricity as described by Surek

---

[30]NSF RA-N-74-073, October 15, 1974 "means" quotation from the report.

[31]Surek T (2003) *Osaka Japan: World Conference on Photovoltaic Energy Conversion.* www.nrel.gov/pv/thin_film/docs/surek_osaka_talk_final_vgs. pdf.

followed the advice of experts:

(a) Finance research to make PV cell cost competitive for utility central power applications, which means to come down to the so called "grid parity"[32]; and

(b) finance demonstration projects to show that PV systems actually work.

The goal was to establish "central ground and/or space power stations" after the cost of a photovoltaic system reached "grid parity" and it was proven that it was dependable. It was believed at that point very large companies, or utilities were going to dump huge amounts of money in building factories to make PV systems, and utilities would immediately switch over to PV.

Somehow, however, we (Lindmayer and I) evaluated the terrestrial PV possibilities differently from what was described in the "Solar Energy Panel's" 1972 December report and what happened was described by Richard Swanson[33] that four, he called them "intrepid entrepreneurs," Dr. Elliot Berman, Power Corporation in 1973 with support from Exxon, Dr. Joseph Lindmayer, and Dr. Peter Varadi who founded Solarex, also in 1973 (sold to AMOCO in 1983), and Bill Yerkes who founded Solar Technology International in 1975 (sold to ARCO in 1978) did not believe in the "prevailing view" that lots of research will be needed to manufacture inexpensive conventional Si solar cells for terrestrial purposes.

It is obvious that the US "prevailing view" was vastly different from what we tried to do. To accomplish anything they believed would need about 7–14 years and we planned to produce PV cells and systems at reasonable price for terrestrial applications in less than only 1 year, and we planned that our company should break even. We totally disregarded the central power systems, instead we tried to achieve low prices by making products and selling them to applications where they were needed, at the price we were able to make them,

---

[32]"Grid parity" is the point at which renewable electricity is equal to or cheaper than grid power.

[33]Swanson RM (2011) The story of SunPower, in *Power for the World-Editor: Palz W*, Pan Stanford Publishing Pte. Ltd., Singapore, page: 532.

and increase the volume to bring prices down, not by conducting long years of research to achieve less expensive PV systems for central power stations.

These "intrepid" people, including the author of this book, came independently from each other to two conclusions, which were completely opposite to the "prevailing view."

The first conclusion was that in spite of the fact that the space and terrestrial solar cells were related, we, the "intrepids," did not accept the "prevailing view" that spending time and money on conducting research would discover how to make space cells cheap to be used for terrestrial applications. Lindmayer, who was very successful in research to improve the efficiency of solar cells used for space, knew that conducting research to make the space-oriented solar cells inexpensive for terrestrial utilization would be a hopeless endeavor, because to make inexpensive solar cells for terrestrial applications required the development of a different, new technology. Also space solar cells were manufactured in small quantities, whereas the mass production of terrestrial solar cells would need the utilization of totally different automatic machinery. Those companies, mentioned by Swanson, developed new technology in the 1970s which was successful in substantially reducing the cost of terrestrial solar cells and modules, and today, about 40 years later it is still the basis of the terrestrial solar cell manufacturing process.

Concerning the second conclusion, Lindmayer and I, and from the actions of the other two "intrepid entrepreneurs" it was evident that they felt the same way, did not believe what the "prevailing view" was proposing, that the goal was to make PV cell cost competitive primarily for residential, ground central station, and space central stations. We believed this was wrong for two reasons.

The first reason was that PV is a decentralized electric energy source, which means the system can be installed near the user or any convenient location and connected to the grid. It does not have to be centralized. W. Palz in 1981[34] made a visionary forecast: "As most of the (PV) systems will be grid connected a

---

[34]Palz W (11 May 1981) *IEEE Photovoltaic Specialist Conference*, Houston, TX, 15.

favorable buy-back rate for the surplus electricity fed by the photovoltaic generators into the grid have to be negotiated with the utilities." Based on the decentralized possibilities of PV, he departed from the centralized PV power station concept. His prophecy became reality about 20 years later with the introduction of the "Feed in Tariff" (FiT).

The second reason was based on the fact that a market for terrestrial use of PV already existed even at higher PV prices. In Japan solar cells were used for terrestrial purposes since 1961, as will be mentioned later. Furthermore, all of the space solar cell rejects were easily sold for terrestrial applications. This indicated that a market existed for the solar electric generator, which could be used in any location and needed practically no maintenance, no refueling or recharging and some of these applications seemed to be not price sensitive. We believed that the development of these markets would increase the volume of production of solar cells, which would decrease prices, which would open up another market and this again increase the volume which would reduce prices again. This meant that the reduction of solar cell prices would be achieved by increased production volume and not by continuous research.

When we started to manufacture solar cells and modules for terrestrial utilization, it was found that PV was not only a small local market, but it was a large global market and there were at least 1.6 billion people without electricity who would require decentralized access to electricity, because they would not be able to be connected to any grid in the foreseeable future. We should therefore, develop a global market for what we could make and for the price at which we could make it.

Paul Maycock, who was managing at the US Department of Energy (DOE) a $900 million applied research and "commercialization" budget under President Carter's program, said in 1980 that "Residential systems (PV) are being sold all over the world right now. Solar cells are already economically viable for off-the-grid, isolated houses, and villages."[35] Paul was right, the terrestrial solar electric business started to fulfill a need in

---

[35]Williams N (2005) *Chasing the Sun*, New Society Publishers, Gabriola Island, BC, Canada.

the world from the North to the South poles to provide electricity, where the electricity from the centralized power stations was not reaching and probably would never reach.

Joseph and I had another fundamental disagreement with the Cherry Hill Conference conclusion, namely that the "government should fund and manage the overall [PV] program."[36] This is perhaps needed for some military-oriented projects, but we both escaped from the communist Hungary and had personal experience about results of the system where commercial projects are funded and managed by the government, where the decisions come somewhere from the top. Not long ago I even wrote an entertaining book about the Hungarian communist government funded and managed projects and systems.[37] We believed in private enterprise, where a person makes the decision. If the decision is good, he will succeed, if not he will go bankrupt. I even went further. I believed that our solar business should rely on many customers of which the government could be one. By that time I knew quite well that relying entirely on government business is dangerous because the direction of government spending can make a 180 degree turn. In the PV area this actually happened in 1981. I summarized this in a very simple manner: If a company has 100 customers and loose one, that is only 1% of the business, but if the company has only the government as its single or majority customer and loses it, the company loses 100% of its business, that means it will become bankrupt.

The conclusion was that if we, the "intrepid entrepreneurs," decided that we were not putting our bets on the government funding to develop the "central PV power station" concept, but betting to develop a totally new technology and sell the PV products. With this ideology we were alone. We also realized that to develop a new technology was only half of the success. The other half was to develop the non-existent terrestrial market where the products could be profitably sold.

The *European* PV research was at that time also space oriented. The European Union's PV program from 1977 under

---

[36]NSF RA-N-74-073, October 15, 1974, 8.

[37]Varadi P (2012) *Comrades, Mistakes Were Made*, Amazon, Charleston SC.

the leadership of Wolfgang Palz believed in trying to reduce cost by research, but he also supported programs to show that PV prices would be drastically reduced if mass production could be achieved. In this respect it is interesting that he realized the issue—as discussed above—that in the future there will perhaps be no need to utilize centralized Utility systems, for one can connect PV into the Utility's grid directly from any location.

Japan was not involved in space programs, but research on PV still existed. Therefore, Japan was the first to use PV systems for terrestrial applications. The Japanese company NEC in 1961 installed the first solar PV system to power a lighthouse. After NEC installed three more lighthouses the Sharp Company of Japan started to install solar-powered lighthouses in 1966. Since then Sharp installed 1500 more solar-powered lighthouses and also produced solar cells to power communication systems. The Japanese "Sun Shine Project" for terrestrial application of PV was initiated in 1974 where the central driving force was Prof. Yoshihiro Hamakawa. As an interesting side effect to the Japanese PV research not being space-oriented, as will be described, one of the applications of PV was for consumer products, including calculators.

This was the beginning of the terrestrial PV age, the "First Part," which lasted little more than a decade.

Chapter 4

The Dawn of the Terrestrial PV—Solarex: The Beginning and …

The major difficulty in accomplishing the opening of Solarex was that we had problems renting a small space of 3500 sq. ft. (325 m²). The problem was that landlords did not want to rent space to a company that was just starting and there was no assurance that it would be in business a month later. I decided that if a bank would give us a line of credit, the landlords would take our company more seriously. Therefore, I started to obtain a line of credit for a company, the address of which was Dr. Lindmayer's home address, had no assets, had at that time no employees, and had only cash. The first bank I tried was the bank in which our lawyer was the Chairman of the Board. They turned us down, as the Chairman probably did not believe that a company involved in solar energy would survive. Our only trump card was that we had at that point a sizable amount of cash from the stocks we sold. I realized that banks prefer to give loans to people who have money and do not need the line of credit, which the bank for various reasons could terminate at any time. After several banks would not even consider our request for a line of credit, I visited the First American Bank in Silver

Sun above the Horizon: Meteoric Rise of the Solar Industry
Peter F. Varadi
Copyright © 2014 Peter F. Varadi
ISBN 978-981-4463-80-5 (Hardback), 978-981-4613-29-3 (Paperback), 978-981-4463-81-2 (eBook)
www.panstanford.com

Spring, MD, and met its president Bob Burke. He was interested in the potential of solar energy and said he would provide us a $100,000 line of credit, but Joseph and I had to guarantee it. I finally talked him out of that idea and walked out with the line of credit. The line of credit from a sizable local bank impressed the landlord and he rented the space for Solarex.

On August 1, 1973, as planned, Solarex opened its doors at 1335 Piccard Dr., Rockville, MD. The group who was at the opening of Solarex was Joseph, myself, Joe, who was my technician at COMSAT, but quit, because he wanted to spend some time to meditate what he should do with his life. I called him and told him to take a temporary job at our new company and when he thought he wanted to continue his meditation, he could quit. After 38 years, Joe was still doing research until BP Solar, which was the continuation of Solarex, closed the facility in 2011. There was also Barbara, our secretary, who worked at COMSAT before, but left because of family reasons. She called me and said she was looking for a job. I told her, "You are hired." We also had several other people working there, but they were workers finishing the construction around us according to the plans I had submitted to our landlord.

In the next 2 months we hired a few more people, technicians, and factory workers, the secondhand equipment, and furniture we purchased was delivered, assembled and put in operation, and on October 5, 1973, a Friday, we started to produce the first terrestrial solar cells. That was the day before Syria and Egypt launched a surprise attack on Israel and about 2 weeks before the famous "oil embargo" was initiated. If you followed Solarex's plans outlined in January 1973, you could see that we adhered to it to the day. But it would be a lie to say the reason our plans were developed was so we could be ready to produce solar cells about 2 weeks before the famous oil embargo started. It just happened that way, with the result that the utilization of renewable energy was recognized, and the spelling of the word PHOTOVOLTAICS became known to many people.

The result of the oil embargo was that Solarex did not have to spend money for advertising as our plant was in a suburb of Washington, DC, about 25 miles from the White House. We received many telephone calls and several reporters and TV crews visited us, so Solarex received a lot of publicity. One of the local TV station's crew came out to film us for the evening news and saw one of our secretaries typing on her typewriter that was powered by one of our few PV modules. While the TV crew set up their equipment, our secretary ran home to put on her best dress. She was filmed for at least 20 minutes, typing on the solar-powered typewriter. After work, she went home and collected her friends and family in front of the TV to see the evening news. In the 1-minute segment the announcer said that what could be seen was a solar-powered typewriter being used at the offices of Solarex, a local company recently formed to commercialize PV for terrestrial use, which, because of the oil embargo, was destined to have a successful future. The TV segment showed the typewriter and our secretary's fingers, but no other part of her. The next day when she came to work she was very upset.

The marketing and selling of PV modules turned out to be much easier than I had expected. The reason was that many locations existed in the USA which required electric power to operate lights or radio equipment for communication purposes, but had to be powered by a battery as no other electric power source was available. These locations were either in the Gulf of Mexico, operated by the various oil companies, or others used by the US Bureau of Land Management, Forest Service, and the National Weather Service[38] in the continental USA and Alaska. In the mountainous part of the USA, state police also used radio repeaters to provide communication between the police cars and their base station. These were also in places having no electric connection. The problem of all of these organizations was that the batteries powering this equipment had to be periodically replaced, which was a very expensive procedure. Some of the equipment used thermoelectric generators, but they had to be periodically maintained and filled up with fuel.

Therefore, the possibility of using PV to power that equipment would have been a great benefit for these organizations. The two space solar cell manufacturers, Spectrolab and Centralab, utilizing the rejects from their space cell production, supplied

---

[38]A part of NOAA—National Oceanic and Atmospheric Administration.

some PV modules to these organizations, but they could not sell more because the space solar cells were prohibitively expensive. Therefore, when Solarex and Solar Power Corporation (SPC) started to manufacture relatively low cost PV modules for terrestrial use, these markets were readily available.

The solar cells produced for use in space were small solar cells; the biggest were 2 cm × 2 cm (4 cm^2) or 2 cm × 4 cm (8 cm^2). By contrast, the cells initially produced by Solarex or SPC were 10–40 times bigger, producing that much more electricity. For this reason, neither of the space solar cell companies could compete.

Elliot Berman, who headed SPC, was very much interested in developing solar business for the oil platforms in the Gulf of Mexico and worked with Tideland Signal Corporation in Houston, TX.[39] Solarex, on the other hand, marketed its products to the US Bureau of Land Management, Forest Service, and the National Weather Service in the continental USA and Alaska, and also to the state police in the mountainous states. For this reason, we were able to sell our production without much competition. The advantage of having our facility just outside Washington, DC was that all of these government agencies had their headquarters located there, and we were able to visit them very easily and tell them about the advantages of utilizing PV systems to recharge batteries in remote locations.

I had never sold anything in my life, but I found that selling PV was actually pleasant. That was for two reasons. This was after the oil embargo and I never went to a single place to sell solar modules where I wasn't enthusiastically received. How many salesmen selling, for example, a vacuum cleaner or a refrigerator can say that? The other reason was that selling new things always faces resistance. As for PV, engineers were at that time among its primary customers deciding what to use, and engineers are understandably extremely conservative by nature. Convincing them to try PV was a challenge. Until that time I was a scientist buying equipment for my projects, and I

---

[39]The Environmental Protection Agency in 1978 outlawed the dumping of batteries in ocean waters. This accelerated the use of PV-powered systems on oil platforms in the Gulf of Mexico.

knew that for an engineer it was a serious responsibility to buy something, especially a new product, because they were putting their reputations at risk. Therefore, I had to figure out how I could reduce their risk and make their project a success.

An example was the Arizona state police, which was one of the first to consider using solar modules to charge the batteries of their repeater station on the top of a mountain located in the Hualapai mountains near Kingman, AZ, about 160 miles northwest of Phoenix, AZ. This repeater station made it possible for the state troopers to communicate in that remote area. Obviously they had problems replacing batteries, as the location could be reached only by helicopter. The Arizona State Police's brave engineer stuck his neck out and was willing to try our solar system for that location. He told me that he would install safety telemetry circuits to monitor the solar system. I encouraged him and told him if he had any problem with the solar system, I promised him that the next day he would receive a replacement. He went ahead and installed our PV system. A few months later he called me and told me that the repeater station had stopped working, and his telemetry unit indicated some problem with the solar module. I told him that he would receive the replacement the next day as I had promised.

I got a call from him the next day and he started to apologize. He said that he received the replacement PV module and went by helicopter to replace the module. He said that the problem they found had nothing to do with the PV module. The contractor installing the entire repeater station had placed the lightning arrester's cable improperly and a lightning bolt hit the station and melted the antenna. That was the reason the station went off the air. He said that because of the cable, the lightning also jumped across the PV module, and burned a hole in it when it went through the PV module to the ground. He said that in spite of the hole, the PV module was working perfectly. I asked him to replace it with the one I had sent him, and send me back the one with the hole in it. I had never have seen anything like that and it would be a good demonstration to customers that our PV module could even withstand a hit from lightning. After they fixed the antenna and properly grounded the system, they had no more problems; he disconnected the special telemetry system and started to buy more PV systems for other locations.

About a year later, I was in Phoenix, AZ and visited the engineer at the state police. He insisted that I go with him in a helicopter to visit the site of the first PV system he had bought, and had the imaginary problem with. I readily agreed as I had never been in a helicopter before. We got into a little police helicopter with two pilots and the two of us. In spite of the ear-plugs we used, this was the noisiest place I had ever been. Either to impress me or just for fun, the pilots flew under and above bridges, and came down over roads almost to the ground. Finally we got to the mountains. This must have been a part of the Mohave Desert, because all the mountains were reddish brown without any vegetation. The pilot was now trying to find the repeater station on top of one of the mountains. It was not easy. He passed by several mountain until he found the one with the repeater. His problem was that the entire repeater structure was painted exactly the same color as the bare mountain. After we landed and inspected the station with our solar system, the engineer explained that they had to paint all of their repeaters and weather equipment the same color as the surroundings, because there were a lot of hunters in the mountains who would use the repeaters for target practice for fun. As the stations are practically invisible, they melt into the mountains and now they have no problems with the hunters. On the way back the pilots did not try to scare me. We only went down practically to the roof of a house. The pilots had some friends there. People from the house came out because of the noise and the pilots waved to them and dropped a small package.

Lindmayer was able to obtain two R&D contracts from the National Science Foundation and one from the Jet Propulsion Laboratory. Because of these contracts we hired Karl Margod, who was a scientist and had worked for Joseph before on research projects. We also added more technicians to our staff.

By the end of 1973, we could predict that by early 1974 or at the latest in April—9 months after Solarex opened its doors—

the company would be profitable and would be able to support itself without additional capital.

With this forecast I told Joseph, that Bill Yerkes's idea to build a PV company that would cover as its business area both space and terrestrial fields would make a lot of sense. I also told him that Centralab, located in a suburb of Los Angeles and which was the older of the two space cell manufacturers (the other being Spectrolab) was a division of a company called "Globe Union" and I had indications that Globe Union would be glad to sell it; so, Solarex would have an opportunity to buy it. Joseph knew the company very well and fully agreed with my idea. This information was confirmed, when a few days later our lawyer received a telephone call from the COMSAT General Counsel's Office advising and suggesting that there might be a corporate opportunity for Solarex to acquire Centralab. COMSAT knew about this situation, because Centralab took a license from COMSAT to be able to produce the "Violet solar cell," invented by Joseph and therefore they were in contact with each other.

After we came to the conclusion that this idea was worth exploring, I called the management of Centralab, whom I knew quite well. I asked them about the situation and they expressed their opinion that the merger of Solarex and Centralab would be an excellent idea. I got the names of the people at Globe Union, whom I should call. I called Globe Union and involved in discussions with them as well as the people working at Centralab. I visited Centralab and these discussions were fruitful. As a matter of fact, Globe Union wanted to transfer Centralab to us as soon as possible and asked for little money to close the deal. The Centralab people were very enthusiastic about the merger with Solarex, as the space business was dependent on a few large orders while the orders for terrestrial systems, if usually smaller, were steadier. We were able to complete the basis of a proposal to be sent by Solarex to Globe Union. In a memorandum dated February 20, 1974, I made a detailed review of the merger and presented it to the Solarex Board. The essence was that Solarex West (Centralab) would continue with the space solar cell business and much unused equipment would be transferred to Solarex East, which we could use to enlarge our terrestrial cell production. Discussing the matter with our Board we came to the conclusion that with a minimum amount of additional investment which was available, we could go ahead with the deal the way I described it in my memorandum of February 20.

Close to the end of March, the deal with Globe Union to acquire Centralab was worked out and it was expected, that it will be completed on April 1 or closely thereafter. This became quite firm and it got to the point, that when the Centralab management had to make a decision they called me to get my opinion about it.

We had several orders in house and the National Weather Service tested the solar module that we designed for them and informed us, that they were in the process of giving us an extremely large order for that system. By the end of March 1974 we were profitable and, with the acquisition of Centralab, ready to multiply Solarex's size. We were very happy to inform our stockholders about our success in a letter.

Chapter 5

... Almost a Premature End

On Friday, March 29, 1974, I was in my office reviewing papers in connection with our merger, when Barbara came in and told me that a gentleman is here and would like to talk to me as he was bringing some papers, which he wanted to give me. A tall gentleman in a dark blue suit came in, greeted me, and when he got to my desk, handed me some papers. I looked at the papers, which were some kind of a legal document. He then handed me another paper and asked me to sign it, acknowledging that I had received the papers he gave me.

I asked him to sit down, looked at the papers, and told him that I was going to call our lawyer Stanley Frosh and ask him what to do. I dialed Stanley's number and got him on the telephone. I told him that I had been served with some papers, which by looking at them looked to me that they were related to some lawsuit which COMSAT had filed against us at the United States District Court for the District of Maryland, on March 22.[40] The gentleman who brought it had asked me to sign a paper and I told him that I would call our lawyer to ask him what to do.

---

[40]*National Archives in Pennsylvania*, Case No. K-74-cv-291.

Sun above the Horizon: Meteoric Rise of the Solar Industry
Peter F. Varadi
Copyright © 2014 Peter F. Varadi
ISBN 978-981-4463-80-5 (Hardback), 978-981-4613-29-3 (Paperback), 978-981-4463-81-2 (eBook)
www.panstanford.com

Stanley sounded surprised and asked me to look at the papers again, as he could not believe that this would be a lawsuit from COMSAT. I picked up the papers, looked at them again, and repeated what I told him before. He told me to sign the paper and said that he would be in our offices early that afternoon.

I signed the paper and the gentleman left.

In the early afternoon Stanley arrived and the three of us, Stanley, Joseph, and I, sat in our "conference room," which had several folding chairs and two folding tables pushed together making one "conference table."

Stanley reviewed the papers and told us that COMSAT was suing each of us, namely Solarex, Joseph, and I, for one million dollar for taking COMSAT's proprietary information and using it in Solarex's production of solar cells.

We told Stanley that we had not taken any papers or anything else from COMSAT and that we were using a new technology for our terrestrial solar cell and module production, because the space cell and module manufacturing process would have been prohibitively expensive. We assured him that we had meticulously made sure that we had not taken any documentation from COMSAT and we had nothing from COMSAT on the premises or even at home.

I told Stanley that business-wise we were doing reasonably well. We had orders and we were expecting a very large order from the National Weather Service, and Joseph was expecting research contracts from several government organizations. Furthermore, Stanley knew as being one of our board members that we were in the final days of completing our acquisition of Centralab. Obviously, COMSAT would probably issue a news release about the lawsuit and notify all of our customers and would-be customers. Because of the lawsuit most of our orders would stop until the lawsuit was settled and we were not going to survive. For this reason, time was very important. We had to settle this idiotic lawsuit as soon as possible.

Our discussion continued late into the night. We had no dinner as both Joseph and I had lost our appetite. The end result was the following. Stanley would answer the complaint, saying that it was ridiculous and we did not use any of COMSAT's proprietary know-how. We were going to ask to start the so-called "discovery" process immediately, which probably could be done, because Stanley knew COMSAT's law firm, which was one of Washington's largest intellectual property law firms located a few blocks from Stanley's office on K Street.

We also discussed why COMSAT had selected this time to bring a lawsuit against Solarex. We all knew that a few days after we opened Solarex's doors, in August of the past year, Joseph and our former boss at COMSAT, Dr. Ed Rittner, had called important persons at NASA, the "Air Force and National Science Foundation," from which organizations we expected to get research contracts, and told them that Solarex "had legal restrictions on the ability to engage in activities in the solar cell field." When we learned about this and notified COMSAT's legal department, Mr. William H. Berman, Associate General Counsel and Assistant Vice President, called the same people and told them—and to confirm it sent them a letter—stating that Rittner's call was a "misunderstanding," because "there was no termination agreement which prohibited us (Lindmayer and Varadi) or Solarex, Inc. from participating in activities related to any application, including space application of solar cells." In the intervening 8 months, we had absolutely no more communication with COMSAT, except in connection with Centralab, and suddenly this lawsuit materialized for reasons not known to us.

The matter probably started, because COMSAT licensed Globe Union's Centralab division to produce the Violet solar cells and COMSAT would provide consultation in connection with the Violet solar cell technology. It seemed that COMSAT was not able to fulfill this requirement and Centralab called Lindmayer, who was the inventor of the technology, to perform consultation. Lindmayer advised COMSAT that he would be glad to help, but all the technical information was left with COMSAT and therefore COMSAT and Solarex had to get an agreement about this. While this went on, Globe Union in late 1973 determined that they wanted to sell Centralab. Solarex's negotiation with Globe Union was with full knowledge of COMSAT. Our assumption was that some people at COMSAT initiated this lawsuit just in time to achieve that Solarex should not be able to acquire Centralab and in the process, as a consequence of a prolonged lawsuit, ruin Solarex. Our assumption was supported by the fact, as mentioned before, that shortly after we opened Solarex, a person from COMSAT tried to interfere in such a way that Solarex should not be able to get research contracts. The Director of COMSAT laboratory personally called Robert Strauss, a department head at COMSAT lab, who was never involved in solar cell research, but being a friend of Joseph and I, invested in Solarex, and warned him, that he was being closely watched.[41] From the timing of the lawsuit, another possibility was that perhaps COMSAT wanted to acquire Centralab and in the process eliminate Solarex as a competitor.

After thinking about this, Stanley brought up the fact that COMSAT was an interesting company, because it was a public firm, with its stock listed on the New York Stock Exchange, but was incorporated by the US Government. He believed three of its Board members were delegated by the government and therefore the US Government was somewhat implicated in COMSAT. His opinion was that because COMSAT's action would probably annul the planned acquisition of Centralab by Solarex, which acquisition was in its final stages, COMSAT's action could be considered to be "relating to interference with commerce," which fell under the Antitrust laws of the United States,[42] therefore, we could bring a civil countersuit under the Antitrust laws. He proposed to sue COMSAT for 25 million dollars and damages of 10 million dollars, which under the Antitrust laws would triple.

Stanley told us that he was going to prepare all the necessary papers, denying COMSAT's allegations and filing the Antitrust lawsuit as a counterclaim. Joseph and I knew nothing about this sort of business as we were never involved in any lawsuit, so we told him to do what he as our lawyer and board member believed was the best for Solarex. Stanley also helped us to write a letter to the Solarex board and to Solarex's stockholders informing them about this enormous problem, which in spite of the fact that we had not, nor were, using any of the solar

---

[41]Information from Mr. R. Strauss.

[42]Clayton Act (October 15, 1914) and Sherman Act (July 2, 1890).

cell "trade secrets of COMSAT," could result in Solarex's running out of money and having to close.

Next day I called our board members, among them Al Nerkin, who after Joseph and I was the largest Solarex stock holder. I told him what had happened and that we had made some profit in the last month and had enough orders in house to be able to keep Solarex solvent, but because quite a bit of Joseph's and my time would be spent on the lawsuit, we probably would run into financial difficulties in June or, at the latest, in July. I told him that Stanley had mentioned that he would pay for everything in connection with the lawsuit and that he would not charge for his time or his associate's time, but expected to be compensated at the end with Solarex stock.

Al asked me if Joseph or I had taken any material from COMSAT or used any of COMSAT's trade secrets. I told him:

"No."

"I take your word, I believe you. I am ready to invest another $50,000 in Solarex, but as you are now profitable, I am willing to pay three times as much per share as I paid when you started almost a year ago. Furthermore, I authorize you to tell COMSAT's lawyers that I have more money to spend on this lawsuit, which you are going to win, than COMSAT's board can spend on such a frivolous lawsuit. Even if they win, they can collect only a fraction of what they are going to spend on this lawsuit, not talking about how much they will spend if they happen to lose due to the Antitrust lawsuit."

I thanked him and was very encouraged because with Al Nerkin's generous investment, I could see that we might be able to finish the lawsuit and have money to continue.

Stanley completed all of the required papers in which Solarex denied the allegations in COMSAT's complaint and COMSAT was served with our Antitrust countersuit on April 8, 1974, filed in the United States District Court for the District of Maryland.

When we informed Globe Union, they terminated the deal that Solarex would acquire Centralab. Also several orders, including the large order we expected from the National Weather Service, were put on hold and also research contracts that Joseph expected were delayed pending our lawsuit. We expected all of this, but we were continuing to manufacture present orders and Joseph and his group worked on our existing research contracts, which had not been cancelled.

Joseph was very depressed, as his credibility and character were implicated in this lawsuit and he knew that the lawsuit was groundless. I knew it too. I just could not understand what had triggered such an action at COMSAT. They could have called us, visited us, and we could have showed them that what we were doing and had nothing to do with any of their alleged "trade secrets."

We had almost daily meetings with Stanley. COMSAT used an outside law firm on K Street to handle the lawsuit, as COMSAT lawyers were not trained to conduct court cases. Stanley had several meetings with them to establish what the procedure should be.

As we learned from Stanley, the first event in a lawsuit is the "discovery" stage, which is a structured exchange of evidence, and includes depositions from persons each party wants to question and also to review documents they want to see.

Stanley told us that Mr. Fred Okash, the lawyer who was in charge of the lawsuit at the K Street law firm, would like to arrange a visit by himself and another lawyer from his law firm, and a lawyer from COMSAT, to Solarex's premises to review our documents and also our production of solar cells. Obviously, they would sign a document promising that such information would not be transferred to anybody else. They also asked for certain documents and if we did not have them, a statement that we did not have them.

We told Stanley that we saw no problems letting them visit Solarex and review whatever they wanted. They should come anytime, because that might finish the entire lawsuit, which was in Solarex's interest, because in a prolonged lawsuit Solarex would not survive. Stanley disagreed and told us that he was dead set against letting them in to look around. We spent some time on this, but could not convince him. As I learned later, lawyers do not believe anything that their clients tell them, because they probably lie. We did not lie, but I do not know if Stanley believed us at all. I think he believed that with the solar energy business we most likely would fail anyhow, but he was sure that as the US Government was somewhat involved in COMSAT, he could win or settle the Antitrust lawsuit and he would make more money from that than with his investment in Solarex.

We had several meetings with Mr. Okash and other lawyers in the K Street law firm's very lavish conference room, but we got nowhere with the discussions between Stanley and them. Joseph was in a very bad shape and many times did not come to the meetings. I was always there and during recesses and after the meetings, I tried to convince Stanley to give them what they wanted, but he told me he knew how to deal with this sort of lawsuit. So, he would not let them onto our premises and he did not give them any documents. With this wrangling, the clock was ticking and April was gone. The depositions started in May.

First they wanted the deposition of our secretary, Barbara. They had no luck with her. They advised her that she was under oath and if she was not telling the truth, it would be a criminal offense. She said that she understood it. When asked, she said that she had been handling the secretarial work from the beginning and that we had a total of one filing cabinet, which was next to her desk. When she started to work, she found a few documents in one of the drawers, which were still there, and the rest of the papers were in two drawers, which were not even half full, and she was the one who put them there.

On the question if she knew whether we had other papers, she said that the researchers had their note books and the production people had sheets about the daily output. The note books were kept by the research guys and the production papers were given to her daily and they were in the filing cabinet. Again when asked, she told them that we had not removed any papers from the filing cabinet. On the question if she knew whether we were using any COMSAT papers, she said she had not seen any papers from COMSAT. Obviously, Stanley had no questions and Barbara was told she could go.

The next deposition was scheduled for the beginning of June. COMSAT and their lawyers were not in a great hurry. We believed that they were trying to drag the matter so we would run out of money and fold. Next they wanted the deposition of Karl Margod, who was hired the previous December from COMSAT where he had worked on research projects for Joseph. Joseph believed he was a very good person for the research projects Joseph was working on and hired him. At Solarex, he had a nice office and spent practically all of his time there, diligently working.

In the meantime we engaged Mr. Walter Ames, another lawyer, whose specialty was litigation in patents, trademarks, and other intellectual property. He was recommended by Robert Strauss, one of our stockholders, because Stanley had no experience in these fields. Walter was a senior partner in a very prestigious law firm, also on K Street, that specialized in these fields. I believed that on Washington's K Street the offices were occupied either by law firms or by firms specializing in lobbying. Walter was an excellent trial lawyer he was a likeable sort, a short, balding guy with a booming voice, which must have been very effective during a trial.

Karl's deposition was scheduled for June 3, a Wednesday, and that would have been the last meeting for 2 weeks, because Stanley had his vacation planned for a long time and could not change it. He and his wife were going to Brazil for 2 weeks. When the day of Karl's deposition came, I was there. Joseph did not come because he believed that Karl was going to say that he knew little as he was not working on solar cells, but on other types of research.

At the beginning of this session, Stanley, who came with us, introduced Walter and said that Walter was joining our team for this lawsuit, since his specialty was intellectual property, which was what the lawsuit was about. He said that this was the last meeting before he would be back from his vacation. It seemed that Okash and the other lawyers knew Walter.

Karl was seated at the head of the conference table next to him; on his right was a stenographer lady, with her machine. Next to her was Walter, and I was sitting next to Walter. On the other side of the table, three lawyers from the K Street law firm and one from COMSAT were sitting. Okash was the second, but he was conducting the deposition. Next to Okash was a tall elderly man, with not much hair and blue eyes. He was probably one of the major partners in the K Street law firm, because Okash seemed to turn to him once in a while for some advice.

Okash started, as with Barbara, reminding Karl that he was giving his deposition under oath and told him the consequences of not giving true answers. Karl assured him that he knew about the system.

Fred Okash was very good. It was like in the movies. If it would have not been such a grave situation for us, I obviously would have enjoyed it. Fred started to build up Karl's image that he knew about solar cells and was an expert in that field. Therefore, his testimony would be extremely important.

He started with Karl's schooling, his jobs, what he did at COMSAT, how long he was working with solar cells, and his knowledge about COMSAT's trade secrets. Finally, his questions focused on his present job at Solarex.

Okash leaned closer to the person next to him, put his hand in front of his mouth, but we could hear him asking:

"Shall I do it now?"

The elderly man shook his head, meaning "No."

Okash then asked Karl if he had access to Solarex's files, to the production records, and to the research lab work. I was surprised when he answered:

"Yes, I had access to everything. At Solarex we have only one filing cabinet and I never had or have any restriction to go anywhere on the premises. I became good friends with the technicians and workers. During breaks and evenings we met and discussed solar cells and many other things."

"Based on your answers, you showed that you are an expert, who is knowledgeable about solar cells, about COMSAT's trade secrets and also what Solarex is doing."

"Thank you for your opinion about my knowledge."

Again Okash leaned closer to the person next to him and put his hand in front of his mouth, but we could hear asking him again:

"Shall I do it now?"

The elderly man had moved his head down twice, indicating "Yes."

"Is it true that you had many telephone conversations with your friends and colleagues at COMSAT, and on one occasion,

when I was also present, said that Solarex was using COMSAT
trade secrets and know-how in the production of the solar cells
they made?"

This was strange. To put this question so bluntly, I realized
that Karl was a key witness and Okash knew the answer to his
question.

"Yes, it is true."

"Based on your knowledge, please state for the record: Is
Solarex using any trade secrets or know-how that belongs to
COMSAT?"

At this point Okash looked straight at me. Walter and I
looked at Karl, who answered:

"I can tell you that Solarex is not using and has not ever used
any of COMSAT's trade secrets or know-how."

At this point, I looked at both of them. Karl was looking
at Okash and Okash suddenly was not looking at me but was
looking at Karl. The difference was that the blood left Okash's
face, which turned white, and the elderly man, who seemed
to be the boss, put his hands on the table in front of him, then
moved his hand to his lips and his eyes seemed not to be
focused anywhere.

Okash, his face white as paper, leaned forward and looked
straight into Karl's face:

"You are telling me that Solarex is not using and has never
used any of COMSAT's trade secrets or know-how?"

"Yes, that is what I told you."

"That is just the opposite of what you told us."

"It is true that I had lots of discussions with friends and
colleagues at COMSAT on telephone or after hours. At that time,
I was under the impression that Dr. Lindmayer must have been
using COMSAT's trade secrets in developing the terrestrial solar
cell, because Solarex came out with its first terrestrial solar
cells extremely quickly. At Solarex I am not involved in solar
cell research or production, my work is writing proposals or
working on existing contracts therefore, I made the assumption
that Dr. Lindmayer used some existing technology. But I
realized that when I was questioned at this deposition, I
would be under oath and would have the consequences you so
clearly explained to me. Therefore, for the last week I wanted

to ascertain for myself that what I was going to testify is the absolute truth and not an assumption. Therefore, I reviewed everything at Solarex. I went through the filing cabinet, through the note books, talked with everybody, and I was surprised but came to the conclusion that during this short time Dr. Lindmayer developed a technology to make solar cells which does not use any of COMSAT's know-how. Therefore, what I am stating here and now under oath that Solarex is not using and has never been using any of COMSAT's trade secrets or know-how, and this statement is based on the truth."

Okash's complexion did not get better. He looked at his boss, who did not say anything. His hands were still on his lips and his eyes seemed not to focus anywhere. He declared a recess so that everybody could get coffee and said the deposition would continue in 15 minutes.

I got up, went to the toilet and Okash walked in a minute later.

When Karl finally looked into what our process was, he realized that what Joseph did was to make inexpensive solar cells. We could not use the "high-tech," "clean room," "batch processing" technology developed for space solar cells. Therefore, he started to use a new process, the "atmospherization" of the terrestrial solar cell production. "Atmospherization" means that for terrestrial solar cells an inexpensive production technology, where all the steps are performed under atmospheric pressure,[43] had to be used.

"Fred, I know you realize after Karl's testimony that your lawsuit is absolutely kaput but we have the Antitrust pending and Stanley said that because three directors of COMSAT are from the government, and also the President of COMSAT, he will have a stellar group to ask questions under oath at the deposition about COMSAT and this lawsuit."

At this point, Okash washed his face with water and I left.

I sat down next to Walter, who said that this was an interesting turn of events from which neither Okash nor his boss had yet recovered. After Okash returned, Walter said that he had no questions. Okash closed the deposition and we all left.

---

[43]Varadi PF (August 1993) *The Photovoltaic Industry: Past, Present, and Future*, Solar Energy Society Meeting, Budapest.

Walter and I went to Stanley's office and we discussed the fact that Karl was the key witness and they had no cards left. My question was what should we do now? Stanley said we would continue when he returned from Brazil. I wished him a good trip. He gave me his whereabouts so I could reach him if something happened.

On the K Street, Walter told me that in his opinion Okash and his boss got an unexpected surprise from which they would not be able to recover. I went back to Solarex, went to Joseph's office and told him what happened. He was very pleased and it was obvious to see that he had started to recover. We discussed what to do with Karl and decided that at that time we would do nothing. It was now evident, that COMSAT foolishly started the lawsuit based on the information that they did not verify. It was an enormous blunder and their lawsuit is actually kaput. On the other hand, our Antitrust lawsuit was more alive than before, as they even used an unfounded lawsuit to try to stop Solarex's acquisition of Centralab and also put Solarex out of business.

Now we faced an incredibly large decision. Continue the lawsuits, the one they started they obviously lost, our counterclaim, however, looked quite promising, because what they did was an enormous blunder, bringing a lawsuit based on hearsay and having good connection with Solarex they did not even attempt to research it to find out if they have a case. If this would be reported in the press, the government, being involved, would not let it go to trial; they would have to settle it out of court. We and our stockholders could make quite a bit of money, especially if they would drag it out and Solarex would have to close. The other side of the coin was that if we offered them to settle, by dismissing both lawsuits, they would with great probability accept it immediately and this idiotic and for Solarex eventually detrimental business would be closed.

The choice to decide became very simple: Should we continue with the lawsuits and we eventually wind up with sizeable amount of money, or settle the matter immediately and continue Solarex. After short discussion, we decided that we started

Solarex to develop the terrestrial photovoltaic business and to be very successful in it and we agreed that was what we were going to do.

I told Joseph that I was to call Okash on the next day and invite him come to visit us and look at what he wanted. Joseph agreed.

On the next day, Thursday morning, I called Fred Okash and told him that we should settle the matter. He said that they have to come to Solarex to inspect everything. I told him they were welcome; they could come immediately or as soon as they could. He told me that he could not talk to me unless my lawyer would give him permission. I told him that, as he knew, Stanley was in Brazil. How on earth could I get permission from him? He said he had to get my lawyer's permission to talk with me.

I told him I was going to try, but in case I could not get hold of him, we were going to fire him as our lawyer and then Fred could talk to me.

I tried to contact Stanley, but was unable. Finally I was able to reach him on the telephone. In those days the overseas telephone lines were not very good, especially to call Brazil. Our connection was OK but not very good. I told Stanley, that until he returned, we were here in isolation and some issues may come up and for this I would need a "Telex"[44] from him that I should be able to talk with Okash. He said that would be OK. I gave him our Telex number and he said that he was going to send me a Telex immediately authorizing me to talk with Okash if it was needed.

The Telex arrived that evening and on the next day I called Okash and told him that I had Stanley's authorization that he could talk with me. He said he had to see when people were free and he would call me back.

Late that afternoon he called me back and suggested that he and two other lawyers, including one from COMSAT, could visit us on June 14 at 2 p.m. I told him that was OK for us, and

---

[44]In those days the Telex machine was the only communication means, as no fax or e-mail existed.

in spite of the fact that that day was a Saturday, I was going to ask our secretary, Karl, and all of the technicians to be there, so they could conduct whatever review, search, or discussion they desired, because I wanted to finish the lawsuit, and if he wanted anything else, he should tell me as I wanted to avoid any delay in their finishing, their investigation, and dismissing the lawsuit. I also told him that damages to Solarex caused by any delay they would cause would obviously be recorded.

He said that if they needed anything else he would let me know. From his voice, I heard that from my offer to produce everything and everybody he understood that COMSAT's game was lost and he would not find anything, as Karl testified, and that we clearly had the upper hand.

The three lawyers arrived at 2 p.m on Saturday. Barbara, who was there, ushered them into our "conference room." Joseph, Walter, whom I asked to be present, and I joined them. Joseph asked them to sit down on one side of the table, where we had prepared three folding chairs. I apologized that we did not have as nice a conference room as they were used to, but if they wanted coffee or water, please let me know and we would serve it to them. At that point Barbara, Karl, and our two technicians entered. I introduced everyone and told our people that they were free to show or tell everything that the lawyers were interested in.

After our crew, except Barbara, had left, Okash handed me a signed document in which they assured Solarex that whatever they learned here would not be disclosed to anybody else, except that which was relevant to the lawsuit.

Barbara sat down at the end of the table and I told Okash and the others that Barbara was going to take notes about our discussions. They said they would like to see our facilities and perhaps talk to some of our people. We said OK and they left the room; we asked Karl to show them around. It was amusing to see the lawyer's faces when Karl distributed safety glasses to them and told them to follow him. Joseph, Walter, and I went to my office and sat down. Barbara returned to her desk.

They spent close to an hour touring our little facility, which by then was already 5000 sq. ft. (465 m^2) in size, since we had added space to our original 3500 sq. ft. When we saw them coming back, we joined them. We all sat down and Okash said that they would like to review what we had in the filing cabinet. I suggested that I was going to ask some of the people to help move the entire filing cabinet into our conference room so that they could take whatever they wanted and review it. I asked them if they wanted to talk some more with our crew and if not, I would let them go, because this was Saturday and they probably wanted to see their families. Okash said they were finished with our people. In the meantime Barbara had organized moving the filing cabinet into the conference room. I thanked everybody who had come in on a Saturday and told them that they could go home.

Barbara offered coffee, which we all accepted. Okash talked with his lawyers how to organize the review of the filing cabinet's content. One of the lawyers was put in charge and removed files and papers and brought them to the "conference table" where all three of them looked at every piece of paper. They made notes and when they had finished, the lawyer in charge of the filing cabinet picked up all the papers and put them back wherever they belonged.

It took them at least 2 hours to come to the last file, which was Joseph's private file. Most of the papers were his CVs and the last one they started to review was a ring binder, with at least 30–40 pages in it. I had no idea what it was. When the folder got to Okash he started to read it page by page with great interest.

Joseph, who was sitting next to me, obviously knew what it was and looked surprised that Okash spent that much time reading it page by page. I started to worry.

Finally Okash came to the last page, turned to Joseph:

"In the preparation of this lawsuit I learned a lot about you, but I did not know that you had so much interest in operas. You know, my hobby is to sing in operas, and I was reading with delight the stories you assembled. Why did you do it?"

This unexpected turn of events caught me by surprise. Joseph answered:

"You know, I worked before at Sprague Semiconductor research center in Williamstown, MA. I knew many people at the local radio station and they asked me if I could assemble a weekly 1 hour show about operas. As I like operas I agreed, and the file you reviewed has the notes of what I said and what part of the operas I played on their phonograph."

"I liked your stories very much especially..."

After this Okash and Joseph had at least a 15-minute discussion about operas, Joseph's notes and Okash's experience as a singer.

Finally I chimed in that it was getting late and was there anything else they wanted to see. I saw the desperation on Okash's face. Joseph's opera radio show had finished him. I concluded:

"Look, obviously you do not have a case. To prolong your lawsuit is going to cause us more damage than what these 2 months have already done. For any unnecessary damages we can file a lawsuit, which we are going to win. I suggest settling the matter expeditiously and completing a draft of the settlement and submitting it to COMSAT. We will submit it to our Board and that will be the end of this idiotic lawsuit."

"Let us talk the matter over between ourselves," Okash said.

"OK," I said, "we are going to be in Joseph's office. Barbara, could you please provide coffee to our guests."

We left. About 15 minutes later Barbara came and said that they want to continue our discussion. After we were all seated, Okash started:

"We decided that we are willing to make an attempt to settle the dispute between COMSAT and Solarex and work out a draft of a Settlement Agreement with you which we would submit to COMSAT management and you to your Board. Obviously it has to start with dropping both lawsuits."

"As you see, COMSAT's lawsuit was completely unfounded. It was based on hearsay and caused large damages to Solarex. To ask us to drop our lawsuit will depend on what are we getting in return. As it is getting late, I suggest that we reconvene this meeting Monday in your offices at 11 a.m. Until then you can discuss the matter with your client COMSAT and explain the situation they are in. I want to settle this as soon as possible, but we have to agree to the terms. I am suggesting the meeting for Monday at 11 a.m. to write the Settlement Agreement. By that time Stanley will be back and if we do not settle the matter the way it satisfies us, we have to issue a news release to the media and Stanley will start the deposition for the Antitrust lawsuit and will have a list of the people whose depositions he wants. Are we invited to your office for Monday 11 a.m.?"

"Yes, you are."

He was clearly surprised, but he had no choice. We stood up, escorted them to the door, shook hands, and said good bye.

Joseph was also surprised about my stern warning. I told him that I had seen enough movies, and I learned that one gets more attention and people listen to you better if you take your revolver from your belt, put it on the table in front of you and put your hand close to the gun. That is what I did. Besides it was late, we could not get to any conclusion that night and I would feel better if we would discuss what we should ask of them. And we could not settle the matter before Stanley returned on Sunday afternoon.

Joseph suggested that the three of us should go over to the Washingtonian Motel's restaurant, the only one in that area, have a drink and a good dinner, because he did not eat much in the last 9 weeks.

The red-haired waitress was there, who usually served us, and immediately brought us a CC Manhattan for Joseph and a dry Rob Roy for me, asked Walter what he would like to have, and we ordered dinner.

We lifted our glasses clinked them and Walter in his booming voice said, "To Solarex, which will become the largest terrestrial solar cell manufacturer on Earth!"

Sunday evening Stanley was expected to be back from Brazil. Joseph called him and told him that there were some developments and we were going to be in his office Monday morning at 9 a.m. and subsequently we were going to have a meeting in the offices of the K Street law firm at 11 a.m.

Monday morning we explained to Stanley what happened and the purpose of the meeting at 11 a.m., to complete a Settlement Agreement. He was surprised, but told us that it was a good development. We started to discuss our proposal for the settlement.

The main points we insisted on were: COMSAT had to dismiss its lawsuit against Solarex, with prejudice, which meant that COMSAT was barred from filing another case on the same claim. COMSAT had to send a letter to Solarex's customers and potential customers advising them that the lawsuit had been settled as COMSAT found that Solarex had never used any of COMSAT's trade secrets. COMSAT was restrained to make any disparaging remarks about Solarex. Both parties would refrain from hiring the employees of the other. COMSAT agreed that if any dispute arose, it had to go to arbitration and not file a lawsuit. In return, Solarex agreed that it would dismiss with prejudice the Antitrust lawsuit against COMSAT.

We set up a meeting for June 13, 1974, in Okash's office, where both parties signed the Settlement Agreement.

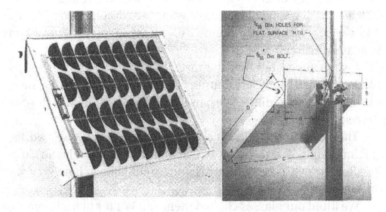

Figure 5.1   The first mass-produced PV system (1974–1979).

Subsequently we got the big order from the National Weather Service, which was for H and HP-type PV modules (Fig. 5.1) with mounting brackets and voltage regulators (6–10 Wp), of which many thousands of units were fabricated and installed worldwide from 1974 to 1979. Solarex received the research projects Joseph expected and Solarex's prestige was established: Solarex had developed new terrestrial solar cell and module technology.

Chapter 6

"Chevron" Solar Cells

COMSAT's "blitzkrieg" and its "lightning speed" settlement exonerating Solarex probably had the beneficial effect that it established Solarex's credibility and its dedication to the terrestrial PV business. I do not know whether it was a result of this, but starting with the large order from the "National Weather Service" more and more orders were received.

Joseph Lindmayer was able to get back to his research to improve the terrestrial solar cell's efficiency and manufacturing technology. He hired Chuck Wrigley, who had worked with Joseph at Sprague Electric in Vermont. During that time he was Joseph's co-author of the classic book, mentioned before, about semiconductors. Chuck joined us in July, 1974; he was a perfect addition to our company: Joseph was an ideas man and Chuck was the person who executed those ideas.

On the side of the silicon (Si) wafer which is exposed to light, which produces the electrons and a metallic grid had to be attached in contact with the Si wafer (as it was shown in Chapter 2) to complete the electric circuit. The efficiency of solar cells, how much electricity they can produce, is very important. In order to increase the solar cell's efficiency a good

Sun above the Horizon: Meteoric Rise of the Solar Industry
Peter F. Varadi
Copyright © 2014 Peter F. Varadi
ISBN 978-981-4463-80-5 (Hardback), 978-981-4613-29-3 (Paperback), 978-981-4463-81-2 (eBook)
www.panstanford.com

design of the front metal contact has to be made. The design problem of these metal contact electrodes are that they are not transparent and they are blocking the light to penetrate into the solar cell. Therefore, the electrode area has to be minimized in order that more light can be absorbed and converted to electricity by the solar cell. On the other hand, if one minimizes the area of the metal contacts that will reduce the quantity of electricity the electrodes can carry. Figure 6.1 shows a solar cell in which the front contact electrode design is very simple but they are blocking too much light and large area is in between the metal grid not making the electric connection to the entire surface uniform.

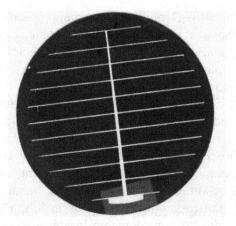

Figure 6.1 "Standard" solar cell.

Joseph and Chuck developed an interesting design in which the electrode's lines were extremely thin, blocking much less light but the thin lines reached everywhere and therefore, were able to pick up more electricity thereby increasing the solar cell's efficiency. This is shown in Fig. 6.2. Joseph always had a flair for names and he called this cell "Chevron," because of its design. Chevron is an English word which means according to Webster's dictionary "a V-shaped bar on the sleeves of a uniform, showing rank." We decided that we were going to trademark it. As soon as we filed the application for registration, we printed brochures describing this new and more efficient "Chevron solar cell," and we were able to legally print the letters "TM" after the capitalized word "Chevron." The patent office notified us that they would accept it as a registered trademark for solar cells, but first it had to be published for opposition. This means that the public, including companies, can object within 30 days.

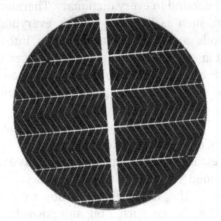

Figure 6.2 "Chevron" solar cell.

Soon after it was published, I got a telephone call. The person identified himself as a lawyer and said he would like to visit me and discuss some matters. We had finished the COMSAT's lawsuit a few months before, during which process I became allergic to lawyers. Responding to my question what the purpose of our discussion would be he said that he represented the Chevron Oil Company and they would like us to drop our application to obtain "Chevron™" as the trademark for our solar cell.

By that time I had enough experience on how easy it is for a large company to start a lawsuit, but at least Chevron had the decency to tell us that before they did anything, they wanted to talk with us, I agreed to have lunch with the lawyer in the Marriott Hotel, which was in Bethesda, MD not very far from our offices.

We met. He was a middle-aged, well-dressed man, with brown hair, combed back like mine was, and he was wearing metal-framed glasses. In spite of him being a lawyer, I found him quite sympathetic. After we sat down at our table he started to explain that the Chevron Oil Company takes it very, very seriously if other companies are trying to use "Chevron" as their trademark. He sort of emphasized the "very, very seriously" part, which sounded like a threat.

I pointed out to him that the word "chevron" is an English word and can be found in every dictionary. Therefore, as it is not a unique word, such as Xerox, or Kodak, everybody can use it and as all goods and services are divided into 45 different International Classes and 60 different US Classes in which one can Trademark a name, the Chevron Oil Company is not able to register the word "Chevron" as a trademark in all of these categories.

He agreed that was a problem, but he mentioned that Chevron is a big company, with lots of resources. I assumed that he meant money, and I think he meant it again as a threat.

So, I explained to him, that he can imagine how horrible it would look if the newspapers wrote stories, that after finishing the oil embargo, such a big and powerful oil giant like Chevron would harass and sue a very small company trying to develop the solar energy business to help reduce our country's dependence on oil, which idea was becoming very popular with the public.

He said, "Look let's make a deal. I would suggest, that you could keep using the word "chevron" to give a distinctive name to your solar cell, but do not use capital C—only a lower case c—and do not put TM after the word, chevron."

"I can see your point, but we spent money to register the trademark, to print brochures, to advertise it, etc. We invested substantial amounts in our "Chevron cell," and you suggest that we just should use a lower case "c" and not put TM after the word, Chevron. You are not serious."

"We would compensate you for all the problems you mentioned. We would pay your expenses."

I was thinking. We are going to use this chevron pattern only as long as we find out something that gives our cells a higher efficiency than the "Chevron™" cell is providing. On the other hand, additional money would provide us with better stability, especially after this stupid COMSAT's lawsuit. So I proposed him that Chevron would pay us a reasonable amount, but I want to have it settled now and receive the check and the signed contract in a few days, by August 26 (1974), which was a Friday.

"OK," he said without hesitation.

At that time the amount I proposed was very large for us, but as he agreed so quickly, I thought that I had left some money on the table, but I had enough of lawyers.

We shook hands. After that we had a very pleasant lunch. During our lunch he told me that he has a lousy job. Practically every day he has to visit a company to convince them not to use "Chevron" in their name or advertisement. Just a few days ago he had to visit a trucking firm in Naples, Italy. They agreed after some "discussion" that the company would drop the word "chevron," but he was not sure if he was going to go home in the passenger cabin of the airplane or in a wooden box in the hold of the plane.

Thursday noon the signed contract and a check for Solarex were on my desk. I called Walter Ames, our patent lawyer, and told him I would stop by his office after lunch, as I wanted to ask him to review the contract before I signed it. He found it satisfactory, I signed it, and Walter's office mailed it. From Walter's office I went to our bank, the First American Bank in Silver Spring, and met its president, Bob Burke. I handed him the check and told him that I would like to deposit it for Solarex and would like to pay back what we owe the Bank on our line of credit. He said that it was not necessary to pay back the line of credit. I told him that we wanted to pay it back as we had a purpose in doing so. My reason was very simple that I wanted to raise our line of credit. We had serious problem getting the first line of credit, but being profitable and needing no money is when one should get a higher limit. He looked puzzled and probably believed that Solarex might go to another bank. I told him that the reason is that, as he can see, Solarex is a very reliable customer, which is able to pay back a line of credit, and therefore, he may want to offer to increase our line from $100,000 to $250,000. He looked at me and said:

"You have a point. We will do it." He picked up the check and asked me what Chevron bought from Solarex. I told him that they did not buy anything. "How is it that they sent you this large check? I never heard that the big oil companies are giving money away."

"Bob, do you know who Harry Houdini[45] was?"

"Yes, of course."

"He was my kindergarten teacher."

---

[45]Harry Houdini was a Hungarian-born, American magician.

Chapter 7

The Silicon Enigma

The development of the terrestrial solar cell manufacturing technology and the introduction of methods to improve its efficiency—like the chevron-shaped grid—resulted in producing them relatively inexpensive. But the basis of the terrestrial solar cells was still the silicon (Si) wafer, the price of which we had absolutely no control of. The purchase of Si wafers resembled the ride on a roller coaster.

The primary source of the single crystal Si wafers came from the manufacturers who produced them for the semiconductor industry. The availability and cost of the wafers depended on how well the semiconductor industry was doing. If it was not doing well, there was no problem in obtaining wafers at the right price. When Si wafer manufacturers could not sell their production to the semiconductor industry, their salesmen would show up in my office and take me out for lunch. If, however, there was a great demand for wafers for the semiconductor industry, the same salesman would not even return my telephone calls. Wafer manufacturers happily sold only that small percentage of the total wafer production, the quality of which was not good enough for semiconductors at

Sun above the Horizon: Meteoric Rise of the Solar Industry
Peter F. Varadi
Copyright © 2014 Peter F. Varadi
ISBN 978-981-4463-80-5 (Hardback), 978-981-4613-29-3 (Paperback), 978-981-4463-81-2 (eBook)
www.panstanford.com

a reduced price to the solar cell industry. For solar cells, even those wafers could be used. But as the solar business increased, more and more wafers were needed and if they were available, they were only at a very high price.

The quality wafers used for the semiconductors were very high priced. The semiconductor manufacturers could afford to pay that price, because on a single wafer they made very many products, for example transistors or integrated circuits. On the other hand, one wafer was for us independent of its size, only one solar cell, depending on its size with one or more W output. The solar modules would not have been sellable if we would use Si wafers purchased at prices transistor or other semiconductor product manufacturers were able to pay for.

It became evident from the beginning that to become successful and profitable in the photovoltaic business depended on the price at which we were going to be able to buy the Si wafers we needed for our growing production.

We realized that at that time there was also a secondary market for the needed Si. Si ingots (rods) (before they were sliced up to become wafers) and wafers were manufactured for the semiconductor industry in a variety of quality, grades, and sizes some of which were for various reasons not used by the semiconductor manufacturers.

To use these "left overs" we needed to develop a solar cell manufacturing technology, which was able to produce good solar cells on wafers of all kind of specifications and also to have the slicing equipment to buy ingots and to slice them. Solarex had an excellent technical staff and under the direction of Lindmayer and Wrigley they were able to develop a solar cell production technology, which was able to use practically any Si material.

Running the operations of Solarex, obviously my department was responsible to make sure that we had enough Si wafers to run our production. After a while, I must have developed a reputation in the Si wafer and semiconductor community that I am buying for Solarex practically all of the Si wafers or ingots, which cannot be used. I would buy it and they at least would get some money for it. This resulted that I got many telephone calls offering me Si wafers or ingots. To give an example about the type of offers I got and that I had to take everything seriously I describe an interesting story.

One day I received a phone call from one of IBM's departments and the caller said they had heard I was buying all kinds of waste Si wafers, and he had 52 barrels full of such wafers that they wanted to sell. I knew that liquids, for example oil or wine was sold in barrels, but I never heard that Si wafers were sold in the quantity of "barrels." I asked what size the wafers were, what type, and what specification did they have? The IBM fellow told me that he had no idea about such things, he just knew there were 52 barrels filled with Si wafers, which he wanted to sell or send to a dump. On my question of how much he wanted for the Si he simply said that if I needed it and transportation could be arranged immediately, we would agree to the price. I said that I would be there the next day and see the material. I was there the next day and saw it, but could not quickly grasp the situation. I asked them to open a few barrels, but they were all the same. The barrels were filled with 2.25 inch (5.625 cm)—which at that time was the standard size—Si wafers, apparently rejects. Mostly not, but several were broken and they apparently were just dumped into the barrels. I called the guy to whom I spoke the day before and said I would buy all of the 52 barrels. He virtually gave them away, but stipulated that I had to pick them up in a few days. I agreed that they would be picked up in 2 days.

And that is what happened. I called Jeannie Rosenbloom, who was the manager of our administration, and told her that by the next day she should rent as many padded trucks as would be able to pick up 52 barrels. Get enough people who know how to drive a truck and send them to the IBM plant in New York State, to pick up the barrels and drive slowly to our plant near Washington, DC, and that the Si content of the barrels should not shake to preserve as many of the wafers as possible. In the meantime, she should set up a bunch of tables and chairs, and get school children to sort the Si wafers from the barrels. She organized it excellently the 52 barrels of dumped silicon wafers turned out to be a treasure house for us.

The school kids were able to salvage a very large number of intact wafers but also many which were broken were useful. We were able to cut them exactly in half or quarters and were able to use them in types of solar modules with smaller power output. Obviously there was a large amount of broken and not useful wafers, but these became very important and used for experiments described later which led to a new and inexpensive casting method to produce wafers for solar cells.

Such cases, providing us with inexpensive Si made it possible for Solarex to become profitable in the first year. However, we anticipated that because of the increase in demand for solar panels, the rods and wafers bought this way, by spot purchases, was only a temporary measure and would not suffice for the future. Time to time we experienced crises during which we could not buy Si wafers or rods at all, not even at a high price, and we had periods when we did not know how we would be able to continue to manufacture solar cells.

We had to take this issue very seriously from day one and find a solution in order to stay in business. Unfortunately we had a handicap. We had very little money and therefore, whatever solution we were going to come up should not cost a lot of money.

Chapter 8

Silicon, the Second Most Common Element in the Earth's Crust

We decided to base our terrestrial solar cell business on silicon (chemical symbol, Si) wafers as the starting material. It is well known that Si is the second most abundant element (27.7%) in the Earth's crust after oxygen (46.6%). But Si wafers were another story. It was obvious that the way to secure the needed Si wafers as described in Chapter) was a temporary solution. We also found Si wafers were an incredibly unreliable base material because its price and availability could wildly fluctuate. With the increasing demand for solar cells and in order to be able to expand and maintain stable production, Solarex needed a permanent solution to have Si wafers at an acceptable stable price and in a guaranteed quantity. In those days, the wafer requirement of the solar business was small compared to the semiconductor needs; therefore, Si wafer manufacturers had no interest to guarantee quantity or proper price for their solar customers. This meant that we had somehow to assure the Si wafers by ourselves.

Si wafer suitable for making semiconductor devices and Solar cells is the last step in the manufacturing process starting

Sun above the Horizon: Meteoric Rise of the Solar Industry
Peter F. Varadi
Copyright © 2014 Peter F. Varadi
ISBN 978-981-4463-80-5 (Hardback), 978-981-4613-29-3 (Paperback), 978-981-4463-81-2 (eBook)
www.panstanford.com

with the source of Si: quartz or quartz sand. In spite of that the Si metal is a very stable material, in the nature Si as a metal is extremely rare. In most cases Si is found in its oxide form, such as quartz or quartz sand. Starting from quartz sand to obtain high purity Si wafers for the production of semiconductor devices or for solar cells, several steps are required which are shown in Fig. 8.1.

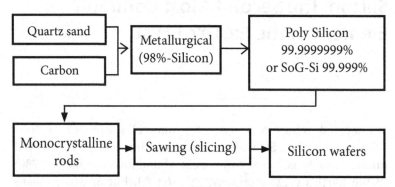

Figure 8.1   From Quarz sand to Silicon wafers.

The first step is to produce "metallurgical grade silicon"[46] (MG-Si) from quartz or quartz sand and carbon in an electric arc furnace. The price of metallurgical Si which contains 98% Si is very low.[47] An extremely large quantity of this material (55%) is used for alloying aluminum or iron. The second largest utilization

---

[46]Definitions:

- *Metallurgical silicon* obtained by reducing quartz (SiO_2) to metallic quartz (Si).
- *Poly (Polysilicon or Polycrystalline) silicon* chunks or rods of solar cell grade (SoG-Si) or greater purity pieces of silicon formed from multiplicity of crystals obtained by the so-called Siemens method or by other process.
- *Monocrystalline silicon* is a block (rod), which consists of a single crystal. This may be produced by the so-called Czochralsky method.
- *EFG ribbon* crystalline growth (**E**dge-defined **F**ilm-fed **G**rowth).
- *Multicrystalline silicon* "bricks" consisting of several large silicon crystals produced when hot liquid silicon is cast into a container the cooling of which is regulated.

[47]The largest producers of MG-Si are listed: Photon International, 1-2009, 140.

of Si (about 40%) is as a raw material in the production of silicones, which is being used for many applications. Only a very small amount of the metallurgical Si is purified to be used for semiconductor and solar cell applications. (Annex. 1 provides the details about the process to produce "metallurgical silicon.")

The second step is to process the metallurgical Si to achieve an extremely high purity Si (99.9999999%) to be used for the manufacturing of transistors, integrated circuits, which are referred to as "semiconductors." For the production of solar cells, a less pure so-called "Solar grade silicon" (SoG-Si) of 99.999–99.9999% is sufficient. These materials called "polycrystalline (poly) silicon" are produced from metallurgical Si mostly by the so-called "Siemens" purification process. (Annex. 1 provides more details about the poly-Si manufacturing processes.)

Buying the poly-Si at that time was not a problem. One could get large quantities on the spot market and the price of poly-Si was stable. In those days one was able to secure yearly contracts.

While in the first decades of the photovoltaic (PV) business, poly-Si was plenty, our problem was the wafer supply. In spite of that the so-called vertical integration of the production should be pursued, i.e., from raw materials to finished products, all the process, every step of the production should be done and controlled by the company. PV manufacturers in general decided that they should not get involved in the manufacturing of either the metallurgical or the poly-Si.

The conclusion was that in order to stay in the expanding solar cell business, Solarex must also participate in the last two steps: production of monocrystalline ingot and slicing them into wafers.

From the outset, monocrystalline Si wafers were used for the manufacturing of solar cells. Monocrystalline Si is produced mostly by the "Czochralsky" (Cz), or by the "Float Zone" (FZ) method. The production of wafers from monocrystalline Si ingots is done by sawing. Small quantities of solar cells were also fabricated on wafers produced by the EFG[48] Si ribbon process.

---

[48]Edge-defined film-fed growth (EFG).

Czochralski Method

The Czochralski method is described in detail, because it has a wide use in manufacturing not only Si but also many type of crystals, including the artificial ruby and sapphire. Jan Czochralski was a Polish chemist, who studied in Germany and in 1916 accidentally discovered how to produce mono-crystals. Monocrystal is the name when the entire piece of a metal or non-metallic material, large or small, consists only of a single crystal. Because of a mistake Czochralski dipped the pen with which he used for writing, instead of in the ink well, into an adjacent dish containing molten tin. Of course, he immediately pulled the pen out, to which a string of tin was now attached. Czochralski found that the string of tin was a monocrystal, which means that the whole string was a single crystal. He repeated the experiment several times with the same result. He published his results in 1918 in a German journal.[49] With this method, Czochralsky opened the way for the production of valuable crystals, such as artificial rubies or sapphires and also Si.

G. K. Tiel and J. B. Little of the Bell Laboratories in the USA, in 1950 has used Czochralski crystal manufacturing method for the production of monocrystalline germanium (Ge), which is one of the elements from which semiconductors can be produced, and thus opened a way for the production of monocrystals.[50] Since then the method was also used for the manufacturing of Si or artificial ruby or sapphire monocrystals. In the case of producing Si monocrystals (Fig. 8.2), the method consists to use the sufficiently pure poly-Si pieces with the addition of the correct amount of dopant[51] (Boron or phosphorus) and mix them in a suitable container (e.g., quartz) and melt

---

[49]Zeitschrift für Physikalische Chemie, 1918.

[50]http://en.wikipedia.org/wiki/Jan_Czochralski.

[51]In semiconductor production, doping intentionally introduces impurities into an extremely pure (also referred to as *intrinsic*) semiconductor for the purpose of modulating its electrical properties. The impurities in the case of silicon are: Phosphorus (P) makes the silicon n (negative)-type and Boron (B) makes it p (positive) -type. www.wikipedia.org/wiki/Doping_(semiconductor).

the mixture under protective gas (A). After the mixture is melted, small Si crystal (germ) is lowered into the molten Si (B) and slowly pulled up (C). The melted Si adheres to the Si germ while it is very slowly pulled up (C), having the same crystal structure as the germ had and a Si monocrystal ingot is being formed (E) (Fig. 8.3).

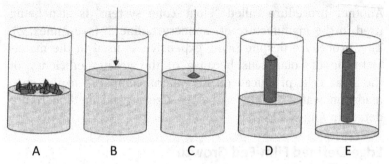

A   B   C   D   E

Figure 8.2 Czochralski monocrystal manufacturing method.[52]

Figure 8.3 Monocrystalline Si rod produced by the Czochralski method.

---

[52]http://en.wikipedia.org/wiki/Chochralski_process.

In 1974, Czochralski crystal pulling machines were already available producing 2.5 cm diameter Si ingots. Now Si crystal pulling machines are available making even 45 cm diameter Si ingots.

Float Zone (FZ) System

Another procedure called "Float Zone system" is also being used in the manufacture of monocrystalline Si (see Annex. 2). This procedure despite being expensive, is used in the manufacturing of solar cells because of the greater efficiency of the solar cells produced on FZ wafers, compared to the ones made on wafers produced by the Czochralski method (more details in Annex. 2).

Edge-Defined Film-Fed Growth

Historically this method has to be mentioned, but for several reasons it is not being used very much today (more details in Annex. 2).

Multicrystalline Silicon

As mentioned before, during the first decade of the terrestrial PV business, solar cell manufacturers were not able to rely on wafer manufacturers to supply Si wafers at the required quantity and at an acceptable price therefore, at Solarex and our competitor at that time Arco Solar came to the conclusion that we have to produce our need of Si wafers by ourselves to assure that we will stay in business. At first, we thought that Solarex should also gear-up to produce our needs of wafers by the Czorchralski or by the FZ method. The Czorchralski or FZ crystal pulling equipment was, however, very expensive and many of these machines would have been needed to provide the requirement of the solar cell production. Arco Solar[53]

---

[53]Arco Solar (a division of ARCO oil company located in Camarillo, CA) having enough money was able to buy a large number of Czorchralski crystal pulling equipment and a suitable number of saws and set them up in a factory in Hillsboro, OR.

having substantial amount of money, solved their problem by purchasing large number of Czorchralski crystal pullers and also slicing machines to produce their own wafers.

The problem at Solarex was that it was a privately funded company, with extremely low capitalization and therefore, was simply not able to buy the equipment. Solarex, unlike Arco Solar, did not have deep pocket, but in spite of that had somehow to come up with a solution to assure the Si wafer supply if wanted to stay in the terrestrial PV business. The fact that a company does not have much money, is not a sin, it only complicates matters. If, however, as in Solarex's case, where the competitor has unlimited amounts of money because the owner is a huge oil company (ARCO), it was already far beyond what one can call a complication, it could be called a "serious problem," that the unstable and unpredictable Si wafer supply could simply put Solarex out of business. The necessity forced Solarex to come up with something new and inexpensive way to produce Si wafers.

The result was the development, which we called multicrystalline Si wafer technology, which became very successful and by now it is widely used. The background of the development of the multicrystalline Si technology was that Solarex, as mentioned before, because of its rapidly expanding business had to use monocrystalline Si wafers whatever quality and crystal orientation could be obtained. The experience was that with slight modification of the production process good solar cells could be produced on a variety of materials. Joseph Lindmayer's opinion was that in trying to find a method to produce inexpensive Si wafers so that the expensive Czorchralski or FZ crystal pulling equipment producing single crystals would not be needed. If one could produce Si wafers, which consist of several large crystals even with different crystal orientations and quality it could also be used for solar cell production. He thought that this can be achieved, by using sufficiently pure poly-Si pieces with the addition of the correct amount of dopant (Boron or phosphorus) and put them into a suitable container (e.g., quartz or ceramics) and melt the mixture under protective gas. When melted, simply pour the melted Si into a large container (e.g., quartz or ceramics), the cooling of which should be regulated, then the cast material will solidify in several mixed size smaller and larger crystals (Fig. 8.4).

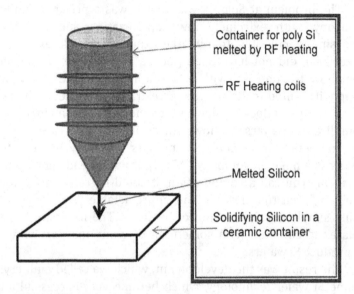

Figure 8.4   Silicon casting system.

Figure 8.5   Cast multicrystalline Si bricks.

This process required inexpensive machinery and could produce large squares, castings, which were called "bricks"

(Fig. 8.5). These square bricks can be cast in very large size and smaller bricks could be made from them by sawing. These bricks then can be sliced into wafers (Fig. 8.6). One benefit of this procedure is that the Si wafers will be rectangular, which when assembled into solar modules provide more active surface, than the round wafers produced from ingots made by the Czochralski crystal pulling method. The disadvantage compared to the monocrystalline wafers is that, the produced solar cell efficiency will be a little lower.

Figure 8.6   10 cm × 10 cm multicrystalline Si wafer. The wafer consists of crystals with different orientations, which in the image can be seen as darker or lighter shaded area.

Lindmayer' experiments were successful[54],[55] and Solarex from the late 1970s started to manufacture solar cells made by slicing cast Si "bricks," called "Multicrystalline" Si wafers.

---

[54]Lindmayer J (November 12, 1976) *Semi-crystalline Silicon Solar Cells*, 12th IEEE Photovoltaic Specialist Conference, 82, 85.

[55]The German silicon manufacturer, Wacker, at the same time developed a similar process to produce multicrystalline silicon wafers, which Wacker brought only later to the market because he was a producer of monocrystalline silicon wafers. (Fischer H, Pschunder W (November 12, 1976) *Low Cost Solar Cells based on Large Area Unconventional Silicon*, 12th IEEE Photovoltaic Specialist Conference, 86, 92.)

Everything was ready for starting the mass production of the cast Si "bricks" and slicing them into wafers. The casting machinery was designed, the ceramic crucible were tested eliminating thereby the expensive quartz material, and wafer slicing ID saw were readily available from companies making them for the semiconductor business. Semix, a separate company, fully owned by Solarex was incorporated. There was only one item missing to start the building of the "multicrystalline" Si wafers production: Money.

Joseph Lindmayer sent his proposals to the US Government organization which was at that time charged to support PV. His proposals were rejected. In spite of our experience with venture capital people, we started to negotiate with two of them, but we could not come to an agreement and postponed the issue for a few weeks, because Solarex's lawyer vent on vacation.

As it will be described later, by that time the French water pump manufacturer Leroy-Somer was a minority stockholder of Solarex with which Solarex had a joint venture in France to manufacture solar cells and modules. When our discussion with the two venture capital companies stopped, I went to Paris to visit our joint venture, France Photon in Angouleme, France, which Leroy-Somer had just started. On the train, going from Paris to Angouleme, Georges Chavanes, president, and Paul Barry, vice president of Leroy-Somer, came with me. During lunch on the train, I told them about the casting process to manufacture multicrystalline Si wafers which we already used to make small quantities of solar cells. I told them we were looking for money to set up a factory to produce these multicrystalline wafers utilizing the Si casting technology in large quantities to supply our and also France Photon's production with low price wafers. I took a piece of paper and drew up the process of what we planned to do and for which we needed some money. I told them that we named this entity as Semix.

They had a very short discussion, after which Paul asked me, "How much money do you need?"

"We calculated that we would need one million dollar."

"We are going to provide you the money. Prepare the agreement and if that is acceptable we will transfer you the money immediately."

A week later Paul arrived at Solarex with Pierre Sanier, their corporate lawyer. We gave them the prepared agreement. Paul reviewed it and accepted the percentage for their money. He suggested that his lawyer and ours, Sam Cross, should spend the day reviewing the details and the next day if everything was in order we could sign the agreement and the money would be transferred immediately to Semix's account. That is how Solarex's Si multicrystalline wafer production line was started. We obviously called the two venture capitalist company and with pleasure informed them, that we already have the money and therefore, no need to continue our discussions. One of the venture capitalists told us that he would like to invest his own money in Solarex. We liked him and told him when the next round of financing will happen we would be glad if he would participate.

In 1982, a 200 kW multicrystalline solar system was installed on the roof of Solarex's Frederick, MD factory. Also in 1982, on the roof of the Intercultural Center at Georgetown University (Washington, DC), a 300 kW multicrystalline system was built. The multicrystalline Si wafers developed by Solarex (and at the same time by Wacker), have been so successful that according to the US National Renewable Energy Laboratory[56] in August 2010 in the world among all the types of solar cells 47.27% were made from multicrystalline wafers, while only 33.9% were made of monocrystalline material.[57] Today several manufacturers are offering so-called "Directional solidification 'Multicrystalline' furnaces" to produce Si bricks.

Sawing to Produce Wafers

Silicon wafers used to manufacture solar cells are made by sawing thin slices from monocrystalline rods or multicrystalline bricks. It is a very important requirement in the sawing process to minimize the kerf loss, and to be able to produce as thin

---

[56]Kurtz S (August 13, 2010) *PV Technology for Today and Tomorrow*, NREL/PR-520-49176.

[57]Wohlgemuth J (June 1999) *Cast Polycrystalline Silicon Photovoltaic Module Manufacturing Technology Improvements* (Final Subcontract Report—8 December 1993, 30 April 1998) NREL/SR-520-26071.

wafers as possible. It is clear that this will reduce the price of the wafers. As mentioned, Arco Solar purchased saws to make wafers from the produced single crystal rods and Solarex equipped itself also to be able to slice (saw) first the mono-crystalline rods, then multicrystalline cast bricks.

In the 1970's, first a multi-bladed oscillating saw was used, similar in principle to what is still being used in shops, slicing a loaf of bread. One advantage of this system was that it was a relatively inexpensive machine and multiple wafers were able to be cut at the same time. But it was a great disadvantage that large amount of Si was lost due to the kerf loss and relatively thick wafers were sliced and the wafer's thickness was not completely uniform.

The other, a more expensive sawing machine was like what stores are using to slice salami or other meat products, i.e., fast-rotating discs, which is cutting off slices. A similar device was used for slicing Si, the difference being that the rapidly rotating blade was like a doughnut, the center was empty and the cutting edge was not on outside, but on inside and the salami, in this case the silicone rod or brick was placed in the middle of the sawing blade. This machine because the inner edge of the blade does the cutting was called ID saw. The kerf loss was much less, than in the oscillating saw, the slices were thinner and smooth, but the sawing process was slow, since it sliced one wafer at a time.

Solarex initially used the oscillating blade saw, and then converted to the ID saws, but ideal would have been a wire saw, in which a thin high-speed wire, e.g., a thin piano wire would do the slicing. Of course, the thin wire used in a piano would not be hard enough to cut the Si, therefore, the wire had to go through a slurry containing fine powder which is harder than Si (e.g., Si carbide) and adheres to the wire. The advantage of such a wire saw would be that due to the thinness of the wire, it would reduce the kerf loss to a minimum and could slice very thin wafers and having a long wire and appropriate mechanization could slice 100 or more slices at the same time.

In 1978, Solarex tried a Yasunaga wire saw. Yasunaga was a pioneer in wire saws to be used for producing Si wafers. The saw indicated that it would fulfill the anticipation of low kerf loss, to be able to slice several wafers at the same time, and produce thinner wafer thickness but the problem was at that time the reliability of the wire saw's mechanization.

When Solarex developed the multicrystalline Si brick (castings) technology, Solarex started its production in Semix Corporation, a fully owned subsidiary in Rockville, MD, later established another factory "Intersemix" (as you remember Joseph's only requirement for a corporate name was that the end letter should be an X) in Gland (VD) in Switzerland because of the relatively low cost of electricity. Dr. Charles Hauser began working at Intersemix in 1982, in the Gland VD factory. Dr. Hauser was given the task to study the slicing technology of the produced Si bricks. The problem was that the cast Si "bricks" have been increasing in size to produce the solar cells cheaper, but the ID saws could only saw what would fit inside of the rotating blade and that would limit the size of the bricks. Thus, the slicing of large Si bricks became a significant problem.

In 1983, when AMOCO the US oil company acquired Solarex, decided to close Intersemix, dismissed the employees and sold the factory building. Dr. Hauser became unemployed but raised some money and near Lausanne founded the company HCT, rented a garage of 200 m² (about 2000 sq. ft.) and continued to study the problems of cutting thin slices of Si rods and bricks. The result was that he developed a reliable wire saw, the essence of which was that a very thin (100–150 microns) metal wire was used for the sawing. In the wire saw, this very strong steel wire at high speed (5–25 meters per second) was running through oil which contained very hard Si carbide powder. The Si carbide particles adhere to the wire and the fast-moving wire was sawing the Si rod, or brick. The continuous wire can be very long, so with a proper mechanical system it could slice hundred or more wafers simultaneously. HCT delivered the first wire saw machine in 1986.[58] The wire saw minimizing the kerf loss, has made possible to produce very thin Si wafers (0.1–0.5 mm) and is able to saw very large (20 inch = 500 mm) rods (or bricks) and speed up the time

---

[58]In 2007, the American Applied Materials Company purchased HCT.

required for slicing. Solarex in the US, manufacturing Si bricks was aware of the advantages of the wire saw and purchased the 2nd, 3rd, 4th, and 5th wire saws[59] produced by HCT. The HCT wire saw was thoroughly investigated[60] and the result was summarized, that one wire saw is able to produce as many wafers as 22–24 ID saws, and 20% more Si wafer can be made from the same rod, or brick. This strongly reduces the price of the wafers and thereby the solar cells, so now wire saw is being used everywhere, to slice monocrystalline Si rods, or multicrystalline bricks.

---

[59]Information from Dr. John Wohlgemuth (presently at NREL).
[60]Cast Polycrystalline Silicon Photovoltaic Module Manufacturing Technology Improvements NREL/ST-520-26071.

Chapter 9

The Evolution of the Terrestrial PV Modules: The Importance of Testing and Quality

After the "intrepid entrepreneurs," as Richard Swanson calls us, started to make solar cells and modules and actually sell them, and after the US Government started to spend some money for development, several entities jumped into the photovoltaic (PV) business and also started to make PV cells and modules. In retrospect, I can see great similarities between the "flying machines" shown in one of my favorite movies *Those Magnificent Men in Their Flying Machines* made in 1965 and the PV modules produced by the various manufacturers in 1974.

So, what is the situation today with PV modules? Again a similarity exists between today's airplanes and today's PV modules. Today's airplanes are very reliable, and so are today's PV modules. The PV module, which is the most expensive component of a PV system, I believe, is one of the few products in the history of mankind for which the manufacturers are giving a 20- or even 25-year warranty that it will deliver at least

Sun above the Horizon: Meteoric Rise of the Solar Industry
Peter F. Varadi
Copyright © 2014 Peter F. Varadi
ISBN 978-981-4463-80-5 (Hardback), 978-981-4613-29-3 (Paperback), 978-981-4463-81-2 (eBook)
www.panstanford.com

80% of its original electrical power output even after so many years. The warranty for such a very long time may sound unbelievable, considering that while the modules have to last for 25 years of outdoor exposure, the manufacturers, many of them are not even 10 years in business; therefore, they could not have waited 25 years to see how their product will perform.

How did this happen when manufacturers are normally reluctant in establishing standards for their products or services. Governments establish regulations or standards usually only after there is a strong public demand. But in general the consumer can only depend on the reputation of the manufacturer of a product, or the service provider.

In the case of many products, manufacturers rely on the so-called "externalization of quality control." This means that the products are not checked by the manufacturer. They are sold the way they come off the production line and the customer using it finds out that the product is good or not good. If it is defective, returns it to the store or to the manufacturer and gets a new one. If the manufacturer is reputable, it will send out a repairman to fix the product. This is cheaper for the manufacturer, but a nuisance for the customer. This system works in the USA or countries in Europe, but not in most part of the world.

For the future of the PV modules and PV systems, it was extremely important, actually a necessity to be reliable and have long life—20–25 years—to become competitive with other energy sources. This required a manufacturing process to achieve a durable PV module and a testing procedure to prove that the product is going to have the required long life. That manufacturers are going to be able to claim a 20–25-year life for PV modules has an interesting background and this, in the world's history almost an unprecedented long warranty provided by manufacturers, could be correct, because the governments, the industry, and organizations such as the World Bank working together developed a quality and certification system which would assure quality and durability. I am using the word "would" to indicate that this would happen "if" the manufacturer voluntarily would do it or the user would only buy PV products which display a Quality Mark.

This interesting story of how this was achieved starts when PV was used for space applications and substantial experience was gathered. PV is used on space crafts, satellites, etc., which are designed and assembled by competent people and the quality specifications are strictly enforced, because the failure of the PV system, the only source of electricity for most of the space objects, would cause the failure of an enormously expensive machine. The experience obtained of how to assure the quality and durability for PV used in space applications provided guidance for PV used for terrestrial purposes.

For the future success of the utilization of PV for terrestrial applications an extremely important event happened, that in January, 1975 the US Government initiated a terrestrial PV research and development project, the aim of which was to help the terrestrial PV industry to reduce prices and produce reliable PV modules. The management of this project was assigned to the Jet Propulsion Laboratory (JPL) in Pasadena, CA. At JPL, John V. Goldsmith who headed JPL's space-oriented PV activities became the Technical Manager also for the terrestrial PV program. Goldsmith, who spent many years working in the field of space-oriented PV systems, knew very well, that one of the most important goals was to achieve excellent quality and reliable PV modules.[61] Utilizing the experience from the space programs, Goldsmith wanted to establish specifications to achieve this goal. The JPL program was simple, but very effective. A quality and testing program[62] was established to buy "blocks" of terrestrial PV modules in meaningful quantities from manufacturers but their products must qualify (JPL Block Program).

In the first "block buy" (Block I: 1975–1976) JPL purchased 54 kW off-the-shelf modules to establish the base line of what is available. The specification for this first Block was required only to verify the electrical output per the manufacturer's rating. This means, how much electricity the module will

[61]https://pub-lib.jpl.nasa.gov/docushare/dsweb/GetRendition/Document-935/html.

[62]A detailed description of the entire JPL *Flat-Plate Solar Array Project, Final Report* by Ross RG and Smokler MI is available: JPL Publication, 86-31. (http://authors.library.caltech.edu/15040/1/JPLPub86-31volVI.pdf).

produce when exposed to the full sunlight. As every module cannot be dragged outdoors and measure its electrical output during the moment when there is "full sunlight" a so-called "solar simulator" had to be fabricated. Also a "standard solar cell" was needed, which was calibrated, so every manufacturer should use the same "sun" exposure. Luckily, this system was already developed for the space solar cells and systems and it just had to be adapted to terrestrial conditions. The environmental tests prescribed by JPL were limited to temperature cycle [repeatedly lowering the temperature to below freezing (−40°C) and raising it to elevated temperature (+85°C), and humidity soak, meaning to keep the modules during elevated temperature also in a very humid atmosphere. These tests were to simulate the terrestrial environment where a solar module has to survive.

Five companies participated in Block I. Even these simple requirements needed for the manufacturers to make several design improvements during production to qualify the product for this program. As an example is the following story:

I happened to be in Florida on a business trip, when I called my office to check back. My secretary gave me information but her normally cheerful voice was not the same as I was used to. I asked her if there is a problem. She told me, that just a few minutes ago she got the information, that our modules failed the JPL humidity test.

I almost got a heart attack. I knew that our modules passed with flying colors the temperature cycling test, which we thought was difficult. After that, we did not think that the "humidity soak" test would be a problem. The solar cells on the module were embedded in silicon rubber, which was water resistant, but as we found out it was permeable to water vapor. I told her I am taking the next flight back. Passing the JPL tests was important, not only because of the prestige, and the customers' confidence in our products, but also because the quantity purchased at that time was comparatively large and not to be able to make those modules would have affected our business.

I got back before the end of the day. Joseph Lindmayer and his people were still in the conference room discussing the matter, but luckily they already found the problem. A step in the production of the solar cells in order to reduce cost was left out starting with the JPL modules. We had to start a new batch for JPL by returning to the original manufacturing procedure and those modules were all able to complete the humidity soak test also with flying colors. This was a good lesson for us to realize how important testing is to achieve good-quality products. In the JPL Block II, modules totaling 127 kW were purchased. (Looking at them today, these quantities seem extremely small, but at that time it was a significant quantity.) The design (e.g., the requirements of interconnect and terminal redundancy) and testing specifications (thermal and humidity cycle, structural loading) were extended. In this program, four companies participated.

It is very interesting to describe the Block II program because it introduced an additional extremely important requirement: the "Quality Management" (QM) System. The requirements of the JPL introduced QM system was very similar to what ISO[63] introduced only 10 years later.[64] The result of the QM is to ensure that all of the manufactured products should have the same quality. This requirement was also used for space solar modules. The JPL program required that the manufacturer had to prepare a Quality Assurance Plan, including inspection criteria. This actually means that the manufacturer has to prepare a detailed plan in writing, how to make the product, and at what point in the manufacturing process the product had to be inspected. The manufacturer's program had to be approved. It was required that a JPL inspector could be in residence in the production area to observe the execution of the QM program. This sounds like a difficult requirement, but it is actually not complicated. At that time it was new, but today in most of the industries very many manufacturers has an ISO certification, for which the requirements are very similar. Now it is also well known that this system is not only resulting in good and

---

[63]The ISO 9000 series of quality assurance-related documents were created by the International Organization for Standardization (ISO) as international requirements for QM Systems.

[64]At that time the ISO Quality Management System now used worldwide did not exist; it was only introduced about 10 years later in 1987.

uniform quality products, but also saving money for the manufacturer.

The JPL program continued with three more Block buys ending with Block V.[65] The specifications were tightened in each Block buy based on the experiences in the previous Block buys. The assumption was based on similar space solar cell specification that PV modules, which passed JPL's Block V tests, would be able to have a 20-year life. The JPL Block program provided manufacturers a means to evaluate quality and the resulting expected life of their PV modules.

The JPL Block I–V program had a big effect on the quality and reliability of the PV modules. One study[66] claimed that Pre-Block V solar module failure rate was 45%, Post-Block V module failure rate was 0.1%. This very low number was verified by the BP Solar/Solarex study.[67]

A combination of accelerated stress testing, field experiments, and statistical data analysis from field returns of commercial products must be used to assess module reliability and lifetime. One of the most important elements is the "accelerated stress testing," which, if the module is able to survive it, will give an indication that it could survive 20–25 years (Annex. 9). An important factual support for this claim is, that the BP Solar/Solarex's data base indicate, that the return of their multicrystalline silicon modules in the time frame between 1994 and 2003 was 0.13%, which represents approximately

---

[65]Block I (1975) manufacturers: M7 International, Sensor Technology, Spectrolab, Solarex, Solar Power Corporation. Block II (1976) manufacturers: Sensor Technology, Spectrolab, Solarex, Solar Power Corporation.
Block III (1976) Manufacturers: Motorola, Sensor Technology, Solarex, Solar Power Corporation. Block IV (1978) Manufacturers: Arco Solar, ASEC, GE, Motorola, Photowatt, Solarex, Spire.Block V Manufacturers: Arco Solar, GE, MSEC, Solarex, Spire.

[66]Kurtz S (NREL) (July 14, 2008) *Reliability Approaches for Photovoltaic Technologies, International Summit on OPV Stability (ISOS)*, and Hipple M (1993) *The Performance of PV Systems, Performance and Reliability Workshop*, 453–460.

[67]Wohlgemuth JH, Cunningham DW, Nguyen AM, Miller J (June 6–10, 2005) *Long Term Reliability of PV Modules*, 20th European Photovoltaic Solar Energy Conference, Barcelona, Spain.

one module failure every 4200 module years of operation.[68] All of those BP Solar/Solarex modules were type-tested utilizing an accelerated stress test, based on the JPL Block V specification, the development of which has a historical importance, because if it would not have been developed, there would not be any confidence in the durability and life of the PV systems as we have now.

The "JPL Block program," its design, and its execution was crucial for the utilization of the PV solar energy to become a success. This was one of the most important and useful government-sponsored PV programs. Without this program the expected failures would have destroyed the image and usefulness of PV.

The "JPL Block program" as described above had three important requirements:

(1) Testing the PV modules to a *Specification* developed during the program.

(2) The manufacturer had to institute a QM system to ensure that all of the products would have the same quality.

(3) The right to inspect the premises and check on the implementation of the QM and select and take samples for testing, if needed.

To easily understand why all of these steps are needed to manufacture any kind of good-quality product, including PV, and to evaluate its quality, usefulness, and safety, we can take the example of manufacturing a life jacket (we all know that the purpose of a life jacket is that if somebody falls into water, it will keep the person wearing it afloat).

Step 1

The first step is to design a life jacket which will comply with an "International Quality Standard" (performance standard/ specification, e.g., ISO 12402–12403) providing the testing procedure that the life jacket should keep a person of certain weight afloat in water.

---

[68]Wohlgemuth J (March 2003) *Long Term Photovoltaic Module Reliability*, DOE Solar Program Review Meeting, Denver, CO.

This will be tested in a Testing Laboratory. If the product (life jacket) fulfills the Quality Standard, which means that the life jacket has passed all the relevant tests and meets all the technical requirements, in this case to keep an object of certain weight afloat, it will receive from the Testing Laboratory a "Type-Test Certificate."

Step 2

A production process has to be developed to make sure that all of the life jackets produced will have the same quality. The manufacturer develops the process of how to manufacture a life jacket and writes a QM plan, describing the steps for the production of the life jacket and establishes the points where some testing has to be done, to assure that all of the manufactured life jackets will exactly have the same quality.

Step 3

As an assurance for the consumer, an independent organization comes periodically to the production area to inspect and verify that the manufacturer's acceptable QM plan is carried out and is being adhered to. If that independent organization finds everything acceptable the manufacturer will receive an ISO 9000 certification. (Step 2 and 3 were the JPL program requirements, the ISO 9000 certification system has identical requirements. It is, as mentioned before, available only since 1987.)

Note: These steps (2 and 3), which are now actually the requirement to obtain the ISO 9000 certification, guarantee that all manufactured items will have the same quality, but it does not care what the product will be used for. (That means the life jacket could be made of concrete, but all of the manufactured product will have the same quality.)

Therefore, Step 1. "International Quality Performance Standard" is also needed to ensure that the life jacket produced was designed to comply to keep a person of certain weight afloat in water.

The example demonstrates that if only Step 1 (testing) is used, there is no guaranty that all of the products will fulfill the requirement. If only Steps 2 and 3 are used, all of the products will have the same quality, but they could be made of concrete, and will be useless. For this reason, all the three steps are essential like it was also mandated in the JPL program.

In life, however, there is one more requirement: safety.

Step 4

Neither Step 1, 2, or 3 will assure that the life jacket is safe, for example, it will not choke the person, or blow up, etc. This will require a "Safety Standard" for the product. Safety standard/specification, is needed which describes the test procedures for (in this case a life jacket's—UL 1177—Buoyant Vests) to ensure the safety of its use. In the USA, the Underwriters Laboratories (UL) established a safety standard for many products also for PV modules (UL 1703) and a safety standard for other products used in PV systems. TÜV Rheinland established a Class II specification for PV modules. Most parts of this TÜV-standard, together with some other international requirements, were accepted as the international standard IEC 61730 "PV module safety qualification."

In view of the above let's see what is the situation regarding PV components, primarily the modules.

Testing Specification Becoming PV Module Standard

The acceptance criteria in the Block I through V program was actually a PV module testing specification developed based on the experiences gained in the consecutive Blocks. An important issue was to develop a PV test and measurement standard by aggregating all of the work in the US Department of Energy (DOE) programs (JPL, Sandia, and MIT etc.). In 1978, Dick DeBlasio of the US Government's Solar Energy Research Institute (SERI)[69] was put in charge to aggregate all of these

---

[69]Solar Energy Research Institute (SERI) began operating in 1977 as the designated national laboratory of the Us DOE in September 1991, its name changed to National Renewable Energy Laboratory (NREL).

specification development efforts into a PV module standard through a consensus process also on international level.

The Block V PV module testing document was later somewhat revised and expanded to include additional tests in the USA. These tests were again validated at the SERI outdoor Test Facility before it became the IEEE1262 standard for PV Module Qualification Testing. About the same time the Block V test specification was established, the European Community was working through the European Union's Solar Test Center, JRC-Ispra on a European PV module standard which had similarities with the Block V document.

Dick DeBlasio's initiative was to establish all of these on an international level under the umbrella of the International Electrotechnical Committee (IEC)[70] located in Geneva, Switzerland, which is considered the globally accepted standard organization for electrical products. IEC established a new Technical Committee for PV named TC 82. DeBlasio with DOE money supported this new organization and established its secretariat in the USA. DeBlasio was chairman of TC 82 in the period of 1996–2002. Finally, all of these specifications were merged and became the presently used IEC standard (IEC 61215 Crystalline silicon terrestrial PV modules—Design qualification and type approval). In 2002, he turned over the Chairmanship to Dr. Heinz Ossenbrink the head of the PV section of the European Union's JRC-Ispra. The Secretariat of TC 82 remained in the USA. Since the IEC Technical Committee for PV named TC 82 was initiated, a great number of PV standards were established[71] (see Annex. 3).

---

[70]www.iec.ch.

[71]Annex. 3 lists all of the IEC photovoltaic standards available as of today (March, 2013). It also lists the PV GAP Recommended Specifications (PVRS) also utilized by IECEE [the Worldwide System for Conformity Testing and Certification of Electrotechnical Equipment and Components (IECEE)] documents to specify technical requirements for conformity assessment purpose for products not yet covered by IEC Standards.

Chapter 10

Dreamers and Sobering Reality

The US Government's first terrestrial-oriented solar activity was initiated under the National Science Foundation's (NSF) Research Applied to National Needs (RANN) program in 1971.[72] As described in Chapter 3, the NSF/NASA panel's conclusion was reported in December 1972. This was followed by the Cherry Hill meeting in 1973. The result was to conduct research to reduce the manufacturing cost of photovoltaic (PV) to be able to be used for either ground or space electric power-generating stations.

In 1975, the Energy Research and Development Agency (ERDA) was created, but its main purpose was nuclear energy. The PV program for this agency was an additional item. The government made a change in 1977 when President Carter created the Department of Energy (DOE) and ERDA became a part of it. In DOE, a division for energy conservation and solar energy was established.

---

[72]Morse FH, 1973.

(http://ntrs.nasa.gov/archive/nasa/casi.ntrs.nasa.gov/19740008685_1974008685.pdf).

Sun above the Horizon: Meteoric Rise of the Solar Industry
Peter F. Varadi
Copyright © 2014 Peter F. Varadi
ISBN 978-981-4463-80-5 (Hardback), 978-981-4613-29-3 (Paperback), 978-981-4463-81-2 (eBook)
www.panstanford.com

Starting with the 1973 Cherry Hill meeting, where it was predicted that the price of the PV cells will rapidly decline from $50/Wp in 1973 to the range of $0.50–$0.10 Wp and this prediction continued, as in 1980 a DOE official predicted that "by 1986 (in 6 years) we expect PVs to be fully economical for residential use in this country (USA) at a cost to consumers of 6 cents per kWh in the sunnier parts of the country."[73]

These predictions of the dreamers had three effects, one good and two bad.

(1) The good effect was that it gave the government the reason to drastically increase funding. The NSF funding for terrestrial solar energy projects (which included besides of PV six other different technologies, for example wind and so on) for the year of 1972 was $1.6 million. This skyrocketed by 1980 to $900 million for PV research and "commercialization."

(2) The first bad effect was that it was a shot into the "manu-facturer's" foot. The customers believing the government's predictions delayed the purchase of PV systems. They said let's buy the PV installation next year as it will be much cheaper referring to the predicted steep price decline.

(3) Another bad result was that the fledgling US PV industry was directed toward a dead end. Major US investors believed the government's predictions and did not believe the management of the PV companies to build up the market brick-by-brick, but directed them to focus on the "big picture" to be able to achieve that PV could be used for central power stations. The problem was that in spite of government predictions, the electric power industry, the utilities showed amazingly little interest in the proposed centralized PV electric power plants.

However, the DOE program also identified an extremely important subject: "communication" of the use of solar energy to the public. Unfortunately nothing came out of that effort.[74]

---

[73]Williams N (2005) *Chasing the Sun*, New Society Publishers, Gabriola Island, BC, Canada.

[74]A detailed description of this effort is in Neville Williams' book, *Chasing the Sun*, New Society Publishers, 2005, pp. 14–21.

The difference in "communication" of the nuclear power stations and PV power station was that nuclear energy was introduced not only to the US but to the entire world by two explosions which wiped out two cities in Japan in 1945, and also in 1953 when President Eisenhower announced to the UN General Assembly the *Atoms for Peace* program (Fig. 10.1).

Figure 10.1 US postage stamp in 1953.

Solar electricity never got such an introduction. As a matter of fact, the curse by introducing it to be used as a central power station degraded it to become one of the many possibilities to generate electricity, without realizing its unique nature, that it can be used as a decentralized electric power source converting light to electricity wherever electricity is needed.

What Is the Difference between the Centralized and Decentralized Electric Power Sources?

We are all used to having connectors on the wall and to be able to get electricity from that connector. If one plugs in, for example a radio or a lamp in that connector, the radio will function and the lamp can be turned on and provide light. One also knows that the connector gets electricity from the wire in the wall, which gets it from wires outside of the dwelling walls mounted on poles or located under the ground. It is connected to other wires, which finally lead to a power station producing electricity, which also provides electricity to a very large number of customers. These are the centralized electric energy sources.

Centralized energy sources belong to the utilities which are operating the so-called power stations, generating a large amount of electricity in a central location and distributing electricity via metal wires—power lines—called the "grid." These power stations are fueled by oil, gas, hydro, or nuclear material.

A decentralized power source, contrary to the centralized one, produces electricity at the location where it is needed and produces it in the quantity needed. One can consider it like a flashlight or a cell phone operated by a decentralized power source, a battery. Electricity generated by solar cells is doing exactly that. Solar systems can produce electricity wherever it is needed and being modular, can produce the needed quantity and be increased if needed by adding more solar modules. Solar systems can be mounted, for example on the roof of a house and its size can be designed to supply the house with the required electricity. A PV system can be, for example, purchased by a store, set up on the roof of the store, and reduce the peak power requirement, which the electric utility is selling at a premium.

The idea that PV should become to be used as a central power station disregarded the PV system's main advantage and unique feature over other electric power sources used for central power stations. It disregarded that it can be also used as a decentralized power source, useful on earth and in space. It can be installed wherever electricity is needed, providing electricity to the satellites for the GPS system, but it also can be deployed anywhere on Earth, not as a large central station operated by the utility but in smaller, being the roof of a private dwelling or larger systems operated by Independent Power Producers (IPP) to produce electricity connected to the grid and sell the electricity to the utility. This idea was around since sometime,[75] but the utilities until recently were fighting

---

[75]Palz W (May, 11–15, 1981) *IEEE Photovoltaic Specialist Conference*, Houston, TX.

the connection of PV to the grid. Instead it would have been advantageous for the utilities to get into that business. Utilities with their centralized power stations would have benefited from a new and environmentally, friendly power generation system.

The utilities may have given some research or study contracts, but they bought practically no serious amount of PV hardware, not even for demonstration purposes. An exception was the Sacramento Municipal Utility District (SMUD). SMUD has been a leader among the utilities in solar power. SMUD's built the USA's first utility-scale solar array at Rancho Seco in 1984 (Fig. 10.2). SMUD owned a nuclear power plant, but shut it down by a vote of the utility's rate-payers in the late 1980s. Although the nuclear plant is now decommissioned, its now-empty iconic towers remain at the site. Solar arrays were installed in proximity to the towers.[76]

Solar array

Figure 10.2 SMUD's "Rancho Seco" shut down nuclear power plant—
utility PV in foreground.

Five additions later, the site generates enough electricity, 3.2 MW, to power 2200 single-family homes. SMUD is also helping local homeowners put solar panels on their roofs.

[76]http://en.wikipedia.org/wiki/Sacramento_Municipal_Utility_District.

Another exception was the German utility RWE, which went in and out of solar cell and module production[77] and now surprisingly realized that PV may provide them with a new and large opportunity.

In retrospect, one can speculate why the electric utilities had not even a little interest in exploring if PV would be useful for them. PV was probably considered a scientific curiosity good for small applications, like satellites. But the utilities' aversion against PV could have been also the result of the utilities' "corporate culture" because PV systems needed no fuel, had no moving parts, and it could be used as a decentralized electric power source, this was totally alien to what the entire electric utility industry was used to.

But at that time the sobering reality was that the dream that the electrical utilities would use PV for central power stations or PV would find its place in their electricity providing system was only a dream as the utilities had no interest to do it as a matter of fact they not only ignored it, but were mostly hostile to the idea. So, who were the people and/or organizations involved and what contributions had they made that the terrestrial use of solar electricity was developed from an idea discussed at the Conference in Paris and during the Cherry Hill Conference in the USA in 1973, to the reality of today, where enough PV systems are globally deployed, which would be sufficient to provide all the electric power needed even for a highly developed country such as Switzerland.[78] The list is very short:

---

[77]Germany's largest Utility, RWE had a nuclear material division NUKEM. In 1979, added to this a solar module manufacturing division (NUKEM's solar division)—probably to give a little green color to the nuclear materials. Under the leadership of Dr. Winfried Hoffmann in 1995 it became a joint venture between NUKEM and Daimler Benz Aerospace and was renamed "Applied Solar Energy" (ASE). In 2002, it became RWE-SCHOTT Solar GmbH and after Schott bought out RWE it became SCHOTT Solar. Schott Solar recently stopped the production of solar cells and modules utilizing Si wafers.

[78]Wilhelm I (May 2011) *President of European Photovoltaic Industry Association (EPIA)* SOLARIS Newsletter.

- governments of the European Union, Japan, and USA
- oil industry
- utilities
- "Determined men who made gutsy calls, took big risks," as Bob Johnstone describes them in his book.[79]

There are a lot of different opinions about the contribution of government, oil industry, and utilities. I was closely involved with them in the first 10 years of the terrestrial PV industry and was an "active" observer for the rest of the time; therefore, I think I am in a position to draw some conclusions. From this group, I believe that the governments and the oil industry made important contributions, while the utilities were rather handicaps.

Government

Government contribution is clouded by the fact that when the government changes, its policy could change drastically. A good example is the USA, where during the Carter administration Renewable Energy, including PV, was very much supported, but when the Reagan administration came to power the government's support was drastically reduced. Companies should be prepared for business cycles, but I believe that it is the company management's fault if the company is fatally hurt by a change. PV is a commercial business in which the government could be a customer. If a company forgets this and relies on the government business only, it deserves to go out of business. It is an old saying: If a business has 100 customers and one changes its mind and is not buying, it is not a problem, only 1% of the business is lost. But if a company has only one customer and that customer changes its mind and is not buying, the company is out of business.

On the other hand, governments provided money for R&D, which nobody else would have done. Governments developed programs to promote the reliability of PV modules, for example the JPL Block buys,[80] without which there would be no PV

---

[79]Johnstone B (2011) *Switching to Solar*, Prometheus Books, Amherst, NY.
[80]Details in Chapter 9.

business today. The government supported demonstration and Pilot Application programs in Europe,[81] Japan, and USA, which provided assurance for the applications of PV.

One more extremely important result of the government contribution was the introduction of the ingenious system of the German Feed in Tariff (FiT)[82] (described later), which is now adopted in many countries. For this reason the government's contribution, in my opinion, in spite of what other people think, was very valuable.

Oil Industry

It is true that the oil industry believed PV would instantly be producing utility-sized central power stations but as this did not pan-out rapidly and because of their corporate culture they missed becoming a major player in the PV business, but in one respect, as it will be seen in Chapter 24, they were crucial to the development of the PV industry.

Utilities

Utilities were and in many places are still trying to prevent the connection of PV to the grid. But the fight with the utilities is more or less over since Hans-Josef Fell and Hermann Scheer, both members of the German Parliament, forced through the German Renewable Energies Law (FiT). Utilities now face the question: Is the Distributed Solar (or wind) going to drive Utilities into Bankruptcy?[83] Obviously the answer is No. Solar electricity is only available during daylight hours, wind electricity is only available when there is wind, but customers require 24 hour uninterrupted electric service. Utilities now have to face the problems of availability, existence, abundant, and reasonably priced solar and wind electricity.

---

[81]Annex. 4, Palz W, Early Photovoltaic Pilot Applications in the European Union, Private communication.

[82]Details in Chapters 32 and 33.

[83]Konrad T (February 28, 2011) *Will Distributed Solar Drive Utilities into Bankruptcy?* Renewable Energy World.

Determined Men Who Made Gutsy Calls, Took Big Risks

The experts looked down on the world from their ivory tower. We "determined men" looked at the world from our trenches. We knew that the business and profitability hinged on the silicon (Si) wafer cost and on the technology for terrestrial solar cells. We had to solve the Si wafer "enigma." We had to solve the technology to manufacture inexpensive quality products. These issues were described in previous chapters. What was left and was equally important to find and develop markets and manufacture products at a price that could be sold to that market. We also knew that this is like building a house brick-by-brick and not only looking at a mirage of a central power station made of PV.

But here comes the sobering reality. The investors and mostly were big oil companies, believed the central utility mirage, the "big picture" which the government experts predicted agreed with the huge business models they liked. But we in the trenches were very happy if some of the roadside emergency phones started to use solar electricity.

This result of believing in the central utility mirage can be described by the following story.

An ENRON executive told me this at a breakfast, when ENRON invested in a production facility to make amorphous silicon (a-Si) solar cells and modules. He said that it would be easier to sell a solar electric power station than one which was powered by oil or gas. The reason was that before building oil- or gas-powered power station three important issues had to be secured. The first was to have a long-term agreement for the supply of oil or gas at a proper price. The second was to raise money to build the plant. The third was to find customers with a long-term contract to buy the produced electricity.

"However, for a solar electric power plant—he said—you need to secure only two issues. One does not have to secure a long-term contract for the fuel, because the nuclear reactor in the sun provides it free of charge therefore, only the money has to be raised to build the plant and customers with a long-term contract had to be found. And this simplifies the matter of building central power stations."

We all knew that what he said was true, but we also knew that in those days "grid parity"[84] was only a fata morgana. The ENRON executive never could have tested the utilization of solar electricity for power plants, because ENRON on account of other "brilliant" ideas went bankrupt and he was probably looking for a job to guide some other company into the simplified establishment of a power plant.

The four of us in the three companies believed in the future of terrestrial PV. We believed that one has to find a market and make the product for that market. This is the first brick, after that the second is to enlarge the production and this has to be done globally. And this was a crucial element for those of us—Bill Yerkes, Elliot Berman, Joseph Lindmayer, and the author of this book—who started the terrestrial PV business,[85] and had to sell the PV products. We found that PV was not only a small local, but it was a large global market. This market grew from practically zero in 1973, in 15 years later in 1988 to 33.8 MW[86] and in 39 years in 2012 to 31,518 MW.[87]

Our idea was the same, but the companies we operated had a different setup.

Solar Power Corporation (SPC) was started by Dr. Elliot Berman but was owned by Exxon. He started to develop markets and started to expand sales to Europe. Exxon obviously did not believe in deviation from the path to reach the big picture. The brick-by-brick and the big picture idea collided and Elliott left Solar Power after which SPC did not reach the goal in 10 years and was closed by Exxon in 1984.

Spectrolab making space solar cells under the direction of Bill Yerkes also started to go after the terrestrial PV market.

[84]"Grid parity" is the point at which renewable electricity is equal to or cheaper than grid power.

[85]Swanson RM (2011) The Story of SunPower; in Palz W, ed., *Power for the World*, Pan Stanford Publishing Pte. Ltd., Singapore.

[86]Surek T (2003) *Osaka Japan: World Conference on Photovoltaic Energy Conversion*, www.nrel.gov/pv/thin_film/docs/surek_osaka_talk_final_vgs.pdf.

[87]Private communication from Paula Mints, SPV Market Research.

As mentioned before, Textron, the owner of Spectrolab, never liked the solar cell business and sold the company to Hughes Aircraft Company in 1975 and Bill Yerkes was immediately let go. He straight away started Solar Technology International (STI) and began to produce terrestrial solar cells. In 1978, ARCO (the big American oil company) bought STI and renamed it Arco Solar. Luckily, ARCO management was not hell bent on the big picture and Bill successfully started the brick-by-brick approach.

Joseph Lindmayer and I started Solarex with a number of stockholders, including AMOCO and were able to successfully develop it on the brick-by-brick and international expansion principle to become the largest global PV manufacturer what Berman tried at SPC. We were not smarter than he was, only probably luckier or maybe we were like the title that Andrew Grove of Intel, another Hungarian refugee, gave his book: *Only the Paranoid Survive.*[88]

---

[88]Grove A (1996) *Only the Paranoid Survive, Doubleday* Business.

Chapter 11

Photovoltaic Systems

Even if we would tell that we were working to achieve the big picture of which we believed at that time it was a "fata morgana," we had to start making solar cells and modules and find an immediate market where they could be used and sold to pay for our expenses. Each silicon solar cell when illuminated produces about 0.5 V electricity. We have seen that by connecting the photovoltaic (PV) cells in series the resulting output voltage is going to add up. Thirty-six cells connected in a series will be useful to recharge a 12 V lead-acid battery. If a higher voltage is needed, PV modules can be connected in series. Obviously the power output, as described, will depend on the size and efficiency of the cell and the intensity of the light. This solar module can then operate light bulbs or other electric systems. Solar cells/modules are very simple to use as compared to most other energy sources. A solar electric system has to be set up outdoors facing south (on the southern hemisphere facing north) it has no moving parts and requires little maintenance, if any.

It was self-evident that if we manufacture PV cells and modules, their sale will require a terrestrial PV market. We

Sun above the Horizon: Meteoric Rise of the Solar Industry
Peter F. Varadi
Copyright © 2014 Peter F. Varadi
ISBN 978-981-4463-80-5 (Hardback), 978-981-4613-29-3 (Paperback), 978-981-4463-81-2 (eBook)
www.panstanford.com

had to come up with PV systems, power sources, providing everything the customer would need to run their equipment. It was obvious that there was a great deal of user education to be done. We realized that the customer's first objection will be that when sunshine or light is not available solar modules do not produce electricity, for example, at night or on very cloudy days. It is a legitimate question how can one depend on a PV system to provide the needed electricity when there are times when it cannot be used.

For the utilization of solar electric power for satellites, this problem was solved by using a battery which is being charged while the satellite receives sunshine and the battery provides the electric power for the satellite when the satellite is in the shade of the Earth.

For terrestrial application of PV one can also use an electric storage system like in space use, for example a battery, but for terrestrial use it could also be water. The system is designed, that the PV array (a number of PV modules connected to each other) supplies electricity to operate the system and if more electricity is produced than it is needed it is stored in the storage system. When the PV system is not producing electricity (for example at night) and the electric system needs electricity, the battery which was charged by the PV system, will provide it. In case water storage is being used, the PV when it produces electricity will pump water up to a reservoir and the electricity can be continuously generated by the water flowing down through a turbine.

The PV system which has a storage media is called "stand-alone" or "off-grid" systems, because the PV system operates independent from the electric power grid, the electric power needed is provided only by the PV modules.

A schematic of a "stand-alone" or "off-grid" system is shown in Fig. 11.1.

During the first 20 years of the terrestrial PV business most of the systems were the "stand-alone" "off-grid" applications. Details of such systems are described in the next chapter.

The other mode of utilization is to connect the PV system directly to the utility's electric grid, "grid connected" systems. Because on many places on Earth there is an electric grid the

PV system can also be connected directly to the grid and supply the electricity supplementing the electric power generated by a central station. The schematic of this system is shown in Fig. 11.2.

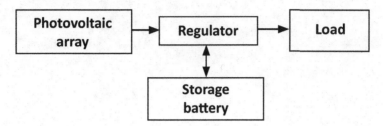

Figure 11.1 Schematics of a "stand-alone" (off-grid) PV system.

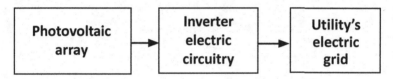

Figure 11.2 Schematic of a "grid connected" PV system.

No electric storage device, such as a battery, is needed instead it uses an electric circuit (inverter) which makes the electric power generated by the PV system compatible with the utility's electric grid. The usefulness of the grid-connected PV is that it provides electric power during the peak power demand, for example at hours before and after noon, thereby helping the utilities to satisfy the needs, without installing more generating capacity, which would be idle in off-peak time.

As mentioned, most of the PV systems were the "stand-alone" types until 1997 when more of the "grid connected" systems started to be used. The table turned around and the great majority of PV systems became connected only to the electric grid. In the USA, this crossover happened in 2004.[89]

Why this crossover happened in Europe only after 25 years and in the USA after 30 years after the terrestrial PV systems were introduced is an interesting question.

---

[89]Price S, Margolis R (January 2010) *2008 Solar Technologies Market Report*, DOE/Go-102010-2867.

Chapter 12

"Stand-Alone" or "Off-Grid" PV Systems

As mentioned in Chapter 11, during the first 25 years of the terrestrial PV business practically all of the systems were the "stand-alone" ("off-grid") applications. They also call them "stand-alone" because the needed electricity is only provided by the PV modules.

There is a very big difference in stability of the business between the "off-grid" and the "grid connected." The "off-grid market" is mostly not subsidized commercial market, selling to customers who need it and can use it at the price the PV system is offered. The "grid connected" was mostly subsidized to promote the utilization of PV for environmental reasons. In Germany as it will be described later, the government established the so-called Feed in Tariff (FiT) system. FiT became very popular and is being utilized in many countries, which made the grid connected PV market very attractive for financial reasons and therefore, it caused a very large demand for PV modules. As expected, this made the PV modules relatively inexpensive, which resulted in an explosion of the demand as described in the later chapters (Act 3).

Sun above the Horizon: Meteoric Rise of the Solar Industry
Peter F. Varadi
Copyright © 2014 Peter F. Varadi
ISBN 978-981-4463-80-5 (Hardback), 978-981-4613-29-3 (Paperback), 978-981-4463-81-2 (eBook)
www.panstanford.com

The drastically lower prices also resulted in the increased usage of PV for the commercial "off-grid" market which, therefore, increased in size. In 1999, when practically the entire PV market was "off-grid," 200 MW was sold[90] for that market. In 2008, the "off-grid" pure commercial market was up to 830 MW[91] and as this market is mostly not influenced by government subsidies one can take its traditional yearly growth of 10–15% and thus can be estimated to be in 2012 about 1.5 GW. The "off-grid" market in 12 years increased about 7.5 times, but as mentioned it is a stable dependable business. On the other hand, the "grid connected" market exploded and grew about 70 times during the same period. But because it is a mostly government-regulated market it is unstable.

The "off-grid" utilization of PV today covers so many applications that it is totally impossible to write down all of them. One can see PV modules to be utilized on city streets, highways, and gardens. One cannot see many applications, because they are in remote areas. The first major utilizations of the "stand-alone" or "off-grid" systems were for three application areas: communications (Fig. 12.1), telemetry, and navigational aids.

Figure 12.1 Telephone repeater station in Thailand.

[90]From PV News, Maycock P, (ed.) (2002) yearly February editions.
[91]Price S, Margolis R (January 2010) 2008 *Solar Technologies Market Report*, DOE/Go-102010-2867.

After the terrestrial PV business was started, we realized the incredible demand and need for the large variety of telecommunication equipment which existed. The telecommunication equipment was needed, but the electric grid did not reach large areas in a huge country like the USA or many other areas in the world and the problem was to supply these needed systems with electricity. The required electric power was usually small and to bring the power line to that place would have been prohibitively expensive. The only way they were able to provide them with electricity was using batteries or thermoelectric generators. The difficulty was that the batteries had to be periodically replaced and the gas for the thermoelectric generators resupplied. In most cases, this was extremely difficult but the need for communication existed and had to be fulfilled. Because of their remote locations and need for a reliable power source, these applications were ideally suited for photovoltaics (PVs). In several countries, telephone companies set up PV-powered repeater stations to cover the entire country. One of these was in Thailand, where the telephone company purchased and installed over 40 repeater stations in 1980.

But at the beginning, the price of PV was high and more markets had to be found to increase the volume of production, which as expected decreased the price of the PV systems drastically. PV became widely used also in countries where 95% of the population had no electricity for pumping water and providing light, powering TV, and one might not expect in those areas, but it is also widely used to recharge cell phones.

PV manufacturers' problem was how to find the users or how the users become knowledgeable that PV existed and could solve their problems. The following story from Nepal actually provides an interesting insight how the need resulted in the utilization of PV for telecommunications and how it solved the difficult battery replacement problem and how the user realized that PV would solve the problem and how they found the PV manufacturer.

We suddenly started to sell large amounts—considering the production of those days—of PV modules to Nepal. The amount was relatively large so during a trip to Asia, I visited Nepal, where our representative, the owner of Greyhound Electronics in Kathmandu, the capital of Nepal, introduced me to the customers. They all worked for the Government's Telephone system. They were excellent Nepalese engineers and several of them to my surprise spoke fluent Hungarian. As it turned out, those got their engineering degree from the Technical University of Budapest, which is one of the best universities. So my discussion with them in Nepal was either in Hungarian or in English. Their story was remarkable, how they wound up starting to use PV in 1978 when it was even in the USA or Europe a new electric power source. Communication equipment was needed in remote areas and PV systems were uniquely useful to provide the electricity for them. After PV proved itself useful and dependable, other applications followed.

The beginning was that an American NGO decided to establish a birth control program in Nepal and for this reason it had to provide for the people radio sets for distraction purposes. The NGO realized that for continuous use of those radios, they needed rechargeable batteries, but as there was no electricity in Nepal, where these radios were intended to be used; they learned that PV modules would be able to charge the batteries. So they purchased small PV systems, including charge controllers and the batteries from Solarex, which we delivered to Nepal. But at that point the NGO ran out of money and was not able to buy the radios. The PV modules and batteries were delivered and stored in a warehouse in Kathmandu.

In the meantime, large number of tourists started to come to Nepal, to visit the Himalayas and the telephone company until then having no need for communication had to establish communications for the tourists. Because of the mountains, they obviously needed a large number of repeater stations to establish connections. Having no electric power lines they could be powered only by batteries. Frequently, they had to replace the batteries so they hired a bunch of people who carried by foot the batteries to the repeaters and replaced the old ones which they had to carry back. Such a round trip meant a minimum of 1 month for that person.

On my remark that this could have been done much faster by utilizing a helicopter, I saw that their opinion that all Hungarians being intelligent people probably changed. They only told me that to some places the helicopter could not get near the repeater station and besides it would have been very expensive compared to the people going by foot.

In desperation they came up with the possibility that if they would try to use the PV modules and batteries which the NGO purchased for birth control purposes and were stored in the warehouse that would maybe solve the problem. They got permission from the American NGO to use them and installed them to power some of the repeaters. When they proved to be reliable and no battery replacement was needed they converted all of their remote repeaters to become PV-powered. That was the reason they were buying many more PV systems. They started to use PV for other applications also. They used it for powering meteorological stations which they also needed for the tourists, but also for remote health clinics having no electricity where they used PV for refrigerators, electric lamps, microscope lamps, and for two-way radios that the health workers would be able to communicate with physicians 50 miles away.

It was also quite surprising to me that one of the first big markets for PV systems was actually in Alaska, where everybody knows a part of the year there is practically no sun. We were not anymore surprised when we started to learn the need for communication, weather stations, and so on where PV was the only solution to recharge batteries, because in Nepal they had the option to send somebody to walk for a month to replace batteries, but in the incredibly large area of Alaska neither this option nor the use of helicopters was feasible.

These applications, some of them small, working only intermittently, and others which were for 24-hour operation, needed a variety size of PV systems. These types of PV systems started to be used not only in Alaska but all over the globe for communication systems, weather stations, and so on and also like the one mentioned for the Arizona police repeater.

When the terrestrial PV business was started in 1973 we had to come up with a way to design these systems, to supply the required amount of electricity to reliably operate the electric load. Obviously that was the time when a "computer" was not a household word and Microsoft Corporation did not exist yet. But when today one goes to an Internet database and asks for information about "PV system design" one gets an incredible amount of sites to obtain "help" to design a PV system.

The situation to design a PV system in those days can be compared, for example, to the Greeks who had no numbers but were able to build edifices that are still around after centuries of neglect. The Romans had numbers, only for counting, because one could not use them to add, subtract, multiply, or divide. In spite of this and the neglect of centuries, their buildings, aqueducts, and so on are still standing in many parts of the world. On the other hand, for the buildings or bridges built with the aid of computers we are lucky if they are still around for a 100 years if they get proper maintenance.

In PV system design in those early days we used the same very simple principles as are being used today in computer-operated programs, except it was done with the handheld calculators and with less accurate information. The question was the same: How large a PV system, including PV modules and batteries, are required to support a given load at a specific site with acceptable reliability? These are the most basic questions in designing a PV system for a specific application.

The difference between the "primitive" design of the 1970s and today's computerized design is that we had to use, like the Greeks or Romans, a high "safety factor," making the system bigger than maybe needed and because of that more expensive than today's sophisticated programs are able to do, minimizing the safety factor using better information and more sophisticated calculations. After years of using only handheld calculators to figure out the need of the load, the first attempt to automate PV system design was introduced by Solarex[92] in 1975, with the SUNRAY I system. The greater precision and capability of SUNRAY II was available from 1981.

---

[92]Solarex Technical Note 8301, October, 1983: A. McKegg and Bill Rever: The SUNRAY II system design program.

To answer the question of how to design a "stand-alone" PV system, first we have to establish that the "stand-alone" PV system has four components: (*In order to make for the reader easier the diagram from the previous chapter is reprinted here.*) (see Fig. 12.2) the "Load" which defines how much electricity is needed; the "PV array," which provides the required electricity; the "battery storage," which as was mentioned stores the electricity and delivers it to the load when the PV array is not producing sufficient amount for the operation of the load. The fourth component is a "charge controller," the purpose of which will be discussed later.

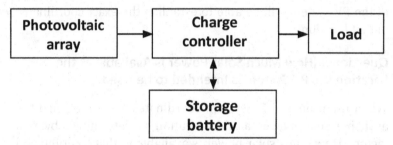

Figure 12.2 Components of a "stand-alone" PV system.

To design the solar electric system to fulfill the needs of the load is very simple. One requires certain information to complete it, but the design can be accomplished utilizing only very minimal arithmetic. One has to answer the following simple questions:

(1) How big is the electric load and how much of the time in a day is it needed?

(2) Where will be the location of the PV system?

(3) How much solar power is available at the location the PV system is intended to be used?

Question 1 (How Big Is the Electric Load and How Much of the Time in a Day Is It Needed?)

- The load can be intermittent, for example lights, radio, and TV used in a house. They are on maybe 5 hours a day. Considering that the load consist of three items: Two 50 W light bulbs and a TV drawing 50 W. Everything is used 5 hours a day, therefore, the total load is: $3 \times 50 \times 5 = 750$ Wh/day.

- If the load is continuous, for example a communication repeater station which needs 50 W for its operation, the total load will be $50 \times 24 = 1200$ Wh/day.

Question 2 (Where Is the PV System's Location?)

This is an easy question. The customer buying the PV system will tell where the system should be exactly located. The designer of the PV system will be able to establish the exact coordinates of that location.

Question 3 (How Much Solar Power is Available at the Location the PV System is Intended to be Used?)

When the proposed PV system coordinates are established the available solar power at that location is determined by two factors. How much solar power is available at that location; and how the atmospheric conditions such as clouds and so on are influencing the availability of solar power?

To find out how much sunlight is available in any given location for the generation of electricity is very simple. This information was available even in the 1970s 40 years ago from government, air traffic, or weather information sources[93] available for most of the Earth.[94] Today, measured solar radiation data are available at a number of locations throughout the world. Data for many other locations have been estimated, based on the measurements at similar climatic locations. The data can be accessed through the Web sites of national government agencies for most countries in the world. Worldwide solar radiation data are also available from the World Radiation Data Center (WRDC) in St. Petersburg, Russia. WRDC, operating under the auspices of the World Meteorological Organization (WMO) has been archiving data from over 500 stations

[93]http://www.oynot.com/solar-insolation-map.html.
[94]http://en.wikipedia.org/wiki/Insolation.

and operates a web site in collaboration with the National Renewable Energy Laboratory (NREL).[95,96]

As it can be seen, in order to find out how much solar energy is available for conversion to electricity by PV systems requires no measurements, it is available for the entire world from government or private (for example airports) sources.

The most useful measure of solar radiation is that which expresses the average number of "peak sun hours"[97] per day for a particular location. As an example is the table in which the average sun hours are listed for two very different locations, Fairbanks, AK (64° 50' 16" N) and Phoenix, AZ (33° 26' 53" N). These two locations were chosen so that one can see that even as far North as Fairbanks AK PV systems can be used. The difference is that they may need to be twice the size that is needed in Phoenix, AZ.

Knowing the average number of peak sun hours at a site and the average daily load requirements in watt-hours, one can quite readily calculate the total watt output necessary for the PV generator. Naturally the system will need batteries that the load, lights, and TV should be able to operate when it is dark. Furthermore, one has to consider that there are maybe days when it will be too cloudy and the batteries will not be recharged at all during those days. For security reasons one can consider that no charging will occur 7 or even 15 days, therefore, the size of the batteries has to be designed accordingly.

The purpose of the "Charge controller" is that in many locations the amount of solar energy available will vary substantially over the course of the year. The example of Fairbanks and Phoenix demonstrates this. As shown in Table 12.1, at a northern location such as Fairbanks, AK, much more energy

---

[95]http://www.worldenergy.org/publications/survey_of_energy_resources_2007/solar/720.asp.

[96]http://wrdc-mgo.nrel.gov.

[97]The definition of "peak sun hour": The equivalent number of hours per day when solar irradiance averages 1000 w/m². For example, 6 peak sun hours means that the energy received during total daylight hours equals the energy that would have been received had the irradiance for 6 hours been 1000 w/m². (http://photovoltaics.sandia.gov/docs/glossary.htm.)

is available in summer than in the winter. As the size of the PV array will be large to assure the system's operation also in the winter months, the PV array will put out more power than is required during some months of the year. To prevent the destructive effect of overcharging the batteries, a voltage charge controller should be used to keep the batteries fully charged, but dissipate the excess power.

Table 12.1 Peak sun hours at two different locations in the USA

Location	Summer average	Winter average	Year average
Fairbanks, AK	5.87	2.12	3.81
Phoenix, AZ	7.13	5.78	6.58

In Fig. 12.3, "Yearly average peak sun hours per day" is included to indicate that peak sun hour charts are available for most of the locations on the Earth. It also shows that one can obtain data very easily, where the radiation figures have been corrected for the solar cell module being tilted 45° and oriented due south. This has been the optimum orientation for modules in the most parts of the United States.

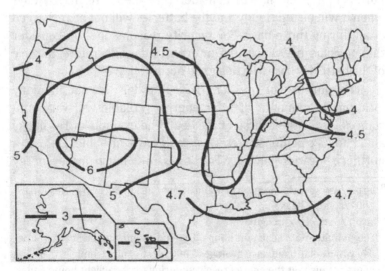

Figure 12.3 Yearly average peak sun hours per day in the USA corrected for the module tilted 45°.

The above only indicates that the design of a PV system to operate a load is not complicated and obviously today many computer programs are available, which can design any size system anywhere on Earth.[98,99]

But, even the best designed system can cause surprises.

Before the introduction of SUNRAY I and the first computer-aided design system, systems were designed by using a handheld calculator and were vastly overdesigned to make sure the system would operate perfectly as we could not afford to have one of our systems fail. In 1975, after design and installation of a 300 W system for a transponder beacon in Kenai, AK, the system worked well until the early part of the winter when a call was received from the Federal Aviation Administration reporting that the system was not working.

One of our technicians was in the process leaving to find out what was the problem and fix it, but before he left we received a picture of the PV system, which we requested as we needed it for an exhibition for the Smithsonian Institution which asked us to provide pictures about the utilization of PV. We knew that the airport was 60.5731° N and the system was designed accordingly, but looking at the picture we realized the beautiful trees surrounding the clearing where the PV system was set up were much taller than we have ever seen. Upon investigation, it was discovered that the modules have been mounted at about 20 feet surrounded by 40-foot-high trees. Naturally, the surrounding trees blocked the sun starting the early part of winter, when the sun is low. When our technician left, his job became very easy. He had to raise the height of the poles the system was attached to over the level of the trees that the sun should be able to shine on it (Fig. 12.4). The modules were raised and the problem was solved.

A problem was also the knowledge of the technical people dealing with the customer's PV systems. Obviously, years ago, in the early days when PV was a novelty this was a much bigger problem at that time than it is today. For example, for the 40 Thai telephone repeater stations in order for them to function well we invited 20 Thai technicians to our factory in the USA to give them a week course to teach them about PV systems including maintenance. We never had any problems with those systems. Today when a large number of PV systems are in operation there are plenty of technicians trained to work, maintain, and repair them.

Figure 12.4 Kenai, AK, 300 Wp PV systems for an airport transponder beacon, 1975. (a) PV system original height. (b) PV system raised above tree level.

Charge controllers are being used also in automobiles. In PV systems their function is to connect the PV system with the battery so all the electricity generated should charge the battery and also connect the battery to the "load" as it should receive electricity when needed. Their function is also to protect the storage battery. Each battery type has its own peculiarity when it is being charged. Overcharging means that the battery is continued to be charged at voltages exceeding the maximum voltage a battery is made for. This is damaging the battery. Charge controllers are also used to protect the battery from total discharging, because that also would damage the battery.

Chapter 13

1975 the Year of Change

As mentioned, in 1973, two companies were started with the purpose to develop and manufacture products for the terrestrial photovoltaic (PV) market: Solar Power Corporation (SPC) owned by Exxon and managed by Elliot Berman; and the privately owned Solarex managed by Joseph Lindmayer and Peter Varadi. In 1973/74, Bill Yerkes was able to expand the Textron-owned space solar cell manufacturer Spectrolab to also add a product line for terrestrial PV cell and module production.

In 1975, the situation changed. SPC was a formidable competitor to Solarex, but Berman left SPC in 1975 and after that SPC's European offices were closed. In spite of SPC having an excellent technical staff, without Berman's vision of how to build up the terrestrial PV business they concentrated on making products which ultimately could be used for central power stations. For Solarex, this opened up Europe and SPC was not a competitor anymore in many of the PV business areas, such as the entire consumer field and smaller systems. Also in 1975, Textron sold Spectrolab to Hughes Aircraft. Hughes immediately let Yerkes go and shutdown the terrestrial

Sun above the Horizon: Meteoric Rise of the Solar Industry
Peter F. Varadi
Copyright © 2014 Peter F. Varadi
ISBN 978-981-4463-80-5 (Hardback), 978-981-4613-29-3 (Paperback), 978-981-4463-81-2 (eBook)
www.panstanford.com

PV business. Yerkes formed Solar Technology International (STI) and continued the terrestrial business, but he had not much capital and had to start on a small scale. In 1978, Yerkes sold STI to ARCO, a large US oil company, and the business was renamed Arco Solar. ARCO invested money and Arco Solar was rapidly expanding. In 1978, 5 years after the terrestrial business started, the total world production of PV modules reached about 1 MW/year.

All this meant that from 1976 Solarex became the largest terrestrial PV manufacturer in the world until about 1980, when Arco Solar under the direction of Bill Yerkes became a formidable competitor. Interestingly, most likely because the oil company's direction Arco Solar also followed the direction toward large-scale central systems, it did not enter the consumer-oriented PV business, where Solarex retained its monopoly. Solarex by 1978 became the dominant terrestrial PV cell and module producer. This is reflected in Table 13.1.

Table 13.1   Shares of PV market percentage of the World in 1978[100]

	Shipped KWp (%)
Solarex	45
Solar Power	17
Arco Solar	12
Motorola	7
Sensor Technology	6
Philips-RTC	5
Sharp	1
Other	7

The reason Arco Solar' percentage was smaller than Solar Power's because ARCO purchased STI only in 1978. Sensor Technology was a semiconductor manufacturer and participated in the JPL block buys only until 1978. Motorola, also a semiconductor manufacturer, started to make solar cells and

---

[100]*Solarex International Newsletter* 2(2), 1 (Based on Strategies Unlimited's findings appeared in the July 30 issue of the "Solar Energy Intelligence Report").

participated in the JPL block buys in 1976 and 1978. Motorola was not able to produce enough solar cells to satisfy the quantity they had to make for JPL and bought what they needed from Solarex. After 1978, Motorola dropped out of the business. In Europe, Philips RTC in France started to produce terrestrial PV modules, mostly for the Australian telecom market. In Japan, Sharp made PV modules for telecommunication and navigational aid purposes.

Solarex, having no restrictions to march only toward central power stations, starting in 1975 developed an extremely diversified PV business, fulfilling requirements for small PV modules, and started a big variety of consumer-oriented business. The consumer-oriented business and producing small PV modules in which Solarex's deep-pocketed competitors did not participate because they were chasing the "big picture" was found to be very profitable for a variety of reasons. Only Sanyo and Sharp of Japan participated in a segment of the consumer-oriented PV business making PV micro modules for watch and calculator businesses.

The consumer product PV business was never taken seriously, "scientists" and "solar experts" probably did not know about it or if they knew they disregarded its existence. It never came up in any government studies. However, in the history of the terrestrial PV business the consumer-oriented business, as will be seen, is definitely extremely important. The importance of the consumer product business of PV was never acknowledged and its history until this book was never written. The importance of the PV consumer business is manifold.

First of all, it helped to introduce the usefulness of PV to people. In the first 10 years of the terrestrial PV business people had no opportunity to see a PV system, because they were all deployed in remote areas. They may have seen it on a roadside emergency telephone if they looked up at the top of the poles they could see a PV module, but they probably would not know what it was because it was never written on the phone or on the module, that it was a solar-powered telephone. On the other hand consumer products starting with watches, calculators, and other products were advertised that they are powered by solar cells, so consumers could notice that.

In addition, consumer products utilizing PV were never subsidized and in spite of that they were very profitable. How profitable they were one can understand from the fact, that the value of its main component the silicon solar cell used for consumer products at that time was zero and even today its value is very low. The reason is that, for example, to make the little modules for watches or other products, one was able to cut them from broken or chipped solar cells or from solar cells which are not able to be used for commercial modules, the value on the books of these solar "scrap" was zero.

In the 1970s, the consumer business was about 20% of Solarex's entire income and probably 80% of the company's profit. I tried to research what is the situation today. How many companies and how many workers are involved in the PV consumer business, how much PV is being used in that business and how much could be the total sales. I was not able to get an answer, not even from people who make their living to make this sort of estimates. Based on various inputs, I assume worldwide there are over 1000 companies manufacturing PV consumer products and about 3–500,000 people involved. The total yearly sale to consumers could reach $2 billion and about 1 GW PV is utilized. It became a sizable business. Many consumer products can be found in many stores as will be described later.

I must disappoint the reader to confess, that Solarex did not enter the consumer PV field because of a market research, it entered by sheer luck and by the management' opinion that if some people need something, somebody should make that product. Chapters 14–16 narrate stories about the consumer PV business which was as mentioned before was maybe rarely but possibly never described, in spite that the events resulted in two very important milestones in the history of PV.

Chapter 14

The Importance of Consumer Business

Before we started Solarex, Joseph Lindmayer knew several photovoltaic (PV) "experts," but I think he never listened to what they were preaching, to conduct research, to decrease the price of space-oriented solar cells, to become inexpensive, and be able to be used for central PV electric stations, because that was the future and that was the business we should go after and this could be achieved in 14.5 years. He did not listen because he was in the process of changing his mind from thinking how to make more efficient space solar cells, how to make inexpensive solar cells for terrestrial purposes, which he had to figure out not in 7 years, the "experts" estimated it could be done, but in a few months before we started Solarex. On the other hand, I never met a PV "expert"; therefore, I did not know what the prevailing wisdom was, that it would take 7 years to develop the terrestrial PV cells.

If we would have known what the prevailing wisdom was, we would probably have written our business plan accordingly to start research to develop solar cells on government money to make central power stations and when we completed

Sun above the Horizon: Meteoric Rise of the Solar Industry
Peter F. Varadi
Copyright © 2014 Peter F. Varadi
ISBN 978-981-4463-80-5 (Hardback), 978-981-4613-29-3 (Paperback), 978-981-4463-81-2 (eBook)
www.panstanford.com

that task we were going to make them for the utilities the same way as the nuclear people did.

But we were ignoramus and I must confess that when we wrote our business plan to raise money to start Solarex, all the products we considered to make utilizing solar cells were not for central power stations, but for consumer products. As an example, one of them was a small solar module for pleasure boats, sail boats, or power boats. I had boats all my life and had a small power boat on the Long Island Sound, in Connecticut. Later I had a sailboat on the Chesapeake Bay, near Washington, DC, I used the boats on summer weekends, and my aggravation was that when I went to the boat and tried to turn on the engine, the battery was dead. I had to remove the battery (not easy and very heavy), take it to a shop, charge it for an hour, put it back in its place, and turn on the engine. Half of the day was gone.

According to the US Coast Guard in 2001 there were 12.9 million recreational boats registered in the United States. One can estimate that in 1973, at least 6 million recreational boats were registered. Based on my experience, I believed that at least 1% of the owners would get fed up spending their weekends charging dead batteries and would buy a solar charger for $100 (which is a ridiculously low amount compared to their 10,000–100,000 dollar boat), to save them from the weekend aggravation. We believed that this business would be a good entry for a startup PV company.

Obviously, when the price of the PV modules will decrease, PV would also be installed supplying electricity to private homes of which there are millions.

As planned, we tried to sell solar modules for boats, but it must be that boat owners like to spend their weekends to lift dead batteries and recharge them and after that recover on deck drinking beer, because that business did not materialize. We installed a PV module on our boat (three of us owned a sail boat) to test if it would do the job and it worked perfectly, the battery on the boat was always charged when we needed it.

Learning from this failure, we derived a few important conclusions which encouraged Solarex to enter the PV consumer business seriously:

- Consumer business is a huge market, but people, the customers, have to learn about PV to develop a need for it. The DOE had a program about this subject: "communication" of the use of solar energy to the public. Unfortunately nothing came out of that effort.[101]
- We learned that we have to develop the PV consumer market, but one should only make products that people need and not what we think is good for them.
- People are not going to be able to see a PV installation, because they are all in remote areas, therefore we have to bring PV to the consumer.
- For Solarex the consumer market would be a very lucrative business, as the existing competitors will never enter it since their owners believe to chase only central power and government research/demonstration business and therefore, they would see the consumer business as an unwanted distraction.

Obviously, the decision was that the consumer-oriented business is very important for the success of PV and for several reasons especially for Solarex it could be an easy way to increase the income with a very lucrative business, since for a time no other company will be in it.

- People have to see and use PV-powered products to support the extensive use of solar energy.
- We have to find out what they need and make it for them.
- The final goal is the utilization of PV systems by the public to produce electricity for themselves and/or to sell them to the electric utilities.

The question was, how could we achieve this?

---

[101]A detailed description of this effort is in Neville Williams (2005) *Chasing the Sun*, New Society Publishers.

Chapter 15

Micro-Generators and ...

The man on the other end of the telephone said that his
name was Roger Riehl and he was calling from Whippany, NJ,
where he was in the process of manufacturing a digital watch
powered by solar cells. He said that he understood we were
manufacturing solar cells for terrestrial applications and he
would like to buy solar cells to power the watches he designed
for production.

This was about a year after the 1973 oil embargo, and I
was used to getting phone calls about strange applications, but
I could not understand why a watch should be powered by solar
cells, which could be easily powered by mechanical winding
as people had done for centuries, or by a tiny battery that
the Hamilton watch company introduced in the USA in 1957.

He explained that his solar-powered watch was a digital
watch, which displayed the time with numbers and not as the
so-called analog watches used since centuries, which had
hands for the hour and for the minutes. The digital watch was
first made by the Hamilton watch company in 1970 and was
called "Pulsar" and the display consisted of light emitting
diodes (LEDs), but he said that the digital circuitry and the LED

Sun above the Horizon: Meteoric Rise of the Solar Industry
Peter F. Varadi
Copyright © 2014 Peter F. Varadi
ISBN 978-981-4463-80-5 (Hardback), 978-981-4613-29-3 (Paperback), 978-981-4463-81-2 (eBook)
www.panstanford.com

display needed more power than a regular mechanical watch powered by battery and therefore, a regular battery would have to be replaced frequently. He designed a digital watch utilizing also the LED display, but in order to eliminate changing the battery, his design used rechargeable batteries and solar cells to keep them charged.

In spite of hearing about the "Pulsar" but never having really looked into what a digital watch was, I told him that this sounded very interesting and we would be glad to make the solar cells for him. He said that he was going to send me the specification for the solar cells he would like to have. He would need 500 pieces now and he would need more thereafter. After I received his drawings, I should give him a quotation. Mentioning 500 solar cells, which at that time was a large number, made me interested and I told him that I was looking forward to seeing his specification.

Mr. Riehl's specification arrived a few days later. It was actually not a solar cell, but a solar cell array about 0.75" × 0.75" (20 mm × 20 mm) size consisting of 8 solar cells each of the size of 2.5 mm × 20 mm.

The required output voltage and current was easy to achieve in our production with a solar cell array assembled in series, like the one his design indicated (see Fig. 15.1).

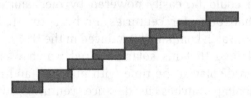

Figure 15.1 Shingled solar micro-generator.

I had a meeting with Joseph Lindmayer and Chuck Wrigley and showed them the specification. They both agreed we could do it. I told them that obviously to make solar cells for watches was a diversification from the goal we started Solarex for. We opened Solarex to ultimately produce PV for electricity power generators for terrestrial purposes. What was needed for watches was a deviation from our goal, because the solar array for watches was only a micro-generator. We were at that time selling all of our production competing with companies with deep pockets, but for watches we were not going to have competitors, as our competitors were part of large companies and would not be distracted from the "big picture" idea. We could set the price and easily get 10–20 times the money for a square centimeter of a wafer, than we could get in our terrestrial business. We had the equipment such as a micro saw, which we were using to quarter or half wafers for small solar modules, and we could produce solar cells with a pattern that we could saw and attach each to the other using solder and a heat lamp. We needed only a tool to hold the solar cell slivers together under the heat lamp.

It looked interesting to Joseph and Chuck, especially to make solar cell arrays for watches, because if one watch manufacturer used them, the others would have to use them too and there were thousands and thousands of watches made. It could become a strange but interesting application for solar cells and a large volume lucrative business too, at least as long as the watch circuitry and display required more power than a small battery could provide.

Chuck said that he could make some samples in a few days. We just did not know how much we could charge for such a tiny "micro-generator." I remembered that Roger Riehl told me that he was expecting to sell these watches to Tiffany. My opinion was that at Tiffany one could not find a watch under $500 and in that case for the first 500 order we could easily get 10–20 times more money for a square centimeter of a wafer, than we could get in our terrestrial business. Both Joseph and Chuck believed that for that price it was worth making the micro-generator, and I could get some samples in a few days to show to Mr. Riehl.

I called Roger and told him that I had received the specification. We could make him some samples and I would like to visit him with the samples and discuss prices, deliveries, and future orders and I would like to see how his watch looked. He was actually surprised at our fast turnaround time and we agreed about the day I was going to visit him.

During the following week, I visited Roger Riehl in his company Ragen Semiconductor, Inc., (RSI) in Whippany, NJ.

Roger was a medium-sized person, with short black hair and a thick black mustache. RSI (which was renamed in 1977 to Riehl Semiconductor, Inc.) was a small factory manufacturing the integrated circuitry for Riehl's digital watch, named "Synchronar 2100" (Fig. 15.2). Roger showed me the Synchronar 2100 watch for which he needed the solar "micro-generator." He designed the integrated circuit which operated the watch and manufactured it at that facility. It was a very advanced circuitry, which provided the time, date, and seconds which could be seen on the LED display. The circuitry was able to correct the watch for leap years until 2100, which year was an exception, as it would not be a leap year. The circuitry was designed to adjust the watch for the hour change due to the daylight saving time. The watch had a rechargeable NiCd (nickel/cadmium) battery recharged by the "micro-generator."

Figure 15.2 Synchronar 2100.

The circuit, battery, LED display on the front, and the micro-generator on the top was contained in a Lexan case, and was filled with silicone rubber. The case was therefore sealed. He was planning to make three versions; stainless steel, gold-plated stainless steel, and solid gold case. I was right estimating the price, as I now found out that it started at $500 and went up to $1200.

Roger tested each of the samples I gave him by connecting them to a load and a meter and turning a lamp on. He said he liked the quality and the output of the samples very much and asked the price. I told him the price for the quantity of 500. He just said OK and proceeded to give me a purchase order for the first 500 pieces and told me that if the quantity of the order increased, he expected lower prices. I said obviously, and we shook hands.

He requested weekly deliveries of at least 50, starting immediately. He gave my samples to one of the technicians and told him that we were going for lunch. By the time we were back the technician should complete the assembly of the solar arrays into watches, to be able to give me a sample watch.

During lunch he told me that he had contacted the other solar cell manufacturers, but nobody was interested in his business. I told him I was not surprised, because some of them made only solar cells for application in satellites or other space projects at very high prices, the very few others were either making only sensors, or did not want to be distracted from the "big picture" of being able to make big power stations in the future. We were also very much interested in developing solar electric products for the "big picture," but we had to work on other applications for solar cells to earn our weekly pay checks.

When we returned after lunch, the three watches with the Solarex "micro-generators" were completed. Roger examined each of them and gave me a gold-plated version. I removed my watch and put the gold-plated Synchronar 2100 on. I checked the time, date, and the seconds, and told him I would be proud to wear one of the first solar-powered digital wrist watches ever made.

I returned to Solarex with the "Ragen Synchronar 2100" on my wrist (I still have the watch, in working condition) and the purchase order for 500 units and told Joseph and Chuck to start producing the micro-generators.

Roger Riehl was a genius designing the integrated circuit and the Synchronar 2100 which was a miracle in 1974, when it was officially announced and went into stores such as Tiffany and other very expensive jewelry places. During the years a total of about 60,000 Synchronar 2100 Mark I to Mark V were made.

The digital display was a tremendous novelty and it became obvious that digital watches were going to have wide acceptance. But in 1974, it was a novelty and people paid for it whatever the price was. The only disadvantage was that it required an LED display, which needed a lot of electricity. The problem was somewhat solved in that the LED display was only turned on when somebody wanted to see the time or date, but it still needed a lot of current and therefore, to use solar cells to charge the battery was a good solution, until a new digital display was found, which required less electricity.

In January 1975, after we started to make the micro-generators for Roger Riehl, I received a telephone call from Mr. Paolo Spadini. He said he was calling me from New York, but he had a watch company in La Chaux-de-Fonds, Switzerland—of which I knew already that it was the center of the Swiss watch industry—the name of his company was "Nepro" and he announced that his company was going to have the first digital watch made in Switzerland. Because of the LED display, which needed a lot of electricity, he wanted to bring a model to the market which was recharged by solar cells. He asked me if I could go to New York to meet him. I said of course and we agreed that I would meet him the next Monday for lunch at the Waldorf Astoria, where he was staying.

I arrived exactly at noon at the entrance of the restaurant and told the person at the entrance that I was looking for Mr. Spadini.

"Oh!" he said, Mr. Spadini just called and asked me to tell you, that he will be a half an hour late and you should please sit down and have a drink until he arrives."

He escorted me to a table, which was set for three. I told him the only thing he could bring me was some water.

The restaurant started to fill up. A little more than a half an hour later, two well-dressed men came practically running. When the first passed the person at the entrance, he pointed with his hand in my direction. His speed increased. When he reached my table, I got up he shook my hand and started to talk in an excellent British English:

"Dr. Varadi, I am Paolo Spadini and he is my brother. We must apologize for being late. I appreciate that you came up to New York to see us. As you probably know my company Nepro is one of the very best watch manufacturers in Switzerland and in the entire world. We are planning now an absolutely superb, beautiful, and trail-blazing project which I hope you can help us to develop."

In the meantime his brother came to me, we shook hands and we were seated, Paolo in the middle and I on his right, his brother on his left.

With this I entered a world which was totally unknown to me, which was started by a Polish watchmaker, Antoni Patek, making pocket watches in 1839 in Geneva. The same company renamed Patek Philippe made in 1868 the first wrist-watch. The global watch business was centered in the French-speaking part of Switzerland with only one world class manufacturer International Watch Company (IWC) in Schaffhausen, in the German part of Switzerland, founded in 1868 by Florentine Jones, a 27-year-old American master watchmaker from Boston. By 1975, there were thousands of watch manufacturers around the world.

The global watch business is very competitive. Some companies sell because of their name which also means quality and service. Others sell because of style, novelty, and quality and others because they are inexpensive. Except the top brand names, for example, Patek Philippe, Rolex, IWC, and Longines, the large group attracting customers with their novelty and design must come up yearly with something new and attractive. It is a hectic life.

Mr. Spadini motioned to the waiter and asked me what I would like to drink. He suggested we should have wine. When the waiter arrived, he told him what he wanted, a Pinot Noir from California. The waiter showed up in a minute, opened the wine, Spadini tasted it and motioned that it was OK. The waiter filled our glasses. Spadini told the waiter to wait, and that we should order. He and his brother selected fish, I asked for a small steak.

After this was settled, Mr. Spadini explained to me that he named his company Nepro, which was started in 1958, from the words New Products. His company was coming up yearly with new products with excellent quality and superb designs. He was planning to come out for the *Basel Watch and Jewelry Fair*, which is the greatest and most important watch fair in the world, with a solar-powered digital watch. He said that he acquired from Ness Time, which was a part of Ragen Semiconductor, the electronics used in the Synchronar, and he was told that Solarex was making the solar cells for that watch. Therefore, he gave me a call as he wanted to buy the solar cells from Solarex and wanted Solarex to say how many cells to use and also to test if the battery he selected would be compatible with the solar cells.

We had almost finished our meal when he suddenly looked at his watch and told us that we had to leave, because he had an appointment at 2 p.m. We should go with him, as his meeting would last only a half an hour and after that we would return to the Waldorf to have coffee and finish our discussion. The way he told me this I had no choice but go with him. I assumed that he did not want to take the chance that I may not wait for him, because he wanted to complete the deal with Solarex. He needed the solar digital watches for the Basel watch show.

I thanked him for the lunch, but by that time he was ahead of us, out through the door, hailed a taxicab and asked us to get in. He told the driver the address, which was near Wall Street. We got in, he next to the driver and his brother and I in the back.

During the trip he apologized and repeated most of the story that he told me during our so-called "lunch." The office he went to was a large law firm. He was escorted to some other area of the office complex and his brother and I waited in the reception area and discussed watches, solar energy, and so on. Obviously his brother was not involved with Nepro, he just happened to be in New York when his brother Paulo showed up.

Paolo reappeared exactly in a half an hour and we went back to the Waldorf, sat down at a table in the bar, and ordered espresso. Paolo explained the project he wanted to carry out. He pulled out three metal cases of the watches from his briefcase and gave them to me. The design was very attractive, truly unique, and excellent for the solar array and for the LED display[102] (Fig. 15.3).

Figure 15.3 "Nepro" digital solar watch.

"Mr. Riehl is a genius, Paolo said. He had the right idea, created a super electronic watch circuit, and digital display, but to a wrong design. The mistake he made was that the solar system was on the top on his design and the LED numbers showed up on the side. Our design corrects this. The solar system is on the top and the LED numbers are also on the top facing the owner of the watch. The other mistake Mr. Riehl made was that he designed a totally and absolutely sealed watch. This may be good if somebody wants to use it at 100,000 meter in the stratosphere or 1000 meter under water, but people do not go there. It is sufficient to have a watch survive 100 meter under water and 10,000 meter over the ocean. One should not encapsulate a watch, because jewelers selling the watch would feel uncomfortable. If for some reason the watch will not work they have to ship it back and get a new one. They cannot fix it. The Nepro design is built as a standard watch, where the rear plate can be opened."

---

[102]http://www.ledwatches.net/photo-pages/nepro-alfatronic.htm.

Paolo added that he would like to use a silver oxide battery made by the Swiss battery manufacturer Renata.[103] They produced a wide range of products such as silver oxide batteries for watches, hearing aids, and so on. They claimed that their silver oxide battery was rechargeable by solar cells. Paolo said he would like if Solarex to do the following. He pulled out a page from his pocket and handed it to me. It was a typed list:

(1) Make tests to make sure that the Renata silver oxide battery could be recharged by solar cells and if so, design the solar cell system, which would fit in the area on the Nepro watch case design and properly charge the silver oxide battery.

(2) Assemble the three cases with a solar array and a battery and test that they perform well.

(3) Send the assembled watches to the factory in La-Chaux-de-Fonds by express air.

(4) The assembled watches needed to arrive in the Nepro factory not later than February 28, 1975 because he wanted to exhibit them with proper literature on the Basel watch and jewelry fair which started on April 28, 1975.

(5) After he received the sample watches he would need 2000 solar cell systems, delivered weekly at the rate of at least 100 per week and hopefully Nepro could increase the order after that.

(6) In case the Renata silver oxide batteries were not acceptable, I should call Paolo immediately and propose an alternate solution. Nepro was also looking for a NiCd battery that Mr. Riehl was using, which would fit into the design.

I was very impressed how organized he was. Reading the list I made some notes:

We could do what Paolo wanted. For the first four items on his list, I had to give him a quotation. For the additional 2000 micro-generators, as we called the solar cell system, as he probably knew from the Synchronar people how much

---

[103]Renata SA CH-4452 Itingen/Switzerland.

we were charging, the price for Nepro would be the same. If Nepro continued to order, for the next 5000 we were going to drop the price.

As the time was very short, Paolo said that he did not need a quotation. He gave me then and there a check, so Solarex could start immediately. If it was not enough I should let Paolo know and Paolo was going to take my word for it. Paolo asked me to telex him the quotation for the 2000 and the additional 5000, but I should take it that he was right now accepting my quotation.

He pulled out his check book and gave me a check for Solarex. I looked at the check and was pleasantly surprised at the amount. Two weeks later I called him and told him that the Renata batteries worked fine. We cycled them many times and found no problem. I also told him that the three samples would be shipped by February 15, 1975. I also told him that after he received the three watches and accepted the micro-generator design, he should inform me by telex. We were going to start to ship him the weekly 100 micro-generators, 2 weeks after he sent me the telex. He thanked me very much for the good news and handling the matter so business like and told me that he was mailing me an additional check as a bonus for our fast and efficient work.

Joseph and Chuck developed a solar cell manufacturing process which made the assembly of those small micro-generators very easy. I handed over the project to Ramon Dominguez, who had just joined Solarex, but with whom I started to work in 1961 and I knew if somebody could do it, he could.

For me it became evident that the advent of the digital watches where the time was displayed with numbers and the requirement that they needed more electricity than a mechanical, now called "analog" watch, opened up the utilization of solar cells in watches. To market our micro-generators for US watch companies, we could do it from Rockville, MD, but most of the watch manufacturers were in Switzerland and some of them in Japan. Low-quality watches were made in Hong Kong.

Joseph and I discussed the European situation concerning the possibilities to expand Solarex's business to sell terrestrial solar products, including the ones for watches in Europe. The European Space Research and Technology Center (ESTEC) located in Noordvijk, the Netherlands, was the focal point of European solar activities.[104] AEG/Telefunken developed the humidity resistant titanium–palladium–silver (Ti–Pd–Ag) contact. Based on G. Seibert's (ESTEC) work and research at AEG-Telefunken, Ferranti Electronics Ltd., and Société Anonyme de Télécommunications (SAT), the titanium-dioxide (TiO_x) anti-reflection coating was developed, which increased the efficiency of solar cells by at least 6%. We agreed that while in Europe a lot of organizations and companies worked on PV research projects and achieved very interesting and good results, there were also companies manufacturing solar cells for space use, and several solar cell manufacturer's cells were even qualified for the US as well as for the European space programs, but until that time (beginning of 1975), interestingly none of them expanded their activity to the terrestrial use of PV (the first was AEG in Wedel, Germany, in 1978).[105] In 1974, Elliot Berman of the US Solar Power Corporation (SPC) started in the UK a subsidiary with the purpose of expanding SPC's terrestrial PV sales in Europe,[106] but our information was that Exxon, the owner of SPC, was in the process of terminating this activity in 1975.

---

[104]Curtin DJ (1972) *European Solar Cell and Solar Array Research*, Proceedings of the 25th Power Sources Symposium, p. 134.

[105]1978, First series production of solar electricity modules in Europe (AEG in Wedel, GER) http://www.schottsolar.com/us/about-us/history/.

In France, Photowatt started operation in 1978. http://www.photowatt.com/fr/corporate/histoire.php.

In Germany, RWE nuclear chemical division in 1979 added the NUKEM' solar division http://www.appliedmaterials.com/news/articles/applied-materials-names-winfried-hoffmann-chief-technology-officer-solar-business-grou.

In the UK, Lucas Energy Systems (in 1980 became BP Solar) was selling the Solar Power Corporation' (US) solar modules starting in 1974.

[106]Private communication from Mr. Bernard McNelis of IT Power Ltd., UK (2011).

Having no competition in Europe it seemed to be a very timely need to expand Solarex's sales activitiesthere for our terrestrial products and also the micro-generators for the watch companies.

Chapter 16

... and a Permanent Bridge to Europe

Mr. Eric Winter visited Solarex in November 1974, and offered to represent Solarex in the German-speaking part of Europe. Joseph and I had a meeting with him in our "conference room." Seemingly he wanted to impress us, because on a sheet of paper he outlined his empire. He owned about eight companies located in the various parts of Germany. But the ownerships of his companies were incredibly complicated. For example, Company A owned a piece of Company B and also C and E. Company B owned a piece of A, C, D, and F. Company C owned a piece in A, B, D, and G, and so on. At the end, the sheet of paper looked like the design of a spider's web in which the various companies appeared like captured flies. He assured us that he was the competent person to be able to sell our products.

We told him we were going to consider his offer and would let him know in a week.

I discussed the matter with Joseph. First of all, I believed that Winter was in touch with Solar Power Corporation (SPC) people in the UK to represent them in Germany, but he was not chosen, and therefore, he visited us. My opinion was that

Sun above the Horizon: Meteoric Rise of the Solar Industry
Peter F. Varadi
Copyright © 2014 Peter F. Varadi
ISBN 978-981-4463-80-5 (Hardback), 978-981-4613-29-3 (Paperback), 978-981-4463-81-2 (eBook)
www.panstanford.com

I could not comprehend the complicated way his empire was structured, probably because I had no business school education, but I thought that except for himself nobody was able to understand his business system; therefore, either he would be extremely successful or he would go bankrupt.

As there was no European company making terrestrial solar cells and the only company was the US Solar Power Corporation (SPC) having representation at least for now in Europe, it would be probably a good time for Solarex to enter the European market; therefore, independent of how good or bad Mr. Winter's activity was going to turn out, he could open for Solarex a bridge to Europe. We notified Mr. Winter that we accepted his offer to represent Solarex in the German-speaking part of Europe and sent him a Representation Agreement, which Stanley, our lawyer, provided for us. This was the situation at the end of November 1974.

I called Mr. Winter from time to time and reminded him to do something. He said that he put two good salespersons on the job to sell our products, but they had no result yet. I reminded him that he had not sent back the signed Representation Agreement. He assured me that he was going to do it immediately.

In January 1975, shortly after the deal with Nepro solidified, I called Mr. Winter in his Frankfurt, Germany, office and informed him that we were expecting an order from a Swiss watch manufacturer located in La Chaux-de-Fonds for a substantial number of solar products to be installed in watches. I told him that in spite that this order was not from the German-speaking part of Europe where he was our representative, as La Chaux-de-Fonds was in a French-speaking Canton of Switzerland which was not on his territory, and we had not received any orders from him, I was still planning to give him a 10% commission on this sale. He was very happy to hear this. He asked me which watch company was involved. I told him that unfortunately I could not tell him, because the company wanted to have secrecy as they were afraid that their competitors were going to find out. This was true, because Mr. Spadini several times told me that we should not disclose what he was planning to do, unless he gave us permission to do so.

I also mentioned to him that I had not received from him the signed Representation Agreement I sent him in November 1974. If it got lost in the mail I would send him one immediately. He said that he received it, but forgot to send it back but he was going to do it immediately.

I was very surprised when 2 weeks after I had called him up in January, I received a call from him. He told me that he had conducted research and if one Swiss watch company was showing at the Basel Fair a solar watch, then all of the others would do the same and it would become a substantial business. Therefore, he incorporated a business in Switzerland. Its name was ESOTRON. He was buying a small factory building in La Chaux-de-Fonds, to be next to the Swiss watch factories, and hired a manager for that factory an engineer, whose name was Michel Joliot.

I told him that this was excellent news and I would send now a Representation Agreement to him made out to Esotron, to replace the one which he did not sign yet, to specify that Esotron was representing Solarex based on the terms and conditions of the agreement. I asked him the address of this new entity, where it was incorporated, and who the owners were. He gave me all the information. He said that he was the sole owner of the company, which was a surprise for me considering the spider's web system he used in Germany. He said that he was waiting for the new Representation Agreement and he was going to sign it immediately.

About 2 weeks later, he called me and told me that the project to set up and equip the building in La Chaux-de-Fonds was progressing well and he would like if Mr. Joliot could visit Solarex to get acquainted with our products and with the people in sales and production. Mr. Joliot was planning a trip to the USA and could visit Solarex in May 1975. We agreed to the date.

I asked him if he received the Representative Agreement and when did he return it, because I had not received it yet. He said he received it. He may want to make a few small changes and his lawyer was reviewing it at that moment.

Mr. Michel Joliot arrived in mid-May 1975 and spent several days at Solarex. Michel spoke good English and spoke obviously French and also very good German. He was a very pleasant person and a good engineer. He seemed to be very competent and had good knowledge of semiconductors and was very enthusiastic about solar energy, which he believed would have a great future and was important for mankind. He spent his time with and was escorted around by Ramon Dominguez, who headed up Solarex's production of cells, modules, solar systems, and also the micro-generators. I was also originally responsible for the production, but with our unexpectedly large sales volume in 1974, which we believed would triple in 1975, we needed a competent person who should be in charge of those areas. My first thought was to offer the position to Ramon, who was a New York University graduate, whom I knew quite well, because he worked with me at a division of Raytheon Company since 1961, when he and his family managed to leave Cuba. I was familiar with his competence and his ability in organization. He joined Solarex in early 1975.

It seemed that the addition of Michel Joliot to the staff in La Chaux-de-Fonds would help to find business for Solarex. He had already found several possibilities in Germany. He believed he found good prospects for our solar modules as well as for the micro-generators. He talked to our research people and also spent time in our production. He asked us to make some samples for him and became familiar with our production method to assemble the micro-generators. Michel suggested that we should produce the solar cells and he could assemble them into micro-generators and also into small modules, for what he believed he would find business. By that time we had several girls assembling the micro-generators and therefore, it was reasonable that he could assemble them also in Switzerland, where he was closer to customers.

I had a telephone discussion with Eric about the Representation Agreement. He said that his lawyer unfortunately had not finished the review, but he would ask him to speed it up.

After Michel returned to Switzerland, he managed to find us business. He sold micro-generators to Witte and Sutor in Murhardt, Germany, close to Stuttgart. They started to use the micro-generators, to charge flashlights. After the first trial order, we received large orders.

The facility in La Chaux-de-Fonds got finished and we shipped the solar cells, and Michel, based on what he learned during his visit, started to assemble the micro-generators for Witte and Sutor and also for other customers he found in the meantime. He also found some small module business for fence chargers. Fence chargers are used to put short high voltage pulses on wires surrounding, for example grassy areas where cows are. When the cow touches the wire, it gets a little high voltage shock and that makes it to avoid going near the wires and not leave the area. Obviously in Switzerland, Austria, and many other parts of Europe, this system was used very much, where cows or other animals were left out for grazing.

Our European business obviously increased and I also wanted to expand it to other parts of Europe. The other possible distributor who called us was Roger Mytton having a company, Solapak, in Newcastle in the UK. I sent him a Representative Agreement. I received it back in a few weeks and he started to sell our larger solar modules primarily in Norway for weekend houses.

I still had not received the signed Representative Agreement from Eric Winter. At the beginning of July, I sent him a telex and told him that I was planning to go to Europe at the beginning of September and if I had not received the signed agreement by that time I would have to make serious decisions whether to continue the business with him.

By that time I started to believe that he was planning to set up his own solar cell and module production in La Chaux-de-Fonds and that was the reason he was not signing the agreement.

I came to the conclusion that in case we had to terminate our relation with Winter, the best would be to start our own "Solarex Europe" and organize our European sales and expansion from there, similarly as Elliot Berman tried to do for SPC from the UK. I discussed the matter with Al Nerkin, who was one of our major stockholders and who ran a big vacuum system company in the USA and had a lot of experience. He asked me two questions: Who are the terrestrial PV manu-facturers in Europe? Is there a great market potential for terrestrial PV in Europe?

The answer was simple. At that time no terrestrial PV manufacturer existed in Europe except Philips RTC in France, but they only made a small module for the Australian market, but because of technical problems it is going to be closed. As of market, there is some market especially at that time in consumer-oriented PV products, which could be assembled in Switzerland; furthermore, the Swiss office would be good because we could reach the French- and German-speaking part of Europe and the Mediterranean area.

He encouraged me telling me that under these conditions he thinks it is a good idea and we should set up our own operation in Europe as soon as possible.

I got acquainted with Georges Didisheim, who lived in Stamford, CT. He was originally Swiss and his family founded the "Vulcain" watch company in 1858 in La Chaux-de-Fonds, Switzerland. He became very interested in solar energy and became a consultant for our European expansion. We decided to plan our trip in September to go to Switzerland and first complete the agreement with Winter and after that to see the possibility to set up an office for Solarex to develop our European business.

I informed Winter that I would arrive in Geneva on Sunday, September 7, and would need the signed agreement before I leave. I received no answer. On the other hand, our bank informed us that a check of $10,000 which we received from Esotron, that is Winter, had bounced. It was not collectible.

I sent a telex to Esotron for Winter and informed him that his check had bounced. Two days before I left with Georges Didisheim for Geneva, I received a telex from Michel Joliot in which he informed me that Mr. Winter was "hardly" injured and he, Michel, would like to meet me immediately after I arrived, to inform me about the details. Reading his fax I immediately realized that English is a very treacherous language. If the letters "ly" are added to a word it increases its meaning, e.g., bad and badly, etc. exception is the word "hard." In this case "ly" minimizes the meaning. Hardly means, that

Mr. Winter was not seriously injured. I, however, having a European origin, immediately understood that Mr. Joliot was tricked by the English language and in reality something very bad had happened to Mr. Winter. Therefore, I answered that I was sorry to understand that Mr. Winter was badly hurt and would not be able to meet me, but as planned, I was going to arrive in Geneva on September 7 and would be glad to see Michel during the afternoon at the Intercontinental Hotel, where I was going to stay.

Michel arrived during the afternoon and Georges and I met him in the bar. I introduced Georges to him. After we sat down and ordered an espresso, Michel told us that Mr. Winter's empire in Germany had collapsed and gone bankrupt, which obviously made Esotron also bankrupt. Mr. Winter, who happened to be in La Chaux-de-Fonds when he received this news, jumped out of his hotel window. Fortunately, the room he was staying in was not on a high floor; therefore, he did not kill himself but broke many bones and was in the hospital in a very serious condition. Michel said that Esotron's building was purchased on a bank loan and he also owed money to banks and to others, including the city of La Chaux-de-Fonds.

I mentioned that he also owed Solarex $10,000 and I showed him the returned check. Georges told him that while in the USA if a check bounced, it is not a big problem, in Switzerland it is practically a crime. Michel agreed and mentioned that neither he nor others working at Esotron had been paid. He said they all hoped that Solarex would take over Esotron and he suggested that the next day, Monday, we should have a meeting with people of the city of La Chaux-de-Fonds and people from the local banks which gave the loans to Mr. Winter. They all agree that Esotron should be offered to Solarex for takeover.

This was an unexpected turn of events. I obviously had to go to this meeting and Georges said he would come with me. Georges said that he knew the area very well as he grew up there and we could be there by 11 a.m. Michel said he would inform everybody who should be there, to complete the deal.

When we sort of finished the discussion I suggested having a drink before Michel drove home. Michel gave a short summary of the business he found. It looked very encouraging, that we had in Europe such a good start. He said that besides himself, Esotron employed one lady in the office and they had two more people for the assembly and shipping, and so on.

I told him that the news he gave us was unexpected. I had to think it over, how this could be done, but for our meeting it would be necessary to have a complete list of Esotron's liabilities.

Next morning I was driving a rented car and Georges gave me directions. We discussed the situation. I said that taking over Esotron would solve the issue that we had discussed, to establish a European branch of Solarex, but my main question was how much liabilities there were and what the banks would do about it.

Georges gave a good description of the situation of the Swiss watch business. He said that the big name watch companies were doing very well. Rich and important people bought the very expensive watches, independent of their price, or they would wear expensive specialty watches. A well to do lawyer and so on would never wear a cheap watch. The Swiss inexpensive watch companies were all practically bankrupt, because of the Japanese watches, for example Seiko and Citizen, flooding the market with analog electric watches. Therefore, in La Chaux-de-Fonds there were many empty buildings.

Georges explained to me how the watch industry started in Geneva and in the Jura mountains. The Jura valley's involvement in clockwork originated in the city of Geneva. Little did puritanical Calvin realize when he banned the wearing of jewelry there in the sixteenth century that he would bring to life a craft that would become one of the most prestigious industries in the world! To ensure their livelihood, jewelers devoted themselves to clock and watchmaking. This new handicraft very quickly spread throughout the Jura region, where local watchmaker-farmers developed unprecedented skills in the field of precision work.[107] We arrived in La Chaux-de-Fonds, fairly high in the Jura mountains (1000 m—3281 ft.),

---

[107]http://europeforvisitors.com/switzaustria/articles/la-chaux-de-fonds-history-architecture.htm.

and very close to the Swiss–French border. Interestingly, La Chaux-de-Fonds was built like Manhattan, with relatively wide straight roads and not like the old European cities, with meandering narrow streets. Georges said that this was because the city was destroyed by fire in 1794 and Charles-Henri Junod created the new city's plan in 1835, and the city is now known for its "modern" grid-like plan. He also said that it was probably the "watch capital" of the world as many watch companies were located there.

When we arrived at Esotron's building, we were surprised that it was a new small building and that Esotron was not located in one of the empty watch factories. When we entered the office, it was very nicely furnished, with two desks, behind one of them a lady. On one side of this office area was a large conference room. Its door was open and we could see a nice big conference table and chairs and several people were already seated. On the other side we saw offices. One big one, and three smaller ones. At the end of this office where we were was a door, which I assumed led to the factory storage and shipping area.

As soon as we entered, the lady got up and came to greet us, shook hands, and told us in English to follow her as the gentlemen were already in the conference room. We entered the conference room. Michel greeted us and introduced us to the six people present. Two of them from the city of La Chaux-de-Fonds, and two each from two different banks, probably the ones Mr. Winter obtained loans from. From each bank one of the people was the president of that bank's La Chaux-de-Fonds office and the other was the bank's lawyer. The city people were one from the Mayor's office, representing the Mayor and the other from the city's financial department.

Michel pointed to one side of the table and asked us to be seated. The four bankers were sitting across from us. Next to us were the two gentlemen from the city and Michel sat down next to me at the head of the table.

The lady who escorted us asked if we would like to have an espresso—it seemed everybody had already a cup in front of them. We both thanked her and told her to get us one. I said I only would like to have sugar. Georges asked for cream and sugar.

I started the discussion:

"I am very sorry that Mr. Winter cannot be present and I hope he is doing well."

"I visited him today in the hospital," said Michel,—"but could not tell him about this meeting because he was sleeping. He is going to recover according to the doctors, but it will take time, probably months."

"I understand," I went on, "that you as creditors would like Solarex to take over Esotron."

They all agreed to this. The bank people and the Mayor's representative gave us a lengthy explanation, that they provided all of those loans, because the city's industry was practically 100% watch-making, which was now not doing so well and they were very interested that a new business would start here, which they believed would have a great future and would provide many job opportunities. Therefore, they would love to have Solarex take over Esotron. I listened to all of that and said:

"I would like to mention that Solarex is also a creditor, because the last payment, a check for $10,000, was bounced by one of the banks, having no funds in Esotron's account."

With this I opened my briefcase and pulled out the check and showed it to the bank people. After they saw it, I put it back into my briefcase. I continued:

"I told Mr. Joliot yesterday that as a first step to consider the takeover of Esotron by Solarex, I need to have a complete list of Esotron's present liabilities and review them. I asked him to put it together for this meeting and," I turned to Michel, "I assume you have it."

"I have the papers here, how much Esotron owes to these two banks and to the city. Unfortunately, I was not able to assemble more, because the files are not well organized."

"If you do not have all of the liabilities now, when do you think the list could be prepared?" I asked.

"I am sure that by tomorrow morning I would be able to prepare it," Michel replied.

"As this is extremely important to proceed, I suggest that we get together tomorrow morning at the same time as today. In the meantime, I think it would be good if Mr. Didisheim and

I could tour the facilities, review some of the papers, such as the present orders, future possibilities, and also get acquainted with the people working here."

The person representing the Mayor stood up and said that as the time was 12:15, the city of La Chaux-de-Fonds would like to invite all of us to have lunch in the "Restaurant— Brasserie Hotel de Ville," which was close to the City Hall. Michel suggested that Georges and I should go in his car and leave ours at Esotron and we all meet in the restaurant.

The restaurant was beautiful and the lunch was excellent. The Mayor's representative selected an excellent wine, and by the time we finished our espresso coffee it was close to 2 p.m. We had a very interesting and good discussion. They were all very upset about the financial condition of Esotron and that they all got bamboozled by Winter and his multitude of intertwined companies. They agreed that without knowing exactly what the liabilities were, no sane person would take over Esotron. We agreed that the next day we should meet at 11 a.m.

Georges and I went back to Esotron. Michel gave us a tour and Georges and I went back to Geneva. I called Solarex that evening and found out what happened during my absence. After we discussed these matters I asked my secretary to connect me to Joseph.

He came on the line and I explained him the situation with Winter and Esotron. After a discussion, we agreed that the best would be to establish Solarex's European division and he said that I should proceed the best way I thought fit.

Georges and I returned to Esotron the next day by 11 a.m. and sat down at the table in the order we had been the day before.

Michel handed out to everybody a sheet of paper, the title of which was Esotron's liabilities. The total of the one page of paper was a very large amount. Included were the bank loans, furniture, and equipment, basically Winter did not pay any of his bills at all.

I asked if that was the complete list. Michel assured me, that according to his review, it was. I turned to the bank and city people around the table:

"By reviewing it you think this is the total liability of Esotron?"

They all said, "Yes."

"Then I think you would sign this and guarantee that there are no more liabilities?"

The way they all said yes before, now they all said, "No."

My problem was, as I discussed with Georges in the car and during last night's dinner, that after the meeting I expected this outcome, that Winter had liabilities but nobody would guarantee how much there was. Furthermore, Esotron's assets were not worth much. The building for example could not be sold at all, because many of the watch companies' buildings were vacant. Furthermore, in spite of what Winter told me, that he was the sole owner, that may not be true and it would make the take over a questionable, lengthy, and expensive procedure. The only asset Esotron had was the customer list and the knowhow of assembling micro-generators that they learned from Solarex. In the entire world only our competitor, SPC, could be interested to acquire this mess, and because of their parent, Exxon, the amount would be a small fraction of a dry well, and they had even in Europe sufficient legal help to complete the transaction. There was a possibility that they may buy Esotron, but if that happened Solarex was going to suffer a big loss.

At that point I suggested having lunch at the nice restaurant we had lunch at yesterday, and this time I invited them to be Solarex's guests. They all accepted very happily and again we had the same arrangement to drive there.

Georges and I got into Michel's car. He was driving and I sat next to him. After we started I asked Michel:

"Do you think that the list of liabilities you presented is accurate and no other liability exists?"

He looked at me as he was driving and said:

"Sorry to say, but I do not think that the list is complete."

"I do not think so either," I said, "and neither the bank nor the city people think that the list is complete, therefore, they said they are not guaranteeing that there are no other liabilities. In this case, however, the only solution is to put Esotron into bankruptcy."

He looked at me, very disturbed. I continued:

"Obviously we cannot hire you before Esotron is dissolved at the end of the bankruptcy procedure, but Solarex is going to pay your salary from the day you have not received it from Esotron and in return we ask you to continue to attend to the present customers and try to get new ones. When you are going to get free from your Esotron obligations, Solarex will hire you with the same salary, as the manager of Solarex's European division, which we are going to form immediately and Georges is going to help you to set it up. Do you accept this solution?"

"What is going to happen about the employees?"

"I am going to stay in La Chaux-de-Fonds tonight. I would like to invite you and your wife for dinner and we can discuss the details."

We arrived at the restaurant.

In the restaurant we were escorted to the same table as the day before. I suggested that we should start with an aperitif. They enthusiastically agreed and suggested we have a glass of Appenzeller. I did not know what that was, but I agreed. After the drinks came, I made a toast for solar energy. I asked Michel to select the wine. He reviewed the extensive wine list and proposed that we should have a wine from the Canton we were in, a rosé from Neuchatel, Oeil de Perdrix. Everybody enthusiastically agreed.

We ordered our lunch. After we started our lunch and had some wine, I told them that they seemed not to believe, that we knew all of the liabilities and outstanding obligations made by Mr. Winter. Even until Mr. Winter recovered, based on the spider's web system Mr. Winter used to organize his empire, we probably would not find out who was or were the owners of Esotron.

They all agreed.

"Obviously under these conditions unfortunately Solarex is not in the position to just simply take over Esotron, I said. The only thing one can do is to put Esotron into bankruptcy. To do this there are two possibilities: either I can do it—I pulled the $10,000 check out of my pocket—as there is no money to pay this check, or you can do it because you are not getting the payments Esotron or Mr. Winter owes you. I am sure that you must have discussed this possibility; therefore, I would like to have an answer from you now, as I am planning to stay in La Chaux-de-Fonds; tomorrow either you or I am going to file the proper papers at the court to start the bankruptcy procedure."

I continued, "As we have finished our lunch, Mr. Didisheim and I are going to sit down at the bar to give you privacy to discuss the situation. Would you agree to this?"

They looked at each other, I think they expected something like this. Finally one of them said, "OK, we can discuss it, but I do not know if we can give you a final answer now."

"As I told you, I said, I am staying tonight in La Chaux-de-Fonds and tomorrow morning either you or I am going to start the bankruptcy procedure; therefore, the only question for you to decide is what is better for you, that you are going to proceed, or Solarex should proceed? Please take your time. We are waiting at the bar for your answer."

With this Georges and I stood up went to the bar and told the bartender to go to our table and ask the gentlemen if they would like to have more wine, an after-lunch drink or coffee. We ordered two espressos. The waiter went to the table and it seemed all of them wanted an after-lunch drink. About half an hour later, Michel came to us and asked us to join them. The Mayor's representative was apparently elected to tell us the decision:

"It was decided, that in the interest of everybody, the two banks are going to start the bankruptcy procedure tomorrow morning and we invite you to witness it and we are going to give you copies of the papers."

That was very decent and fair of them.

"In return, Solarex will not become a creditor and will not request any part of the bankruptcy for the unpaid $10,000 check."

We agreed that we are going to meet the next day at 10 a.m., at the court. We all stood up and shook hands. They left, I went to pay the bill and Georges and I got into Michel's car and left. I told Michel that I was going to take Georges back to Geneva, but would return by 7 p.m., and if he would please make a reservation for dinner at the same restaurant for him, his wife, who I hope will be available, and me, and also to make a hotel reservation.

I arrived in the restaurant somewhat before 7 p.m., and waited at the bar. At 7 p.m. Mr. Joliot and his wife, a very good-looking blond lady arrived, followed by a small black poodle. I was surprised to see a dog in the restaurant, which is prohibited in the USA. As I found out later, however, in Switzerland and France, dogs are welcome to restaurants. If in the restaurants there are chairs around the table and one of them is padded and the others are not, the padded chair is where the dog is going to sit. After shaking hands, Mrs. Joliot—Denise—put a little rug on the floor and the dog immediately occupied it and lay down. We sat at the table, the waiter arrived immediately, and we ordered an aperitif. I complimented Denise on her excellent English and started to discuss the day's events.

I repeated what I told Michel in the car, that I would like for him to come to Geneva to our Hotel the next day at 2 p.m. Georges recommended a young lawyer in Geneva, Dr. Laurent Levy of the Geneva law firm Brunschwig & Biaggi, who would be Solarex's lawyer to help us to set up the company. He will be there too. He, Georges, Michel, and I are going to discuss the plan for the immediate establishment of "Solarex Europe." I was going to open up a bank account before I left and transfer money to pay for expenses.

While I was talking, I looked out the window and saw that snow was falling. It was September 9! I turned to them.

"I hope that you agree that Solarex Europe should not be set up here in La Chaux-de-Fonds, where it is snowing already at the beginning of September, but somewhere near the Lake of Geneva, near the Geneva Airport."

They both looked at me, finally Denise said: "I like your idea very much."

The next day everything was agreed, who was going to do what. Georges stayed there for 2 more weeks. The bankruptcy case was completed in December and Esotron was dissolved. Michel became officially the Manager of Solarex S.A. (Europe), which was established in Gland, Canton Vaud, Switzerland, not far from the Geneva Airport, with the program of developing Solarex's sales and manufacturing of terrestrial PV products in Europe.

By December 31, 1975, a permanent bridge from the USA to Europe for the explicit purpose for the promotion and manufacturing of terrestrial PV, the *first such company in Europe*, had been established.

Chapter 17

The Micro-Generator Mushroom Resulting in a Solar Thin-Film Solution

(A classical business school case study:
The effect of new technology)

The year of 1976 started at Solarex with big increases in the production of micro-generators manufactured for the Ragen, Nepro, and Uranus (Greenwich, CT) watches as well as for flashlights and other consumer items. The display on the Ragen, Nepro, and Uranus watches was of the light-emitting diode (LED) type.

During the beginning of the 1970s, another digital display started to be used for watches. This was the liquid crystal display (LCD). I had a friend, who was a professor at a university, teaching Einstein's relativity theory. He told me the story that he could explain to anybody the relativity theory very easily. When he was asked to explain it, he told the person that if he had only one hair on his head, that was too few, but if somebody found one hair in the soup, that was relatively too many.

Based on this story, I gave much thought to how I could explain simply to somebody, how the LCD worked. I could not

Sun above the Horizon: Meteoric Rise of the Solar Industry
Peter F. Varadi
Copyright © 2014 Peter F. Varadi
ISBN 978-981-4463-80-5 (Hardback), 978-981-4613-29-3 (Paperback), 978-981-4463-81-2 (eBook)
www.panstanford.com

come up with a simple explanation, only a somewhat complicated one: One should consider a special liquid put between two glass plates, which are coated with a transparent electrically conductive layer on the side they have the liquid. The light passes through this sandwich. But if an electric field is established between the two conductive layers, depending on the strength of the field the liquid changes from transparent to gray and if the electric field gets stronger, to black.

This effect was discovered in 1888 but it took a long time, until 1970, when the Swiss pharmaceutical company Hoffmann La Roche identified a proper liquid which made this change reliable. Finally, the Japanese electronics industry produced the first digital wrist watches with LCD.

The LCD, unlike the LED display, was visible all the time, the problem was that while the display required less electric power than the LED display, the electronic circuitry driving the watch still required quite a bit of electric power. At that time, the watches required two batteries and finally some of the manufacturers decided to install solar cells to recharge the batteries.

One of the first solar LCD watches was made by Sicura, located in Grenchen, Switzerland (Fig. 17.1, in the Jura Mountains. Sicura produced not only an LCD digital solar watch, but because the solar watches became very popular it also manufactured a solar analog watch. The watch cases for the two types were identical. There were other watch manufacturers in Switzerland, for example Mondaine in Zurich and also others and in the USA, Optel, and Chronar in Princeton, NJ, and Uranus, as well as manufacturers in South Korea and in Hong Kong. There was a very big demand for Solarex's micro-generators.

Solarex's micro-generator production's crown jewel was when the Swiss Patek Phillippe asked Solarex to make the solar cells for an analog clock that Patek Philippe was designing. The solar cells had to have perfectly an identical blue color. The solar cells were mounted on the top of the clock (as seen in Fig. 17.2). The clock was manufactured in Geneva, Switzerland, in 1977. The price of the hand-made clock at that time was about $15,000. (The cost of the solar cells was less than 3% of the clock.)[108]

The demand was not only for watches, but also for flash lights, cigarette lighters, and many other battery-operated gadgets.

Solar Micro-generator

Figure 17.1 Sicura digital solar LCD watch.

Figure 17.2 Patek Philippe solar clock with Solarex's micro-generator, 1977.

---

[108]The Patek Philippe solar-powered clock is exhibited in the Patek Philippe Museum in Geneva, Rue des Vieux-Grenadiers 7, in Switzerland. The pictures are shown with the permission of Patek Philippe, Geneva, Switzerland.

The electronics and LCD power requirements decreased and smaller and smaller micro-generators were needed. Solar LCD watches became so popular that large numbers of the inexpensive digital LCD watches were manufactured in Hong Kong. At one point, even "solar look" digital LCD watches were manufactured, where not a real micro-generator but only the picture of it was printed on the face of the watch.

In addition to the demand for solar micro-generators for watches was growing, the era of small handheld calculators began, mainly in Japan. From the beginning, the calculators had an LCD display. The electronics and the LCD required quite a bit of electric power. Calculators were made with batteries, but calculators using only solar power were also manufactured, with micro-generators and without batteries Fig. 17.3. The demand was so high that in Japan, Sharp Corporation also started to make micro-generators using crystalline silicon (c-Si) solar cells.

Figure 17.3 Teal (Japan) solar calculator without battery with Solarex (c-Si) micro-generator.

With calculators it happened as with watches. The electronic circuitry required less and less electricity, but on the other hand, while watches needed to operate 24 hours a day, calculators operated only when they were needed and only when there was light; therefore, because a battery needed replacement and the solar micro-generators needed no replacement, the calculators with micro-generators, without batteries, became popular.

The idea was that the LCD could only be seen when there was light, and when there was light the micro-generator provided enough electricity to power the calculator. The problem was that if the micro-generator was covered even for a short time, the calculator forgot the number it was on before the covering happened. This was rectified by using a small inexpensive capacitor which stored electricity for a period (for example 10–20 seconds), therefore, a short coverage of the micro-generator did not erase the result of the calculation. This made the product very successful and the simple handheld calculators were all "solarized" without a battery. More and more micro-generators are needed to be manufactured from c-Si solar cells.

In the early 1980s, however, the c-Si micro-generator business collapsed. One reason was that the need for c-Si micro-generators rapidly decreased for watches, when at some point a single small battery was able to provide enough electricity for the digital LCD or analog watch to be able to operate more than a year without solar assistance.

The other reason was that in 1974 Dave Carlson at the RCA Research Laboratory in Princeton, NJ, invented a new type of solar cell. This was a thin-film solar cell made of amorphous silicon (a-Si) layers[109] (more details in Chapter 22). The Japanese company Sanyo Corporation started the development of a-Si solar cells in 1975 and in 1980 started to produce a-Si solar cells for their own calculator production and also marketed them for solar calculators. It turned out that a-Si solar cells did a better job to power calculators than the micro-generators assembled from c-Si solar cells.

---

[109]Carlson DE, Wronski, CR (June 1, 1976) Amorphous silicon solar cell, *Appl Phys Lett*, **28**(11), and Carlson DE (December 20, 1977), US Patent # 4,064,521.

The a-Si solar cells required a bigger area than the c-Si micro-generators, because their efficiency was not as good. They probably were not cheaper at that time than the mass produced c-Si micro-generators, but they were superior in one respect: The a-Si solar cell worked much better in an office or home environment than the c-Si one.

The reason was very simple. Dr. Yukinori Kuwano of Sanyo Corporation tells it in the following way[110]: "It was late at night when a traditional c-Si solar cell was compared to a-Si solar cell. Daytime performance of the c-Si solar cell was better, but at night the a-Si characteristic was found better. But ghosts come at night and playing this trick? At day time it is measured by sun light but at night of course are measured under fluorescent light." He assumed that this important difference was due to the collection efficiency (or spectral response) of the a-Si and the c-Si.

Dr. Yukinori Kuwano was right. The efficiency to convert light to electricity of the a-Si was by far not as good as of the c-Si solar cells, but a-Si cells were much more efficient in a narrow range of the visible light (between 400 and 600 nm of the optical spectrum), but that is the spectrum of the indoor light, for example fluorescent light illumination.

It is clear that c-Si is much better in dark places, such as bars lit by candle light, but calculators are rarely used there, they are used in offices or at home with indoor light sources and there the a-Si is much better. That resulted in the c-Si micro-generators being less and less used for calculators, while Sanyo's production of a-Si solar cells for watches and calculators for the years starting with 1980 was several tens of kWp.[111] Now, more than 30 years later, all of the millions of the simple handheld calculators are still powered by a-Si solar cells, and solar-powered watches utilizing a-Si solar cells are still on the market.

These were the causes of the c-Si micro-generator market collapse during the first 3 years of the 1980s. This is what is taught in business schools as the effect of new technology.

---

[110]Private communication from Dr. Yukinory Kuwano and also http://us.sanyo.com/Solar/SANYO-Solar-History.

[111]Private communication from Dr. Yukinory Kuwano of Sanyo Corporation.

Sharp started to manufacture a-Si micro-generators in 1983. Solarex also realized the usefulness of the new a-Si thin-film solar cells and acquired in 1983 the capability from RCA, where Dave Carlson worked and made the invention.

The two milestones in the terrestrial PV history resulting from the story described in these three chapters (Chapters 15, 16, and 17) were

(1) Europe

As a result of the micro-generator business, in 1975 Solarex established Solarex S.A., in Gland, VD, in Switzerland. This was the first terrestrial PV product manufacturing, marketing, and sales operation in Europe. Later a multicrystalline wafer manufacturing factory (Intersemix) was also set up there.

This also resulted in the establishment of three joint ventures for the manufacturing of solar cells, modules, and systems, utilizing Solarex technology in Europe. These joint ventures were

- The Leroy-Somer/Solarex joint venture—"France Photon" in Angouleme, France. This was later merged into Photowatt.
- The Holec/Solarex joint venture—"Holecsol" in Eindhoven, the Netherlands. This was later acquired by Shell.
- The ENI/Solarex joint venture—"Solare"—outside of Rome, Italy.

(2) Thin-Film (a-Si) PV

The description of the history of the micro-generators also included the first commercial use of the thin-film a-Si PV system in 1980 by Sanyo, replacing c-Si technology. A viable and successful alternative solar cell system was established besides the c-Si cells.

Chapter 18

PV for the People

This story started a few weeks before I got the telephone call from Roger Riehl, which got us into the watch/calculator business.

Joseph's secretary came to my office and said that a lady was calling Joseph Lindmayer, who was not yet in, but she sounded that what she wanted to discuss was very urgent, so she would connect her to me.

I picked up the phone and the lady said she wanted to talk to Joseph. I told her that he was not yet in and I did not know when he would arrive. This was on a Friday, a nice fall day in mid-October, 1974. I told her that maybe I could help her as I was the co-founder of Solarex.

She said she just would like to ask me to tell Joseph Lindmayer that she enjoyed his statements the day before at the congressional hearing and she briefly talked to Joseph after his talk and he told her that Solarex could provide her with a small box which would demonstrate the utilization of solar energy by having a solar cell, and when it is exposed to light, a little fan, a propeller, would start to spin.

I did not know what Joseph had told her, because we did not have anything like that. I thought I will ask Joseph what he

Sun above the Horizon: Meteoric Rise of the Solar Industry
Peter F. Varadi
Copyright © 2014 Peter F. Varadi
ISBN 978-981-4463-80-5 (Hardback), 978-981-4613-29-3 (Paperback), 978-981-4463-81-2 (eBook)
www.panstanford.com

had told her and postponed the issue by telling her that I was going to tell Joseph about her telephone call and either Joseph or I was going to call her back on Monday. She also said that she would like to have five pieces and how much would that cost? I told her that we were going to call her back.

Shortly after this telephone call Joseph arrived and came to my office and said to go out for lunch, which we did a few times a week, where we were not disturbed and could discuss our business matters. After we sat down, I told Joseph about the lady's telephone call and what was that he mentioned she could buy to demonstrate the utilization of solar energy.

After thinking it over he said that he vaguely remembered that from the crowd a woman asked him how could one demonstrate the electricity generated by a solar cell and he answered that one could take a solar cell, connect it to a little electric motor to which one attaches a propeller. One can see that the propeller starts to spin when the solar cell is exposed to light, and its speed will vary according to the intensity of the light, but he did not tell her that we had anything like that in the office and certainly not that she could buy one.

After we finished our discussion about current business matters I told him:

"You know, one thing I learned since I left science and became a businessman, for example, with our attempt to sell solar modules to boat owners, is that the 'customer is always right.' If this lady misunderstood you and thinks that a small demonstration of the utilization of solar cells is a good idea and she wants to have one and give it also to other people, then we have to make it, because if one customer thinks it is a good idea, then many other customers are going to buy it too."

That gave me the idea that it is important to educate the public. Why would anybody install a PV system on their house if they know nothing about it? These people are not attending scientific meetings they do not go to congressional hearings, or read government-sponsored studies. They are only going to believe that PV is good for them, if they can see it, touch it, find it in stores, see it in mail order catalogs, or read about it in popular magazines or books.

"We have to go into consumer business, sell PV products made for them for their education. This will have three results: The consumers will be educated, Solarex's name will be known, and Solarex will have a profitable product line because none of our competitors are in this business at this time and probably they will never get in, as they are chasing the 'big picture,' to produce PV systems for utilities."

"We should get into the consumer business. Step 1. We make for this lady what I call a 'solar fan cube' for demonstration. Step 2. I am going to collect a number of catalogs, where they are selling technical products, and going to convince them to put our products directed to consumers in the catalog. Step 3. Try to convince not scientific but popular technical magazines to carry articles about PV."

Joseph, who usually was smoking continuously, got a cigarette, lit it, took a few puffs, and said:

"You have a very good point. We need to educate the consumers, to install PV systems on the roof of their houses."

After we returned to the office I asked Jeannie to find Charlie and ask him to come to my office and she should also come with him.

Shortly after they both arrived, I briefly explained them the telephone call I received from the lady and that I think it would be a good idea to make a "solar fan cube." Both of them looked at me as if I just had escaped from some asylum. I explained:

"I think the matter is very simple. Jeannie, please go to a photo shop, I have seen that they are selling a plastic cube, each side about 4 inches and it is open at the bottom. The purpose of this clear plastic box is to put pictures on each of the four sides of the box. Please buy 10 of those boxes."

"Charlie, I would like to ask you to select 10 of our good-looking chevron solar cells, they must not be the best ones and please go to Radio Shack or some airplane model store and find a little motor, which a solar cell would be able to operate and also get a little propeller, which can be mounted on the shaft of the motor."

"When we have all of that, Charlie please ask one of the technicians to assemble it. Attach the solar cell to the top of the box and drill a hole on one side. Mount the small motor inside so that its shaft would go through the hole and connect it to the solar cell to the motor and mount the propeller on the shaft. That's all. When light will hit the solar cell the propeller will start to rotate."

"One more thing. I have to call this lady Monday morning and I am going to tell her that she can pick up the five 'Solar Fan Cube' on Tuesday morning. Oh, we also need a box in which this contraption would fit in."

At this point the two of them looked at me as if they were now convinced that I just escaped from that asylum.

Monday, after I returned from lunch and entered my office, I saw 10 small white boxes on my desk and in the window I saw a small propeller spinning on the side of a plastic box. I went to the window picked up the plastic box and as I picked it up from the sunny window sill the propeller stopped. On the inside of the top of the box was not one large, but three half good-looking "chevron" solar cells, and inside of the box was a small electric motor, as I described last Friday afternoon (Fig. 18.1). I put it back, the propeller started to spin, and I started to go to the door of my office, when Jeannie entered with a smile on her face, like a cat, which just ate a canary. I must have looked very surprised, because she said:

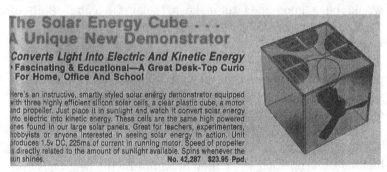

The Solar Energy Cube . . .
A Unique New Demonstrator

Converts Light Into Electric And Kinetic Energy
Fascinating & Educational—A Great Desk-Top Curio For Home, Office And School

Here's an instructive, smartly styled solar energy demonstrator equipped with three highly efficient silicon solar cells, a clear plastic cube, a motor and propeller. Just place it in sunlight and watch it convert solar energy into electric into kinetic energy. These cells are the same high powered ones found in our large solar panels. Great for teachers, experimenters, hobbyists or anyone interested in seeing solar energy in action. Unit produces 1.5v DC, 225ma of current in running motor. Speed of propeller is directly related to the amount of sunlight available. Spins whenever the sun shines. **No. 42,287 $23.95 Ppd.**

Figure 18.1 Edmund Scientific catalog, 1977.[112]

"Isn't it what you described to us Friday that you needed for that lady who called?"

---

[112]I would like to thank Mr. Robert Edmund giving permission to reproduce from pages of Edmund Scientific catalogs.

"Yes, it is. How on Earth did you get it together so fast?"

"You told us it would be very easy, described where to go to get the parts and you would be surprised if we could not do it. And now you look surprised that the first solar consumer product is on your desk."

"I am not surprised that you and Charlie produced these Solar Fan Cubes, I am only surprised by the speed and the quantity. But you are wrong, this is not the first solar consumer product, it is the third one. The first was invented many years ago and it was a light meter for photography. That had a selenium PV cell in it. The selenium PV cell efficiency to convert light to electricity was a small fraction of what this Si cell is able to do. The second one was the solar module for boat owners, which was not successful. This will be the first Si cell used for consumers, which will be successful. But we are going to make more gadgets to educate people how effective PV is. Could you please find Charlie and both of you come to my office as I would like to thank both of you."

Before they arrived I took one box to Joseph's office. He was not there, so I put it on his desk. When I returned to my office, they arrived, I thanked them and they described to me how they were able to make these Solar Fan Cubes. Charlie told me that instead of one large solar cell he had to use three half cells, because he could not find in a hurry a motor which would operate with about 0.5 V of one cell. He could only find a motor which was operating with three cells, about 1.5 V. Now the only question was left, how much we should charge for this. Adding up the cost and time to make it and adding a nice profit they came up with $20. I picked up the telephone, called the lady, and told her that she could pick up the five Solar Fan Cubes any time, their price is $15 each, but that she should pay only for four pieces, one is a gift for her, and when she comes I would like to give her that personally. She deserved it, because by misunderstanding Joseph she launched Solarex's consumer business.

More and more of this "Solar Fan Cube" were sold. By 1975, the need was so much that we had to hire several people to assemble them. The quantity increased and we decided to hire a product design specialist to make an attractive design and packaging. After we selected the design we started to make the Solar Fan Cube and its box as shown below in Solarex's Hong Kong factory (Fig. 18.2).

Figure 18.2 "Solar Fan Cube."

The Solar Fan Cube which was started in 1974 and in 1983 was still in Solarex's *"The Solar Wonder Book"* consumer product catalog for $19.95 and also in kit form for those who wanted to assemble it for $14.95.

As a measure of how much people were fascinated and interested in the utilization of PV to generate electricity from this simple demonstration gadget, many, and many 10,000 were made and sold. As an example, Saudi Arabia ordered 10,000 Solar Fan Cubes for their school system.

With this, PV entered the consumer business and as I told Joseph the next step should be to sell PV through mail order catalogs.

I found several mail order catalogs. Only a very few had some PV in them. I found in one a selenium solar cell with very little conversion efficiency and in another a few small cells from the American IRC company. These solar cells were quite expensive. Nothing was close to our excellent PV cells. The next project was to assemble 10 items to offer to those companies selling through their catalogs, and prepare a nice data sheet about them.

Edmund Scientific was the first organization I visited. Edmund Scientific was a very reputable company, mostly concentrating on optics and other not electronic scientific products. It had also a selenium solar cell in the catalog. Besides the catalog, Edmund Scientific had also one large store in New Jersey. It was a family owned company. I called Robert Edmund, who owned and managed the business, and arranged a date to see him. Robert was a young very nice man, who had quite an interest in solar energy. He immediately said that he was very interested to carry the line I offered and introduced me to Bob Edgerton (Fig. 18.5), who was responsible for a part of the catalog in which the solar energy would fit in, at present represented by some selenium cells.

Bob looked at my list and said he would put all of them in the catalog. He explained me how the system works. They are going to order not a large amount to have it in stock so when an order comes they can ship it immediately. When the inventory decreases to a certain level they reorder it and they expect to get the shipment from us immediately. Obviously any item which is not selling well will be removed from the catalog. He said if we have more solar products, he will be glad to see them and if he thinks it is interesting he would put it in the catalog. He also said that he has to select items which sell well, because each page has to result in a certain minimum amount of revenue, if not, he has to change the slow moving items.

Bob's selection included a number of solar cells and also solar modules. An example is a part of a page from Edmund Scientific's 1977 Spring catalog (Fig. 18.3):

The Edmund catalogs were issued quarterly, so I walked out with a purchase order to ship them the items immediately and they will be in their next catalog.

It was surprising that not only the "Solar Fan Cube" and similar products were consumer products, but our entire line of solar cells and solar modules became "consumer products." In a short time, we wound up having three pages in the Edmund Scientific catalog. I visited Robert and Bob periodically. I liked Robert's store, with many fun things. When I went there I could not resist and always bought something. Bob was also very pleased, because the Solarex pages were selling well, more than 3 times what the minimum would have been for a page. This contradicted the theory that PV's future is only the utility-oriented business. It showed that the consumers were also a market, even at the retail prices shown in the Edmund Scientific catalog.

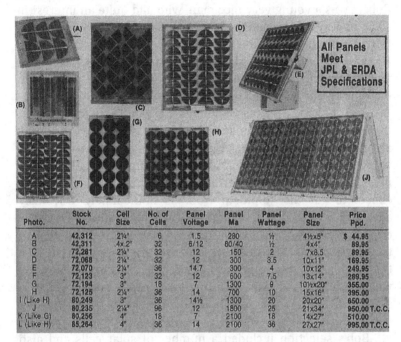

Photo.	Stock No.	Cell Size	No. of Cells	Panel Voltage	Panel Ma	Panel Wattage	Panel Size	Price Ppd.
A	42,312	2¼"	6	1.5	280	½	4½x5"	$ 44.95
B	42,311	.4x.2"	32	6/12	80/40	½	4x4"	89.95
C	72,281	2¼"	32	12	150	2	7x8.5	89.95
D	72,068	2¼"	32	12	300	3.5	10x11"	169.95
E	72,070	2¼"	36	14.7	300	4	10x12"	249.95
F	72,123	3"	32	12	600	7.5	13x14"	289.95
G	72,194	3"	18	7	1300	9	10½x20"	355.00
H	72,125	2¼"	36	14	700	10	15x16"	395.00
I (Like H)	80,249	3"	36	14½	1300	20	20x20"	650.00
J	80,235	2¼"	96	12	1800	25	21x34"	950.00 T.C.C.
K (Like G)	80,256	4"	18	7	2100	18	14x27"	510.00
L (Like H)	85,264	4"	36	14	2100	36	27x27"	995.00 T.C.C.

Figure 18.3 Solar modules in 1977 Edmund Scientific catalog.

At one time when I visited Bob he told me that one of our competitor's salesmen visited him and offered three types of solar cells to replace ours, and he offered them at almost half the price we were selling them.

I told Bob that is a very good price, we could not sell it for that price and I understand that he would buy those and replace ours. He was very pleased and smiled. After that I added:

"Bob, you will have only a small problem."

He looked surprised at me.

"You will have to remove the three pages having Solarex products, because we cannot lower our standards and have our excellent quality products appear mixed with the low-quality products made by our competitor."

"OK, OK, OK," he smiled and said, "you convinced me not to mix PV products."

I was thinking that I am going to tell Joseph, that I was right when I predicted that Solarex will have a profitable consumer business. Our competitors probably will never get in as they are chasing the "big picture," to produce PV systems for utilities. Maybe now they want to get in, but they do not have the right products.

At the beginning of 1976, I visited Bob again. I invited him for lunch. We had a glass of wine to start with.

"Bob, as you can see Solarex has several consumer-oriented products; furthermore, it seems that PV products in general such as PV modules, systems, and so on are also of interest for the consumers. We also have a factory in Hong Kong, which is making these consumer products and also is assembling our small solar modules.

"As you can see, from the success of the solar pages in the Edmunds catalog and from the others also, people are interested in solar energy. But we have to extend the education to everybody about what solar cells can do, because that is the only way they are going to support it and use it themselves. Therefore, we want to put in much more effort to develop our consumer business, which is now important, bringing in revenues, but it is very important for the future, that people should know about it and accept it as a replacement of nuclear and later fossil fuel."

"PV should be in every catalog, in several stores and we should also have a mail order catalog for solar products and ultimately I would like to have solar stores that people could see PV modules and house models, touch them, and want to have them on their roof."

He agreed.

"Well, we need somebody, who is building this up, and I hope you would accept this job."

He looked at me surprised, but 2 months later I walked with Bob around at Solarex and introduced him as the Manager of Solarex's Consumer Product Department.

Solar modules and their application for a multiplicity of uses were our primary interest to sell to consumers. The other interest we had was the education of people, including children. For this reason Solarex published books. The first book was: *The Solarex Guide to Solar electricity*. It was first published in 1979 and was revised and reprinted in 1980 and 1983. This book was so popular that several hundred thousand copies were either distributed or sold.[113]

Bob created and also found new products. One of them was a solar hat.

One day Bob came to my office and had a "safari hat" in his hand.

"Dr. Waters, a medical doctor, visited me and brought this hat. He said he has a patent on a solar-powered hat. He said that a small solar module powers a tiny motor spinning a propeller which blows air over a water-soaked sponge, generating a cool, steady breeze to the person's forehead. In the forehead is the temperature sensor of the people and if it is kept cool the heat is more tolerable."

I really did not know what to say. I did not wanted to hurt Bob's feelings by telling him what I thought about the idea, but as he was successful with many other products, I only told him that it sounds interesting and asked him who would buy a safari hat with a motor and a propeller sticking out in the front of it?

He was very reassuring, "This is a cool product (in those days the word "cool" was used frequently), we can call it the "Solar Cool—Safari hat". I thought it over and I thought we could bring this out also in a baseball hat and that would be a large success (Fig. 18.4).

Making a long story short, Bob made them in our Hong Kong factory and several ten thousands were sold. One very

---

[113]New and/or used copies of all editions are still available at http://www.amazon.com/s/ref=nb_sb_noss?url=search-alias%3Dstripbooks&field-keywords=The+Solarex+guide+to+solar+electricity.

large order came from the soccer team in Holland which was playing in the Soccer World Championship in 1978. The official color of the Dutch soccer team was, and is, orange. Because the Dutch team was hoping to advance and expected to play in the final of the Championship, they ordered 10,000 orange-cultured solar baseball hats for their fans to wear. The hats were made in Hong Kong and delivered to Argentina, where the championship game was played. The fans were wearing the orange-cultured baseball hats and the team successfully advanced and played in the final. In spite of the solar hats, they lost 3:1 on 25 June 1978 to the Argentinian team.

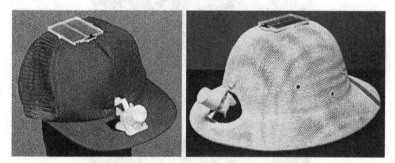

Figure 18.4 "Solar Cool"—Baseball and Safari hats.

Bob (Fig. 18.5) introduced Solarex products to several mail order businesses and also started Solarex's mail order business, by producing *The Solar Wonderbook*, a catalog featuring mostly PV products, and components to build PV systems. He also arranged that popular science magazines should also run articles about PV and its utilization.

As the consumer business was growing it was needed to separate Solarex's commercial and utility-oriented business from the consumer business, and therefore, the consumer business was carried out under the name of "Energy Sciences."

In 1981, we decided to take the next step to bring PV to the public that people should be able to see and touch PV products and systems, by opening "Energy Science" stores. The first Energy Science store was opened in 1981 in Rockville, MD. It was a store adjacent to other stores selling modern electronic equipment. The next store we decided to open was in a shopping mall. We selected the shopping mall called Tysons Corner, which was the newest and one of the best shopping malls of the Washington, DC, area located in Virginia. The Tysons Corner store was opened in the fall of 1982 and Virginia Governor Charles Robb cut the ribbon to open the store.

Figure 18.5 Bob Edgerton.

By 1984, Energy Sciences had opened five stores in the Washington, DC, area and one was opened in a new shopping mall in Baltimore Harbor.

A counter to monitor how many people visited was set up at the entrance of the store. The store was a great success also financially, but more important, it provided an opportunity for people to get acquainted with PV and learn what it is good for. Around Christmas about 10,000 people visited the store daily.

In the Solarex catalog, there were pages offering experiments for children and a page to describe and teach people how to design a PV power source for their home or business. To install PV in houses considering the infancy of the PV business seems to be early, but it was not.[114] By 1980–1983 many houses were equipped with PV, assembled and built by the owner, or by a local electric shop.

These were some of the reasons for the high demand for our PV modules. Obviously as the demand grew, larger quantities of PV modules were manufactured and as a consequence prices went down (Table 18.1).[115]

Table 18.1 Catalog retail prices for PV modules (1974–2011)

Year	1974	1977	1982	1983	2011
$/Wp	$35.00[a]	$27.60[b]	$13.85[c]	$12.48[d]	$2.13[e]

When in 1983, AMOCO acquired Solarex, as described in another chapter (Chapter 24), Amoco made changes in its management team overseeing Solarex. The new people believed in the "big picture" and therefore eliminated everything which was not related to the "core business," meaning not in the direction to achieve "grid parity" and therefore, the entire consumer business, catalogs, and stores were discontinued. All the stores were closed by 1987/1988. The Hong Kong factory was also shut down by 1989. Not only the consumer business was closed, but also everything considered to divert manpower and attention from the "core business" was discontinued. These were Solarex's Aerospace and Corrosion protection divisions and all of the European activity including Intersemix, the multicrystalline manufacturing factory was dissolved, leaving only sales offices.

But the PV consumer business was already unstoppable. It continued in the USA, Europe, and Asia, and also in the developing

---

[114]See in it Chapter 12 Paul Maycock's statement made in 1980 that "Residential systems (PV) are being sold all over the world right now. Solar cells are already economically viable for off-the-grid, isolated houses, and villages."

[115](a)Estimate.

(b)Edmund Scientific Co. catalog 1977.

(c)Solarex, Energy Science catalog 1982.

(d)Solarex, Energy Science catalog 1983.

(e)Home Depot Web site 2011 (www.homedepot.com).

countries. Consumers realized the advantages of PV and if there is a market, somebody will cater to it. Several companies were started to cater to the existing consumer business, for consumer products used in the industrialized countries, for example garden lights, illuminated house numbers, recreational products, and products used for automobiles and for the developing countries, small home systems, lanterns, and so on. The unfortunate consequence was that as the major PV manufacturers were mostly interested in large projects, the consumer product's manufacturing was done mostly by small manufacturers without any quality control and a lot of product ended up in failures.

However, PV consumer products survived this period and today the shelves of major stores are packed with a lot of consumer PV products. To show that solar is now available in the main stream of the US consumer business, one can review the offerings of several chain stores. In Europe, in spite of the success of the FiT program and the deployment of many larges size PV systems, interestingly the consumer PV offering is not as popular as it is in the USA.

One example is a very large store chain in the USA, a retailer of home improvement, construction products, and services, which operates (in 2011) 2248 big stores across the United States, including all 50 US states, Canada (all 10 provinces), Mexico, and in China, with a 12-store chain. The web catalog of this company has more than 120 solar products priced from $10 to $1500. It also offers to install PV systems on the roof of homes.

Another example is a similar chain, a retailer of home improvement, and appliance stores, which operates (in 2011), 1710 stores in the US, and 20 in Canada. The web catalog of this company has more than 230 solar products priced from $10 to $1500.

An electronic retail store chain that has 4675 company-operated stores in the USA and Mexico carries 80 solar products from $5 to $600.

In Europe, one of the chain stores has more than 390 stores and carries 36 solar products in the price range of £5–£4000.

Another chain that has over 500 stores carries only 10 products in the range of €5–€50.

An interesting niche PV market exists; a lot of PV camping gear is available for those who love the "great outdoors."[116]

---

[116]Photon (2010) *The Photovoltaic Magazine* (6), p. 60.

Chapter 19

Story about Success and Cash Flow

As planned, and in spite of the COMSAT's lawsuit, Solarex broke even sooner than its first year of operation which ended in mid-1974. After that, Solarex finished the next 5 years with at least 10% profit of which about 5% was paid to the government as income tax. After 1975, the year of change until about 1980 when Arco Solar became a strong competitor, Solarex had an easy life because as described before, Berman left, Solar Power Corporation closed its European offices, and in the same year Hughes took over Spectrolab and did not pursue the terrestrial PV business strongly. Yerkes started STI but with very little money, so in the first years STI was not a big competitor and ARCO acquired it only in 1978. In Europe, Photowatt was started, but that company was still in its infancy, and Philips RTC, after technical problems, discontinued its terrestrial solar business. Sharp was mostly concerned with their domestic market. These events made it easier for Solarex to expand sales for PV modules, systems, and consumer products and become profitable.

However, one does not need a business school diploma to realize that 5% profit is an excellent achievement, but it is

Sun above the Horizon: Meteoric Rise of the Solar Industry
Peter F. Varadi
Copyright © 2014 Peter F. Varadi
ISBN 978-981-4463-80-5 (Hardback), 978-981-4613-29-3 (Paperback), 978-981-4463-81-2 (eBook)
www.panstanford.com

highly insufficient to finance the business if the business is very successful and doubles every year. That was the problem we were facing at Solarex. At the beginning we agreed that Joseph was going to do the technical matters and I was going to run the operation. I was doing sales, which went very well, but I was also responsible for accounting and so on, which meant also to pay the employees and the bills. This meant that to use our line of credit for this was necessary. Now you can understand why it was so important for me to raise our line of credit, which is described in Chapter 6.

In those days, to raise money on the stock market for a company such as Solarex was impossible. In spite that the very nice guy, Don at J.H. Whitney & Company assured me that our plan looked very interesting and he would consider to invest if I came back 2 years later, now 2 years later both Joseph and I got still an allergic reaction if somebody mentioned to us the words "venture capitalist."

I do not know what I would have learned if I had gone to a business school, but to solve our financial problem I came to a simple solution. I realized that if I would be the banker and my client took 99% of its line of credit, I would be very nervous. On the other hand, if the customer pays back its line of credit when it reaches 75%, the banker will be impressed and as the customer is not tied to the bank by owing money it may go to some other bank, and therefore, the customer could be retained if the bank would immediately offer a much higher line of credit.

That is what I did. When we were near 75% of our line, I went to Al Nerkin and asked if he would invest some money. Al was on our board and knew everything about the company in which he originally invested money because he considered solar energy as a charity, but at that time he had seen that it became a fair-sized profitable outfit. This time he considered it not as charity anymore, but as a business investment. He took a piece of paper and a pencil and after a few minutes of calculation he said how much he would invest and what share price he would pay. He wanted to explain to me how he calculated it but I told him I accepted it as his word was enough that it was a fair value. I contacted our other stockholders and also few more people who contacted us that they would be glad to invest, and in a few days I had the money deposited in the bank and told Bob Burke that I was paying back our line. He was surprised because he saw that even if we pay him back we still have quite a bit of money. I told him that we would be glad to stay with his bank, but we would need to increase our line of credit. We agreed that he was going to double it.

I was able to do this one more time, but then the numbers got quite big, and I was sure that we would not be able to do it again. That was in 1978. By that time Solarex had a very good name, we were the largest terrestrial PV company in the world, but we still could not get on the stock market and our opinion about "venture people" did not change.

What changed was that large companies wanted to get at least their finger into the PV business. Of the oil companies, Exxon and Mobile were already involved. Shell was involved before, but started to get involved again. One day we received a telephone call from ARCO that they would like to visit us.

Two people from ARCO visited us and expressed interest to buy Solarex. I think we had two meetings with them when we told them, that if they want to invest in Solarex we can discuss the matter, but we are not selling the entire company. Both Joseph and I liked the solar business. He was very much involved in the development of the Si casting project to produce multicrystalline wafers and I loved the incredible opportunities to develop both the commercial and the consumer PV business. They said no, they are only interested in a buyout, we said no, it is not for sale. The other possibility ARCO had was Bill Yerkes's STI. Bill sold it to ARCO and it became Arco Solar and Bill was managing it.

By that time Solarex was well known in Europe. Our company in Gland, Switzerland, was selling PV modules to a French company Leroy-Somer. Leroy-Somer was a very large company and one of their divisions was manufacturing water pumps. They bought the PV modules to provide electricity to drive the pumps in places where there was no electricity available. They were very much interested in the French areas on the western side of Africa, where the water table is very close to the surface and where the inhabitants needed water for drinking, for animals, and also for irrigation. So, Leroy-Somer approached us to buy a minority interest in Solarex as well to form a joint venture in France to produce solar cells and modules for their pumps. This sounded very interesting. We were selling a few systems for water pumping, but to tie up with a major pump manufacturer would open up a new segment for our PV business, furthermore, to establish a joint venture for manufacturing PV products for the French market was also interesting.

At the same time, a Dutch company, Holec, N.V. of Utrecht, the Netherlands, also approached us. Holec was a large company in the field of electrical equipment. They wanted to invest in Solarex. Their representative Mr. van Zwet visited us and expressed his interest to buy a minority position in Solarex.

Both companies decided to become minority investors in Solarex and we agreed that it would also be beneficial for Solarex. There was a third investor, Renzo Ghiotto, who owned a pump manufacturing company in Vicenza, Italy, which was loosely associated with Leroy-Somer. Leroy-Somer people recommended that Mr. Ghiotto should also be able to invest in Solarex. It was decided that the final discussion to agree to the terms would be in Paris during the last part of December, before Christmas 1978. van Zwet of Holec, Paul Barry of Leroy-Somer, Renzo Ghiotto, Joseph, and I participated in that meeting. We agreed on the terms very quickly and told our decision to the six lawyers, two representing Holec, two representing Leroy-Somer, one representing Solarex USA, and one Solarex S.A. Switzerland. The problem was that the six lawyers did not agree with each other about the various parts of the agreement. They discussed the matter all day. As we were in Leroy-Somer Paris office, they had plenty of secretaries who would write and re-write the text. The five of us, van Zwet, Barry, Ghiotto, Joseph, and I, discussed the plan about what Solarex should do in the future. I liked those investors, because their philosophy about the solar business agreed with mine. Expand the commercial and consumer business, do some but not very much government study or demonstration work and beside of agreeing that the central power stations, the "big picture" maybe the future, we should do nothing about it. We agreed on everything, had a very nice lunch, in the evening dinner, finally around midnight we told the lawyers still discussing the matter because they were not finished, that they should give us the signature page, we sign it and they can finish the agreement whenever they want. About a week later we received the 1-inch thick agreement with the signature page attached to it. The relation between Leroy-Somer, Holec, Ghiotto, and Solarex was so good, that nobody ever opened the agreement and we never found out what the lawyers' discussion was about.

After these investments were completed and the money was on our bank account, we were contacted by AMOCO, the big oil giant expressing interest to visit us to discuss their interest to invest in Solarex. Gordon McKeague, Manager of AMOCO corporate development, and John Johnson, who worked for Gordon, visited us. We liked them and their ideas about PV and its future and they liked what they had seen, so after a few meetings and their due diligence we announced on June 29, 1979, that Standard Oil (Indiana) "AMOCO" purchased a minority interest in Solarex Corporation. AMOCO's investment in the company was on terms similar to recent investments in Solarex by two European corporations, Moteurs Leroy-Somer of Angouleme, France, and Holec, N.V. of Utrecht, the Netherlands. AMOCO, in recognition of the potential of PVs as a major energy source, made this investment to help foster rapid development of the technology. The purchase of this minority interest by AMOCO provided the final increment of investment capital needed by Solarex to construct an advanced production facility. In addition to this direct equity investment, AMOCO was also considering funding of specific research and demonstration programs.

In 1979, with the biggest market share in terrestrial PV, a substantial amount of cash, and three large investors, Solarex became for the next 3–4 years the world leader in terrestrial PV's commercial and consumer business, by establishing factories in the USA, Switzerland, Australia, Hong Kong, and the UK, and joint venture factories first in France after that in Italy and in the Netherlands. After a few years Arco Solar became the world leader.

Solarex also established factories in the USA and in Switzerland for the production of multicrystalline Si wafers utilizing Solarex's casting process. Solarex acquired from RCA the a-Si technology, hired all of the technical people responsible for the development of a-Si and started a factory near Princeton, NJ, to manufacture a-Si modules.

Chapter 20

Water Pumping with PV

From the beginning of my involvement in PV, I always believed that solar water pumping should be one of the first important applications of PV in developing countries, where water was needed for people and animals and for irrigation. It is very much needed where rivers or lakes were far away or the area was arid. Also in developed countries where electric power was not available, water was needed for people and animals. I was very surprised that the application of solar water pumping had a very slow start and how this idea was started in reality.

Besides myself, several other people also believed this, but the beginning of the utilization of solar water pumping I am describing here is a story which never happened, but the reality was not far from it. I assumed that the first application could be a demonstration project in the western part of Africa, where the water table is close to the surface and not much electric power is needed to pump up sufficient amount of water for the people and animals in an area, which has no electricity, is very poor and badly needs the water. I believed

Sun above the Horizon: Meteoric Rise of the Solar Industry
Peter F. Varadi
Copyright © 2014 Peter F. Varadi
ISBN 978-981-4463-80-5 (Hardback), 978-981-4613-29-3 (Paperback), 978-981-4463-81-2 (eBook)
www.panstanford.com

that one of the organizations involved to solve developing countries' needs may initiate such a program. It really happened. A Request for Proposal (RFP) was initiated by one of those organizations to study the need and possible utilization of solar water pumping in Western Africa and also demonstrating PV water pumping.

I was very glad to read the RFP and expected that we could submit a bid to supply the PV system when requested by the company which would be selected and receive the contract. The organization selected a consulting group and awarded the contract. We did not receive a request to bid on a PV water pumping system. As it turned out, some employees of the consulting group made several trips to West Africa to study "the need and possible utilization of solar water pumping" and assembled their report. The report clearly described that the utilization of PV to supply the demand for water in the Western African region was a very good idea.

The trips and the work to assemble the study basically exhausted the money allocated for the contract and there was only money left to "demonstrate the PV water pumping" which was sufficient to buy a small aquarium, a water pump used for small aquariums, a lamp, and a small PV module to power the aquarium pump. When the lamp was turned on, it was demonstrated that PV could be used to pump water.

This demonstration system with an artistic rendering of a PV system and a reservoir in which one could see the water from the pump pouring in, surrounded by admiring natives and some animals, was exhibited in the sponsor organization's reception area. The receptionist turned on the lamp every morning and turned it off in the evening when she left and during the working hours every visitor could admire the utilization of PV for pumping water from the aquarium back into the aquarium and its planned application in developing countries.

The reality was, however, somewhat different. One can find an excellent description of the reality in Bernard McNelis' paper "Promoting PV in Developing Countries.

[117]I would like to quote only one sentence from that paper[118]: "My trip to initiate the project there (solar pumps in Mali) was a total disaster (and not because of my poor French)."

But luckily, besides some of the world's bureaucracy trying to "help" the poor people, there were people in some other part of the bureaucracy and in the industry who envisioned the need and the market for solar water pumping in that part of the world. The first solar-powered water pump was actively supported by Wolfgang Palz of the European Commission and deployed in 1974 on the island of Corsica, France.[119] This project was implemented by the French company, Pompes Guinard, and the University of Lyon under the direction of Dominique Campana.

Earlier, in 1958, Georges Chavanes[120] took over the control and Paul Barry became the Managing Director of the company. Leroy-Somer, a subsidiary of which was the above mentioned Pompes Guinard, a manufacturer of pumps. They both were visionaries. They learned from the success of the Guinard solar water pump installed in Corsica. They believed that the solar water pumps could not only become a business, but they could install them in the western part of Africa and help the poor people there. They decided to establish the Guinard solar water pumping line. They also wanted to secure the supply of PV systems and establish its manufacturing in France.

By that time, Solarex was the only privately owned PV company which was also well established in Europe. As described in the previous chapter (Chapter 19), Chavanes and Barry approached Solarex and offered to invest money and obtain a minority position. This offer was a great benefit for Solarex, strengthening its financial and market position and having a large European company with great interest in commercial PV

---

[117]Bernard McNelis's paper: *Promoting PV in Developing Countries*; W. Palz (2011) *Power for the World*, Pan Stanford Publishing Pte. Ltd., Singapore.

[118]Bernard McNelis's paper, p. 437.

[119]Palz W (April 2003) *My Life with the Renewable Energies*, Paris.

[120]Georges Chavanes became the Minister of Commerce of France (1986–1988).

sales as a minority stockholder. Paul Barry in 1978 was elected to be on the board of Solarex and also on the board of Solarex SA, Solarex's European operation in Switzerland.

Leroy-Somer developed Guinard pumps to be used with PV and started to use Solarex PV modules for its Guinard pumps. Solarex started to market solar water pumping systems utilizing the Guinard pumps. The cooperation continued in 1978 when a PV manufacturing joint venture was started in Angouleme, France. The joint venture's capital came from Leroy-Somer and Solarex provided the knowhow and the training of the French personnel. For this, Solarex received 30% ownership of the joint venture which was named "France Photon."

Leroy-Somer's marketing PV water pumping systems in the villages of the West African countries was helped by the "White Father" Verspieren.

Leroy-Somer[121] in 1982 created a fully owned subsidiary "SolarForce" with the purpose to develop, market and install all over the world, first PV-powered pumping systems (Fig. 20.1) and later PV-powered refrigerators and also systems for telecommunication equipment. This operation was managed by André Zanetto and Jean Posbic. It had several technical people from Guinard as well as from other Leroy-Somer operations.

The Guinard pumps were developed and designed for PV modules so that no electronics was needed to convert the PV modules direct current to alternating current. The pumps required only 4–6-inch boreholes. The output of these systems was 5–90 m^3 water/day from a depth of 5–90 m. The requirement for PV modules was 0.5–3.7 kW.

SolarForce sold thousands of these systems in Africa (Mali, Niger, and Zaire) and these and larger systems in Pakistan (UN and EU funded) and Mali (Saudi funded).

Solarex also sold these PV-powered water pumping systems utilizing Guinard pumps which were able to convert 90% of their electrical input to pumping power.

---

[121]A part of the story described below is from notes made by Dr. Jean Posbic and provided to me.

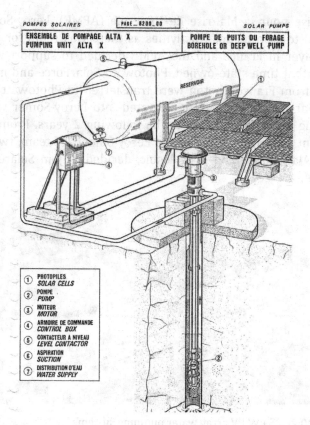

Figure 20.1 Schematic drawing of the Guinard ALTA X type solar pump system (1979).

In 1979, a 25 kW PV array, driving a 15 hp pump, the world's largest (at that time) PV water pumping system, funded by the US Department of Energy, was installed at Mead Field Laboratory in Nebraska (Fig. 20.2). The PV system was used to provide water to irrigate 80 acres of corn during the growing season. During the off-season, the arrays' power operated fans to dry the crop.

When in September 1983 AMOCO became the sole owner of Solarex, Leroy-Somer received all of Solarex's shares in France Photon. France Photon and its related company SolarForce by that time had an excellent technical crew. It was unfortunate that in 1985, with diminishing support from the French Agence

Francaise pour la Maitrise de L'Energie (AFME), Leroy-Somer decided to stop its solar activities as there was room for only one player in France and the AFME decided to support what was at that time state-owned, Photowatt. SolarForce and many assets from France Photon were transferred to Photowatt and the SolarForce team was reintegrated into Leroy-Somer while finishing all solar contracts in the following 2 years. From the excellent technical people of the Leroy-Somer PV team, two of them, Alain Ricoh and Jean Posbic, decided to join Solarex in the USA.

Figure 20.2 25 kW PV array water pumping system.

Sahel and Sahara

Two interesting and useful PV solar water pumping scenarios have to be reviewed, where the sun represents an abundant source of energy, and where the water pumping can be used for the improvement or development of life in that part of Africa.

Sahel: European Union (EU) Solar Water Pumping Project (1990–2009)

A major European Union (EU) project to provide water for people in developing countries in a semi-arid part of Africa was the solar water pumping project in the Sahel. Sahel forms a belt spanning Africa from the Atlantic Ocean in the west to the

Red Sea in the east a 5400 km (3400 miles) long belt that varies from several hundred to a thousand kilometers (620 miles) in width, covering an area of 3,053,200 sq. km (1,178,800 sq. miles). It is a semi-arid grassland, savannas, steppes between the wooded Sudanian savanna to the south and the Sahara to the north.

In spite the rainfall in the Sahel area being only between 150 and 600 mm/year, the water table is at most 100 m (330 ft.) down and in some areas is as close as 2 m (7 ft.) to the surface. It is practically independent of the amount of rain and it seems to be rising and not dropping. This means that the pumping of water by PV is very feasible.

The Sahel region comprises the West African countries of Burkina Faso, Cape Verde, Chad, The Gambia, Guinea Bissau, Mali, Mauritania, Niger, and Senegal. At least 68 million people live in the region, with some 30% in urban areas. Forecasts put the population at over 100 million by the year 2025, half of whom will live in urban areas, according to the Permanent Inter-State Committee for Drought Control in the Sahel (French abbreviation CILSS).

A large number of this population needs to transport the water from a distance. The average African women and young girls spend a considerable amount of time carrying up to 60 L (60 kg equivalent to about 130 lb.) of water 10 km (about 6.25 miles) every day.[122] The EU implemented a Regional Solar Program (RSP) by installing a large number of solar water pumping systems. The RSP is a vast, ambitious, and innovative program which was launched in 1986 by the Heads of States of the CILSS countries and financed by the EU. The importance of water can be seen from the information that 39% of the inhabitants of the Sahel do not have access to drinking water (more than 50% in Burkina Faso, Guinea Bissau, and Mauritania). The EU's RSP-1 was carried out from 1990 to 1998 and produced excellent results. It was continued by RSP-2 which was completed in 2009 and provided more than a million people with drinking water. Under RSP-2, 1000 solar water

---

[122]Campana D (2011) *PV Power Systems for Lifting Women Out of Poverty in Sub-Saharan Africa*; Palz W, *Power for the World*, Pam Sanford Publishing Pte. Ltd.

pumps were established, as well as 16 pumping systems for irrigation in the Sahel region.[123]

Based on the previous experience, the EU solved the problem of the after-installation maintenance of the solar water pumping systems. What unfortunately was not easy to solve was the theft of substantial number of solar modules (up to 30% in some of the countries) and inverters from the pumping sites. Each country or region developed their own strategies and introduced safety measures.

The stealing of PV modules indicated that there was a great need to provide electricity also for other purposes. Most likely the stolen PV modules were used for lighting and powering radios or TV receivers. At this time the situation is probably changed, because the price for PV modules decreased very much, which probably makes stealing not a lucrative business anymore. Furthermore, recent solar water pumps are the more efficient DC pumps and not AC pumps and therefore, the solar water pumping system does not require anymore an inverter.

Sahara—The Last Frontier

The **Sahara** is the world's largest desert. At over 9,400,000 sq. km (3,600,000 sq. miles), it covers most of Northern Africa, making it almost as large as Europe or the United States. The Sahara stretches from the Red Sea, including parts of the Mediterranean coasts, to the outskirts of the Atlantic Ocean. To the south, it is delimited by the Sahel, described above, a belt of semi-arid tropical savanna that composes the northern region of central and western Sub-Saharan Africa.

Very interestingly, under the Sahara are three major aquifers that hold water in pores of stones like a wet sponge. One of these, extending over 2 million sq. km, underlies in Egypt west of the Nile, under the western part of Libya and Sudan. This aquifer has about 400,000 km^3 of water—the equivalent of 4000 years of Nile River flow. It is called the Nubian Sandstone Aquifer System.

---

[123]http://www.cilss.bf/prs/article.php3?id_article=43.

Due to the abundance of the incoming solar irradiation, many studies were made about using the Sahara desert area to install PV systems to generate a large amount of electricity and export it, for example to Europe. A solar generating facility covering just 0.3% of the area comprising North Africa could supply all of the energy required by the EU. This indicates how much solar energy is available potentially, for example in the Sahara, but to export it via wires to Europe, as shown by W. Palz, is not a feasible idea.[124]

There were also studies made to utilize the water from the aquifers for drinking water for people and animals and also for agricultural purposes. This is a very good idea because of the immense quantity of water available, in spite that this water will not be replenished. In the extremely arid Sahara region, this large amount of water is not much below the sand dunes and could be pumped up by the solar energy. The utilization of this water has been started and soon it will be needed to set up some control in spite of the quantity of the water lasting for at least 1000 years.

I believe one of the first solar water pumping system on the "real" Sahara (which is not close to inhabited areas) was the one Solarex installed in the East Owainat area of the Sahara, where the Egyptian Government was studying desert reclamation. The East Owainat part of the Sahara is in the south west corner of Egypt, close to Libya to the west and Sudan to the south. It was established that the soil (sand) in Owainat has the ingredients which makes it suitable for agriculture if there is a sufficient amount of water.

Striking water while drilling for oil or gas may be a disappointment or annoyance in any part of the world, but in the Egyptian desert it's a significant consolation. That's just what happened to the General Petroleum Company (GPC)[125] as it was searching for oil in Egypt's vast and desolate South-western desert. GPC's agreement with the Egyptian Government

[124]Palz W (2011) *Power for the World*, Pan Stanford Publishing Pte. Ltd., Singapore.
[125]The GPC is the exploration subsidiary of the Egyptian General Petroleum Corporation (EGPC), which is the national oil company of Egypt, founded in 1956.

addresses such a possibility. It provides that the company test found no oil but volume of water available than if warranted, develop the site. This particular well site, when it was discovered that instead of oil it was providing water, was selected to provide water for an experimental desert farm in the East Owainat area. The water in this well site was only 25 m (80 ft.) below the surface.

The selection was because in the late 1970s the distinguished Egyptian scientist, Dr. Farouk El-Baz, who was the scientific advisor to the late President Anwar Sadat, started his research about the availability of water in this zone. El-Baz, who was a consultant to NASA's Apollo program, in addition to being the initiator of the new field of remote sensing, was able to determine the amount of water available in this zone. By using very advanced remote-sensing techniques (satellite images and radar), he was able to reach the conclusion that under the Egyptian Western Desert is a large water reservoir. As we know today, this region is located above the Nubian Sandstone Aquifer System which could provide a plentiful supply of water.

The Egyptian Government, based on the suitable soil and the possibility of abundant water, wanted to reclaim land from its desert areas—possibly as many as 3 million acres (12,000 square km/4600 square miles) in East Owainat. Previously useless, the reclaimed land could mean more crop production and relocation of some of the nation's nearly 45 million people (in 1985, in 2010 it was already 81 million). Most of Egypt's population now lives on a very small part of the land—where the water is—on the shores of the Nile, and along the coastlines of the Mediterranean and Red Seas. The success of such reclamation projects rested upon developing efficient, economical means to bring the plentiful underground water to the surface.

The problem was that there was no infrastructure connecting the Owainat desert with anything. The nearest point from where the electrical line could be brought to this area was more than 500 miles (800 km) away. Utilizing diesel generators to provide the electricity for water pumping, irrigation, and for the people working at the farm was also impossible, because the infrastructure of suitable roads to truck the fuel for the diesels was at that time non-existent. An automobile trip from Cairo to East Owainat in those days required 3 days on roads which going from Cairo were less and less developed as they approached the site. Getting there faster required going by plane or helicopter. A PV system to produce electricity at East Owainat overcame one of the major obstacles to desert reclamation—fuel supply—for neither the cost of fuel nor the expense of transporting it was a factor. Remoteness and isolation provide further complications.

The desert presents special challenges to this task. Its climate is notoriously harsh, with the corrosive effects of blowing sand, high temperatures, and daily fluctuations of up to 60°F (16°C). The PV system was chosen to test and demonstrate the applicability in remote and environmentally hostile areas as a means of conserving conventional sources of energy, especially oil and the water pumping system had to be able to meet the irrigation requirements of the 10 acre experimental farm.

GPC selected Solarex for this project because Solarex by then had completed several other PV projects in Egypt, including the PV systems for the navigational aids in the Suez Canal. The contract specified that a turnkey system (completely designed, installed, and ready to operate) be provided. Though the requirements of the system were defined, the methods of fulfilling them were to be engineered by Solarex. The most important contract requirement was a simply stated performance specification: The system was to deliver 350 m^3 of water a day during the summer season (approximately 35 m^3 an hour over a 10-hour duty day). In addition, 350 m^3 of storage capacity were to be provided.

Among other requirements were that the system should operate automatically and have built-in safety features to protect itself. This was a very interesting requirement, which was needed because it was expected that people working at the desert farm would not have the skill to select the various functions. Examples were: System control had to select the duty cycles of the system's two pumps (the submersible pump, which brought water up from the well, or the centrifugal pump which distributed water to the storage or for other use). The pumps had to start, operate, and shut-off automatically at programmed times. This was an important feature in the desert, where irrigating at specific times of the day was critical to avoid huge evaporation losses. The system's extensive internal protection protected not only the system battery bank but also the pump/motor units, inverters, and storage reservoir.

The installed pumping system was powered by an array of PV modules which produced 21.6 kW at peak power. It also included a sizeable battery bank (352 kWh capacity) to provide electricity for lights and irrigation at nights.

The system was installed at the East Owainat site by IMF Solar of Cairo. Bill Hufnagel of Solarex spent 6 weeks at the site supervising the installation and training the operators at the desert farm. The array support structure, like many of the other system components, was designed specifically for the conditions at the site. Sand replaces the more typical cement as ballast, an innovation which saved substantial transportation expense. Because of the lack of roads it was impossible to provide a steel storage tank having 350 m^3 of storage capacity, therefore for storage reservoir a reinforced synthetic rubber "bladder" was used, which Hufnagel referred to as "the world's largest waterbed" (see Fig. 20.3).

Storage tank of 350 cubic meters capacity

Figure 20.3 Owainat—Egypt, PV water pumping system.

When the evaluation and acceptance tests were conducted in April 1984, the system delivered 41 m^3 of water an hour, some 17% above its required performance.

The Egyptian desert was in most ways a harsh and inhospit-able place, but it could support both agriculture and people providing that water was available and could be economically brought to the surface. PV was proven as an obvious choice for that job, for it suited the very nature of the desert—an isolated and inaccessible land of abundant sunshine.

Today, almost 30 years later, the Egypt State Information Service advertised *East Owainat*[126]:

"East Owainat project has turned the barren desert into a green paradise in which trees, flowers and fruits have grown for changing the face of the Egyptian history. It is one of the Egyptian mega national projects that intend to increase the cultivated area in Egypt to create a new residential community in South the Valley. East Owainat is located in the southwestern part of Egypt's Western Desert. The planned area to be cultivated amounts to 220 thousand feddans[127] that depend on subterranean water available in the project area."

Solar Water Pumping—Today

In those early days, solar water pumping was a novelty. Today, solar water pumping is already well established. It is being used all over the world more and more. There is probably no country where the infrastructure of selling, installing, and maintaining solar water pumps is not available. One can go to the Internet and try countries in every continent (except Antarctica) to find organizations offering solar water pumps.

The primary utilization is in agriculture, animals, and drinking water. There are reasons why solar water pumping is being used more and more:

- One obvious utilization of solar water pumping is where the central electric power is not available or its extension would be very expensive.
- Places where diesel engines could be used, PV is getting a strong foothold. The maintenance of the diesel engines

---

[126]http://www.sis.gov.eg/En/Story.aspx?sid=2667.

[127]When Egypt adopted the metric system, the *feddan* was the only old unit that remained legal. Currently taken 1 feddan = 4200 m^2 = 1.038 acres.

and the continuous supply of oil especially in agricultural applications could prove to be a problem. Furthermore, to have a continuous supply of fuel may require the installation of a large tank and the purchase of oil to fill it up. The price of this investment approaches the price of PV, which decreased tremendously in the past years. An excellent comparison between the "Unit Water Cost" (UWC) diesel water pumping and solar water pumping can be seen in a study commissioned by Deutsche Gesellschaft für Technische Zusammenarbeit (GTZ) and Lorentz GmbH.[128]

- Because in the past few years the combination of the Dollar/Wp prices of PV modules became drastically cheaper and more efficient water pumps were also developed, which utilize less power to pump water, and also because the price of electricity supplied by utilities is increasing, solar water pumping became attractive for new applications also in developed countries. One example is the solar water pumping for swimming pools. It is needed so the entire water of a swimming pool should be circulated through a filter once a day. The present swimming pool electric water pumping system is the second largest user of electricity of a household. Obviously the air conditioning is the largest. The efficiency of the AC pumps used for the electricity provided by the utilities is about 50%, while the efficiency of the DC pumps developed for solar water pumping is over 90%. This reduces the cost, because a smaller size of the PV module system can be used. The disadvantage of the solar water pumping system is that its cost has to be paid at installation, while the electricity is paid monthly. But at present the cost of the solar pump can be recovered in 3–4 years and thereafter there is no electric cost. In Florida alone there are about 1,200,000 swimming pools.

It is very hard to estimate how many solar water pumps are sold and installed in each year, however, information from the various solar water pump manufacturers indicate that there

---

[128]www.gtz.de or www.lorentz.de.

were probably 40,000–50,000 pumps sold and installed in 2010, which would mean that about 50–100 MW PV modules were used.

The description of the major manufacturers of water pumps made specifically to be used for solar electricity gives also the history how solar water pumping became from its infancy in 1974 to a standard product sold all over the world. As mentioned earlier, the French manufacturer, Pompes Guinard, was the first in 1974 to make and deploy solar-powered pumps. Guinard, however, discontinued promoting solar-powered pumps in the 1980s.

The Grundfos company[129] was established in 1945 in the small Danish town of Bjerringbro. The company has now (2011) an annual production of more than 16 million of various types of pumps which makes Grundfos one of the world's leading pump manufacturers.

In 1980, Grundfos started manufacturing and marketing water pumps to be powered by PV modules. Solarex by that time was marketing and selling PV systems utilizing the Guinard pumps. Grundfos, which had by 1982, an office established in Clovis, CA, started to market and sell solar water pumping system utilizing Arco Solar PV modules.

When the changes occurred in France and Guinard pumps were not readily available, Solarex started in 1983 to also use Grundfos pumps. At that time, Grundfos offered six basic types of solar water pumps and four different power levels (560, 840, 1120, and 1400 Wp). The pumps could be used to pump water from 5 to 120 m (16.5–340 ft.) depth. The amount of water delivered obviously depended on the PV array power and the average insolation available at the site where the pump was located.

Grundfos estimates that at the beginning (1980s) they sold about 100 pumps per year[130] powered by PV. By 2000, the pumps were redesigned to be more efficient, having an integrated control system, less expensive, and also required smaller PV arrays. The PV array size for these pumps was 200–1500 Wp

---

[129]www.grundfos.com.

[130]Estimate from Mr. Jan Heegaard of Grundfos Co. (www.grunmdfos.com).

and the pumps were able to pump water from as far down as 300 m (700 ft.) about 5 m^3 (1300 US gallons) per day and from 40 m (100 ft.) down about 75 m^3 (19,800 US gallons) per day. With these improvements, and because of the lower PV module prices, Grundfos from 2000 on was selling thousands of solar water pumps per year. The quantity rapidly increased and now (2011) they are selling solar water pumps in the several 10,000 a year.

From the middle of the 1990s, because of the increasing fuel cost for diesel engines and rapidly decreasing PV module prices, the emergence of a substantial solar water pumping business became evident. Realizing this, several pump manufacturers started to develop pumps tailored to be operated by PV modules.

One of these manufacturers was Bernt Lorentz & Co.,[131] founded in 1993 in Hamburg, Germany. The company decided in 1996 to manufacture only solar water pumps. Lorentz now is represented in more than 100 countries. Their solar water pump's power range is from 0.15 kW to 21 kW, covering lifts of up to 350 m (1100 ft.) and flow rates of up to 130 m^3/h. Lorentz sold more than 80,000 solar water pumps in the last 5 years.

SHURflo is part of the Flow Technologies Group of Pentair, Inc. (PNR), a diversified company listed on the New York Stock Exchange. SHURflo was founded in 1968. The recreation market was searching for a dependable fresh water pump to deliver water stored in tanks to faucets, showers, and toilets installed in the vehicles. SHURflo started to make water pumps for the solar water pump market approximately in 1996 and is selling solar water pumps all over the world. The solar water pumps are submergible and operate at 24 VDC. The maximum depth they can pump water is 70 m (230 ft.).

Today, PV-powered water pumps are manufactured on every continent by several manufacturers. One can go to any search engine (Google, Yahoo, Bing, and so on) and find one which is the most suitable for the application. In the USA there is, for example,

[131]Information received from Mr. A. Honey of Lorentz (www.lorentz.de).

Dankoff (www.dankoff-pumps.com) and SunPumps (www. sunpumps.com). In Canada it is SunPetra (www.SunPetra.ca).

The reason is why so many pump manufacturers are offering PV water pumps, some of them even switched their entire production to PV water pumps, because on huge continents, like for example Africa are vast areas where there is no electricity and people can live without it, but neither people or animals nor vegetation can live without water. But water is practically everywhere including under the Sahara available not very deep underground and with solar electricity and a pump manufactured to be efficient with PV electricity is not expensive anymore to pump it up. Pump manufacturers are also making water pumps for swimming pools, which according to them are cheaper than using electric power from the utilities.

One can buy today solar-powered water pumps or a system from manufacturers and installers even in most countries in Africa on the telephone, on Internet, by mail order from a catalog or from a garden or building supply stores. One can even buy a solar-powered pump for your aquarium.

To put today's situation in perspective below are two examples which can be found in Africa.

Figure 20.4 shows a simple utilization of a solar water pump one can see everywhere.

Figure 20.4 Simple utilization of a solar water pump.

But to show the incredible changes in utilization of PV water pumping in Africa lets repeat Bernard McNelis's observation in 1978 mentioned at the beginning of this chapter.[132] "My trip to initiate the project there (solar pumps in Mali) was a total disaster (and not because of my poor French)." I did not ask Bernard if he believed at that time that 35 years later, today, he could buy in Mali a solar water pump system on the telephone or on the Internet (Fig. 20.5).

Multi Solar Pump System
1 × PS1800, 1 × PS4000
Location: **Bamako, Samanko**
Solar Pump System 1: PS1800
    Vertical Lift: 82 m
    Flow Rate: 28 m³/day
Solar Pump System 2: PS4000
    Vertical Lift: 78 m
    Flow Rate: 40 m³/day
Installation: April 2010

> **Bamako**
> **Mali**

> **Rue Nolly dérriere Malimag Imm Sylla Porte N° 29**
> **Centre Commerciale**
> **PO Box E877**
> **Bamako**
> **Mali**
> **Telephone:**
> **+223-6-669 7273**

Figure 20.5 Example of the global availability of PV water pumping systems.

Solar water pumping is today a global unsubsidized large and rapidly growing market segment for PV.

---

[132]Bernard McNelis. Promoting PV in Developing Countries, in Palz W (2011) *Power for the World*, Pan Stanford Publishing Pte. Ltd., Singapore.

Chapter 21

Navigational Aids

> *Of all the edifices constructed by man*
> *the lighthouse is the most altruistic.*
> *It was built purely to serve.*
>
> —George Bernard Shaw

Today, using GPS receivers[133] to go from one place to another on water or on land and surly arrive there one cannot imagine how people could have navigated from one location to another on oceans or on big lakes, utilizing manpower or wind and sails, without knowing exactly where they were, not having accurate or even inaccurate maps, and not knowing where the rocks, or shoals are? But obviously they did, and in most cases successfully and in few cases they perished.

The first attempt of mankind to guide ships through dangerous waters into a safe harbor was the Lighthouse of Alexandria (Egypt), built between 280 and 247 BC on the ancient island of Pharos. This lighthouse was considered one of the seven wonders of the ancient world. The height was estimated to have been between 393 and 450 ft. (120 and 149 m). We can

---

[133]See Chapter 26, *Space Applications in Everyday Life*.

Sun above the Horizon: Meteoric Rise of the Solar Industry
Peter F. Varadi
Copyright © 2014 Peter F. Varadi
ISBN 978-981-4463-80-5 (Hardback), 978-981-4613-29-3 (Paperback), 978-981-4463-81-2 (eBook)
www.panstanford.com

consider this the first lighthouse utilizing solar energy, because it had a mirror on its top and it was claimed, that during the day, the reflection of the sunlight could be seen from as far as 35 miles (50 km).

About 2300 years later, we have the GPS system to tell us where we are, accurate maps, and several thousands of lighthouses and hundreds of thousands of buoys, many of them lit up during the night or operating horns to warn and direct sailors at night and during foggy days. Around the turn of the twentieth century many lighthouses and buoys used acetylene gas, generated in situ from calcium carbide and water, which came into use to produce light.[134] It was very expensive and on the top of that dangerous method to operate the navigational aids. Obviously, when diesel generators and batteries were available the systems were switched to utilize electric bulbs. One can imagine the expenses to supply diesel fuel to extremely remote lighthouses or periodically exchanging batteries in thousands of buoys many of them located in remote waters.

In 1961, about 2200 years after the Alexandria Pharos "solar" lighthouse started its operation, NEC Corporation of Japan installed the world's second lighthouse utilizing solar energy.[135] NEC in the following years installed seven more solar-powered lighthouses in Japan. The Sharp Company of Japan started to install solar-powered lighthouses in 1966. The first one was installed on the Japanese island of Ogami in Nagasaki Prefecture, utilizing PV modules of 225 Wp, eliminating the need to ship fuel or change batteries. Since then Sharp installed 1500 more solar-powered lighthouses.

The number of boats and ships in ancient times were relatively few. Even comparing the end of the nineteenth century to today the increase of watercrafts is mind boggling. In the USA, in 2001, there were 12.9 million boats registered. One cannot even judge how many watercrafts exist in the entire world. If we

---

[134]Ross A (1907) *Report on Use of Acetylene Gas by the Canadian Government as an Illuminant for Aids to Navigation.* Government Printing Office, Washington, DC.

[135]Hirokane J, Saga T, Muramatsu T, Shirakawa I (2010) *History of Contribution of Photovoltaic Cells to Telecommunications.* The second region 8 IEEE Conference on the History of Telecommunications, Madrid.

make a distinction between boats and/or ships sailing further than 100 miles from the coast, the number of those boats are relatively few. "Few" probably means more than a 100,000.[136] The captains of those boats are well-trained people. But the large majority of the boats/ships operate near the coast, which is the most dangerous place, but the majority of those who operate those boats are untrained Sunday boaters. This is the reason that the navigational aids are so important. If somebody has a GPS instrument, they can easily tell where they are. But one cannot tell what is under the boat or ahead of the boat; for this they have to rely on navigational aids, lighthouses, lights, horns to direct and/or warn people on boats, and/or ships, to mark the waterways they should use.

There is one big difference between a way on land and on the water. On land in the majority of cases electricity is available to power the warning lights, while on the waterways that is not available. As mentioned, during the end of the nineteenth and the beginning of the twentieth centuries for lights acetylene light sources were installed. After that, electric lights and horns were used, and to operate them the electricity was provided by either diesel engines or batteries. Servicing them was expensive. Navigational aids were an ideal, important, and instantaneous application, and market for solar electricity. How big is this market?

At present there are about 13,000 lighthouses and many tens of thousands of navigational aids [the US Coast Guard (USCG) has probably at least 5000] and a large number of structures, most of them related to oil, which have to be marked.

When in 1973 the terrestrial PV industry started in the USA, Elliot Berman of Solar Power Corporation (SPC), because of their affiliation with Exxon, believed that the number of offshore oil platforms and other oil-related offshore facilities would rapidly increase. All of these had to be equipped with lights and also other marine navigational aids, which all needed electric power. Many of these offshore facilities had no electrical generating systems and, therefore, batteries were used to provide it. These batteries needed periodic replacement and

---

[136]On this web site, http://www.marinetraffic.com/ais/ one can see some of the ships at any time sailing on the oceans.

servicing which was an expensive program. SPC proposed to replace them with solar-powered navigational systems, which needed practically no servicing. SPC teamed up with Tideland Signal, a company in Houston, TX and started equipping offshore platforms, which had no electric supply, with solar system for lights and also to sell solar-powered buoys.

When Tideland Signal's competitor Automatic Power Inc. (now its name is "Pharos Marine Automatic Power"—where Pharos is a reference to the world's first "solar" lighthouse) also in Houston, TX, realized the success of their competitor, they started to buy solar modules from Bill Yerkes' new venture Solar Technology International (STI). Both of these companies did very well and SPC and STI were able to sell their products in the navigational aid market.

Besides navigational aids, solar modules were also used for many other purposes on the rivers and on the oceans. One early example was when a German navigational aid firm contracted Solarex in 1977 to design a marine panel to the German firm's specifications. Solarex made six watertight, trapezoidal panels to fit on the rim of each of six round buoys. The buoys were located in the Elbe River, Germany where they have been monitoring pollution (Fig. 21.1).

Cadet Lloyd Ralph Lomer USCG in 1960 received the Coast Guard Academy's annual award to the "Midshipman who is adjudged best in those courses which are most basic to Naval Engineering." By 1973, at that time already Lieutenant Commander Lloyd Lomer had the idea that by using solar cells to power navigational structures, buoys, lighthouses, and other navigational aids would save a tremendous amount of money. At that time, the Coast Guard was using boats, called "Buoy tenders" to service those buoys on which the battery went dead. In this case the buoy tender had to go and locate the buoy, lift it out of the water and replace the dead battery with a new one. One can imagine how much such a battery replacement cost. Consider the price of the ship, the fuel of the buoy tender, the time and crew, and the new battery. The dead battery had to be brought back and discarded. This was not simple either, because the battery contained large amounts of lead and sulfuric acid.

Figure 21.1 Pollution monitoring buoy—Elbe River, Germany (1977).

After studying the situation clearly, Lomer understood that solar cell modules would solve the problem to save quite a bit of money compared to the battery replacement cost. However, while solar modules are working extremely well for a very long period of time, for example, in Arizona's dry climate, in the Coast Guard's applications they have to work in an unforgiving environment. The buoys and the solar modules on top of them are practically in hot and/or cold salt water and salt water spray, but at the same time also under extremely hard UV radiation, windstorm, extreme rain and humidity, snow and ice, hit by hail, and the birds, which love to build nests on buoys and making them also dirty. I think one of the most unforgiving places on Earth is the Gulf of Mexico, which has everything enumerated above, except snow and ice.

Lomer's problem was that the solar system in the Gulf of Mexico as well as in waters where there is ice and snow, are all parts of the area where the USCG must also provide solar systems to be deployed and which cannot fail, because ships in treacherous waters rely on the navigational aids and if one of them fails to function it could even cause serious damage to the ship and loss of life. Lomer took this very seriously. He talked this over with John Goldsmith, who was the head of the group at the Jet Propulsion Laboratory (JPL) in California working to establish quality standards and test procedures for solar modules. He also managed to conduct tests in the Coast Guard Laboratory and also testing existing solar modules.

When Admiral Yost in 1977 became the Commander of the Coast Guard, Lomer finally got funded and had people to work on this project. The main issue was to come up with a testing and design specification which would make sure that if a solar module, when tested and survived it, would have a long life under any conditions at any place the Coast Guard was going to use it.

The final specification prescribed a glass front, metal backing, and metal frame and the structure had to be tested under extremely rigorous conditions, far more rigorous than JPL's Block V test specification which at that time was used for commercial solar modules which with some modification later became the International Electrotechnical Commission (IEC) Standard (IEC 61215).

When Lomer was satisfied that the Coast Guard requirements would insure durability and success for the solarized navigational aids under the harshest conditions anywhere where the USCG would use them, it was decided to convert the Coast Guard navigational aids to solar electric power. On November 30, 1981 USCG—Solar Photovoltaic Power Equipment Test Specification No. 368[137] was issued.[138]

When that happened, Solarex was ready to supply the Coast Guard the solar module which was able to survive Lomer's testing requirements. The reason was that while SPC and SPI

---

[137]US and Canadian Coast Guard Design Criteria and testing requirements are listed in Annex. 6.

[138]The International Association of Lighthouse Authorities (IALA) on their meeting on September 20, 1982 adapted the US Coast Guard specification as guidelines for the construction of a specification for a solar photovoltaic array for aids to navigation.

was supplying solar modules for the oil platforms in the Gulf of Mexico, Solarex concentrated on the market for solar modules for navigational aids deployed by governments all over the world. The difference between the requirements for oil platforms and navigational aids, for example buoys, was that the buoys were in the water and the solar modules were drenched with salt water, while the solar modules used on oil platforms were much above the salt water and that meant that they were deployed like all other solar modules which were somewhere on the land, while the modules on buoys were seeing harsh environmental conditions.

Figure 21.2 Solar module with multicrystalline solar cells used by the US Coast Guard.

The US Coast Guard selected Solarex to design and deliver a photovoltaic flash tube beacon system to be installed on major aids to navigation on the Florida coast. This was the first solar panel built to the Coast Guard specification (Fig. 21.2). In its November 12 issue, the Coast Guard's "Commandant's Bulletin" reported that solar was chosen because of "cost savings, elimination of the battery disposal problem, and improved aids for the mariner because of longer operational ranges." The Coast Guard had been using and testing Solarex panels in smaller systems for many years. That contract called for six flash tube beacons. They were installed as reef lights in the

Florida Keys. One parameter was that the PV module be capable of operating after 100 immersion and temperature cycles in sea water. The project was completed at the end of 1980. Soon after 1980 practically all of the USCG's navigational aids were converted to solar.

Figure 21.3 shows a navigational aid, a flashing light which is located in Chesapeake Bay on the East Coast of the USA. It demonstrates that the cost saving is not only due to the battery not having to be exchanged periodically, but also that, because a small battery is sufficient, the entire structure can be much cheaper. It also demonstrates that the birds like it too. One bird family built a nest on this aid to navigation, but the solar modules were equipped with "bird wires" which discourages the birds to go near the modules.

Figure 21.3 Flashing light in Chesapeake Bay.

Before the USCG's specification was completed, the Canadian Coast Guard approached Solarex with their specification of solar modules to start a program to convert all of the Canadian navigational aids to solar power. Mr. Sunny Leung, who by now is retired, was the architect of the Canadian Coast Guard PV program. It should be mentioned that Canada started deploying solar-powered navigational aids before the USCG began. The reason was that the introduction of PV to power the USCG's navigational aids was technically more complicated because of the environmental conditions: the waters with which the Canadians were concerned posed much less difficulties than the USCG's waters. They had to consider only cold salt water, snow, ice, windstorm, hail and rain but no hot salt water, no hard UV radiation, and no extreme humidity. This made a big difference in specifications (see Annex. 5). As failure was not an option, the PV modules for the USCG required very protective encapsulation for the solar cells and their interconnections. The Canadian Coast Guard established its own PV module specification in 1982 (MA 2055 Issue E., November 15, 1982—Annex. 5).

Solar modules

Figure 21.4 First solar-powered lighthouse in North America—False Duck, ON—installed in 1981.

The very first solar-powered lighthouse in North America was the one at False Duck, ON (Fig. 21.4). The history of this lighthouse is very interesting. It not only shows the durability of the solar modules, but also how much the electronic technology improved through the years. Originally when this lighthouse was commissioned in 1981, it needed a hybrid system with 2 diesels and twenty-two 30 Wp Solarex modules. The load was reduced in 1995 and the diesels were removed and 18 out of the 22 modules were transferred to other locations. At present after 30 years of service the original four modules are still operating the lighthouse. Figure 21.4 shows the lighthouse as it is today operating with the four 30 Wp Solar modules.

In the years that followed, as far as I know, practically all of the Canadian lighthouses and navigational aids were converted and are powered by solar energy. There is information available about the enormous savings resulting from the fact that the batteries do not have to be frequently replaced and recharged. The 30 years of successful operation of the PV modules is a testimony of the durability and reliability of PV systems.

Figure 21.5 Navigational aid near the Earth's Magnetic North Pole, Canada—1981.

The northern-most navigational aid installed by the Canadian Coast Guard, powered by PV is at Resolute Bay, 150 km from the magnetic North Pole of the Earth. This navigational aid was installed in 1981 and is still in operation[139] (Fig. 21.5).

The savings utilizing PV to charge the batteries in navigational aids with PV for the Canadian and the USCGs were substantial. President Ronald Reagan commended Captain Lloyd Lomer[140] for "saving a substantial amount of the taxpayer's money through your initiative and managerial effectiveness as project manager for the conversion of aids to navigation from battery to solar photovoltaic power." The conversion of navigational aids to be powered by PV was and is being done all over the world.

Figure 21.6 Suez Canal—Buoy with PV system—1982.

---

[139]Varadi PF (November 2006) *CanSIA Solar Conference*, Ottawa, ON.
[140]President Ronald Reagan's letter to Captain Lloyd R. Lomer dated September 29, 1986.

The Suez Canal was reopened in June 1975. The navigational aids used at that time were replaced in 1982 by 628 modern buoys produced by the Resinex Corporation of Italy (Fig. 21.6) All of them were equipped with PV-powered lights. The utilized PV system, modules, and charge controllers were supplied by Solarex Corporation, providing also engineering supervision of the installed PV systems. This is also a good example that even at the beginning, when PV was still expensive there was a market for it, where PV was less expensive than the alternative.

A description of the utilization of PV for navigational aids in England is given by Trinity House: *"Lighthouses (there are 66 in England), two unattended light floats, for use in exposed deep sea locations, and an increasing number of light vessels have also been converted to solar power. All lighted buoys have now been converted; solar power replaces dissolved acetylene gas which needed regular and costly refueling."*

Figure 21.7 Ramon Dominguez being introduced to Prince Philip, Duke of Edinburgh.

By 1983, the utilization of PV for navigational aids was worldwide fully accepted and the US manufacturers, Automatic

Power, Solarex, and Tideland Signal were invited to participate and exhibit at the meeting of the prestigious "International Association of Lighthouse Authorities" (IALA) in April 1985 in Brighton, UK, organized by the British Trinity House, on which meeting HRH Prince Philip Duke of Edinburgh, Master of Trinity House was also present. Mr. Ramon Dominguez represented Solarex's Navigational Aid Department (Fig. 21.7).

The usefulness of utilizing PV cannot be better described than by Roger Lockwood CB, Chief Executive, Northern Lighthouse Board, as he did in his speech in February 2011[141]:

"The Northern Lighthouse Board (UK) today operates over 200 lighthouses. They are all automated and monitored from a single Monitor Centre in the Board's Headquarters in Edinburgh and many of them are solarised and solarisation, in addition to being a very efficient source of renewable energy, has allowed the removal of maintenance heavy generator equipment. The Board also manages over 150 buoys which are all solarised."

---

[141]Roger Lockwood CB (February 4, 2011) *Emerging Issues in Aids to Navigation*, Conference at the Royal Society of Edinburgh.

Chapter 22

Amorphous Silicon (a-Si) the First Successful Thin-Film Solar Cell

I am looking at this text on my computer's 23 inch (57.5 cm)-wide and less than 1 inch (about 2 cm)-thick color monitor and later I am going to watch my color TV set which is 46 inch (1 m 12 cm) wide and also about 1 inch (2.5 cm) thick (today even 70 inch (1 m 75 cm) TV sets are also available). I just realized that this marvel of technology exists because of the existence of its "brain" the so-called "thin-film transistor" (TFT). What very few people know that it was the result of the continuation of the research first resulting in the development of the "amorphous Silicon" (a-Si) thin-film solar cell. This simply means that the flat panel display which is being produced in bigger and bigger sizes and purchased and watched by millions of people was the result of the continuation of the development of thin-film solar cells.

Dr. David E. Carlson[142] was a research scientist at the world famous RCA Sarnoff Research Laboratory in Princeton, NJ. His research project was to use gases, which in an electric field,

---

[142]A part of the story described in this chapter is from notes made by Dr. Carlson and provided them to me.

Sun above the Horizon: Meteoric Rise of the Solar Industry
Peter F. Varadi
Copyright © 2014 Peter F. Varadi
ISBN 978-981-4463-80-5 (Hardback), 978-981-4613-29-3 (Paperback), 978-981-4463-81-2 (eBook)
www.panstanford.com

called "plasma glow discharge," would alter the surface of glass. After the oil embargo in 1973, he was thinking about solar energy and solar cells as a good, long-term solution to replace oil, and with the same technology he was experimenting with, using silane (SiH_4) gas, he could probably fabricate low cost thin-film polycrystalline Si solar cells. He knew that some people already using the same technology deposited polycrystalline Si layers on crystalline Si wafers. He believed that this might also be done on an inexpensive substrate, such as aluminum, steel, and glass coated with a conductive layer.

After some experiments in October 1974, Dave Carlson was able to produce the first "thin-film Si solar cell." He did this by depositing thin layers where he introduced into the silane gas dopants, first phosphorus[143] (n) and for another layer boron (p) separated by an undoped layer (i) and forming three layers (which he believed were polycrystalline Si) on glass, coated with a transparent conductive layer and applying silver electrodes on the other side of the thin films. He was able to observe an electric current with an extremely low efficiency of about 0.1%, when this sample was illuminated. This triple layer thin Si film was actually a "solar cell," meaning that it was able to convert light into electric current. In further experiments, he was able to achieve better efficiency.

Dave's first problem was similar to what happened to Columbus. Columbus believed that if he was sailing west for a long time, he was going to wind up in Asia. Instead he wound up in a different continent not known to mankind of that age. Dave found out that in spite of other people depositing polycrystalline Si on a crystalline Si wafer, what he deposited was not polycrystalline Si, but was a material lacking crystal structure. These were "amorphous" thin film Si layers, verified by the X-ray analysis by the RCA analytical group in December 1974. To simply explain the difference, diamond has a crystal structure and an amorphous material is like a frozen jelly.

---

[143]Phosphorus in the form of phosphine and boron in the form of diborane are both compounds and gases and can be mixed with silane which is also a gas.

Dave knew from other people's work, which was reported[144] that, a-Si could be deposited from silane in a low temperature (<400°C) "plasma glow discharge" and the thin film produced had unusual properties, but his colleagues at RCA Research Lab were skeptical that solar cells could be produced from a-Si, because it was believed and was also stated by Sir Neville Mott (who won the Nobel prize) that amorphous semiconductors for various reasons cannot be doped therefore, the p-i-n layers cannot exist and so the a-Si solar cell cannot exist. Miracles are very rare, but here was one. Dave Carlson made a-Si solar cells and with further experiments improved their efficiency. Dave found out several years later that what he did, producing a-Si solar cells in a low temperature glow discharge system, where the gas was silane, was not a miracle. If he would have used other deposition techniques, for example evaporation, he would not have been able to produce solar cells. The reason he discovered he was able to produce solar cells was due to the fact that the hydrogen present in the glow-discharge he was using produced a new type of amorphous semiconductor, namely hydrogenated a-Si:H. Similar results on doping a-Si: H were obtained by Walter E. Spear and Peter G. LeComber of the University of Dundee (Scotland)[145] about the same time.[146]

In November 1974, Dr. Chris Wronski of RCA laboratory started working with Dave Carlson to characterize the "thin-film Si solar cells" (a-Si) that his experiments produced. Dr. Wronski established that these solar cell samples were producing electricity mainly in the visible range of the solar spectrum as can be seen from Fig. 22.1. There was no significant response in the red part of the spectrum, beyond about 700 nm. This was different from what crystalline Si solar cells produced. On the other hand, the difference as shown in the figure indicated that in the blue range a-Si produced more electricity than the c-Si (crystalline silicon) solar cells. The

---

[144]Chittick RC, Alexander JH, Sterling HF (1969) *J Electrochem Soc*, **116**(1), 77.

[145]Spear WE, LeComber PG (1975) *Solid State Commun*, **17**, 1193.

[146]As a result of a Patent infringement law suit in 1986, it was established that Dave Carlson's laboratory notebook showed that his work with doped a-Si preceded the work at the University of Dundee.

importance of this for devices used indoor and powered by solar cells as later observed by Dr. Yukinori Kuwano of Sanyo Corporation in Japan resulted in the first successful application of the a-Si thin film solar cells for handheld calculators as described in Chapter 17.

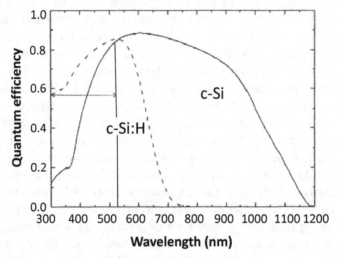

Figure 22.1 The spectral difference a-Si and c-Si solar cells convert light to electricity.

At this point, Dr. Carlson encountered a problem. The problem was that all this work that Dave was doing was a bootlegged project (this means it was not a project authorized by the management) and his manager at that time thought that he was not devoting enough time to his prescribed projects, so he received a relatively poor performance review that year. However, he did receive good support from the RCA patent department since the patent attorneys thought his solar cell work might be important and were willing to file patent applications. Carlson's first patent application was filed in 1976 and the Patent was granted in 1977.[147] The first publication to describe Carlson's and Wronski's result was in 1976.[148]

---

[147]Carlson DE, US Patent # 4,064,521, Filed: July 30, 1976; Issued: December 20, 1977.

[148]Carlson DE, Wronski CR (1 June 1976) Amorphous silicon solar cell, *Appl Phys Lett*, **38**(11), 671–673.

In July 1977, Dave's new boss encouraged his solar cell work and he was able to assemble a team of about 12 very good scientists and technicians to work on the a-Si project.

The success and the need for the thin film a-Si solar cell was demonstrated by the fact, as described before and also in Chapter 17 that, in 1980 the Japanese Sanyo Corporation started successfully the production of a-Si solar cells to be used in handheld calculators. I was obviously informed by our (Solarex) Hong Kong company about Sanyo's starting to market a-Si-powered calculators, as we started losing some of our micro-generator business.

In 1982, when it became evident that the a-Si thin-film solar cells can be used for various applications, RCA management did not know how to commercialize it and hired Booz Allen Hamilton, a leading provider of management and technology consulting services to help commercialize the a-Si solar cell technology. They proposed a research partnership comprising four companies: RCA, an oil company (AMOCO and Phillips Petroleum were interested), a utility (RCA talked to several) and an engineering company (RCA talked to several). AMOCO was interested in RCA's plans, to be able to advise Solarex, in which by that time they were a major investor. AMOCO knew by that time from Solarex that Sanyo was successfully manufacturing a-Si solar cells and using them in handheld calculators and that I was very interested to get into the thin-film consumer-oriented solar cell business which was pioneered by Sanyo.

As a result, Solarex and AMOCO were jointly looking into thin-film solar cell possibilities. I received permission from AMOCO that John Johnson of AMOCO and John Goldsmith of Solarex should help to evaluate the situation. One of the possibilities was the cadmium telluride (CdTe) system, the disadvantage of which was that Cd is a very poisonous material. John Johnson and I visited in mid February 1982 a company in the Los Angeles area, experimenting with CdTe. But the Sanyo success shifted our interest in the direction of a-Si. On July 15, 1982 Johnson and I visited the RCA Laboratory in Princeton, NJ and received a tour of the a-Si laboratories and talked to the scientists there including the inventor of the a-Si solar cell, Dave Carlson and his boss Browne Williams. My conclusion was that RCA's main asset was the scientific staff working on the a-Si development.

In early 1983, it became apparent that the Booz Allen concept was too difficult to implement, and a new RCA management decided to shut down the solar program despite the fact that Sharp Corporation joined Sanyo in manufacturing a-Si solar cells for handheld calculators, replacing their c-Si products. Carlson was able to convince the RCA management that the a-Si technology was an asset which could be sold and a few people were allowed to work as a team while they tried to find a new home for the technology.

In the spring of 1983, RCA entered into discussion with several venture capital groups. The result was the same that I had with similar groups 10 years before in 1973. The people at the ventures learned how to spell PHOTOVOLTAICS, but RCA did not get money.

At Solarex by that time, we were fully aware of the potentials of the thin film a-Si as by 1983 we lost our very lucrative and fairly large handheld calculator micro-generator business, because Sanyo and Sharp switched to a-Si solar cells. We had quite an interest to get into the manufacturing of a-Si solar cells and we knew that at least for consumer products in which Solarex was heavily involved it was an excellent possibility, but it could also be a possibility for commercial applications.

The issue, however, was complicated. By the end of 1982, we knew that sometime in 1983 AMOCO was going to acquire Solarex, because we were unable to finance the rapid growth of the company. The complication was that we needed AMOCO also agree to this transaction because we lacked available funds to buy the a-Si capabilities from RCA and because of Solarex's upcoming buyout by AMOCO.

I had several meetings with Gordon McKeague, who was the AMOCO representative on Solarex's Board, and with John Johnson, who worked for him, and they fully agreed that I should get a free hand to negotiate, and if needed, AMOCO would provide the cash. They arranged that I should meet Larry Fuller to whom Gordon reported. At that meeting, Fuller agreed that Solarex should go ahead to buy the a-Si capabilities from RCA. Accordingly, a meeting was set up at AMOCO's headquarters in Chicago, where Larry Fuller, McKeague, Johnson, and myself sat on one side of the conference table and three RCA representatives sat on the other side.

It was a very funny meeting. The RCA people did not look at me at all, because Solarex represented a zero company compared to the giants RCA and AMOCO. They were facing mainly Larry Fuller or Gordon and talking to them. At which point Larry turned to me to answer. I did. When the RCA people talked again the procedure was the same. When they left, I told my opinion that the value was in the scientific people and that the patents had some value, but if we would start manufacturing the a-Si by ourselves, we probably could get a license the same as Sanyo received. The value of the equipment we would receive from RCA was not worth much. On the other side of the balance sheet, if a large company like RCA decides to shut down an operation, the speed with which they accomplished it was a bigger value than the money they were asking. Furthermore, if we hired their people, they would save money in severance pay. Therefore, the money they were asking for was way out.

RCA got really anxious to shut down the a-Si project and they already knew that Solarex was the only graceful exit for them. So after sending the proposals back and forth, finally they accepted our offer, to sell everything to Solarex at the price we offered, but we made it contingent that we were going to be able to hire the key people. By that time it was July 5, 1983.

Now the problem was to hire the key people. The obvious handicap was that they were working for the big and stable RCA and why should they join a fledgling small company like Solarex? Solarex was at that time already not fledgling and a fairly large size, 10-year-old company. What we could not tell them was that Solarex would become a division of AMOCO very soon after they would have joined Solarex. What none of us had even an inkling about at that time was that the great and stable RCA would go kaput 3 years later in 1986 and would be dissolved by GE. RCA's famous research laboratory, which was renamed the David Sarnoff Laboratories in 1951 to honor the founder of RCA, would be spun off after RCA was purchased by the General Electric Corporation. The RCA Sarnoff Laboratories was renamed the Sarnoff Corporation and became a subsidiary to SRI International, the non-profit company that was once the research institute of Stanford University in California.

By promising that Solarex's thin-film division would be located somewhere near Princeton, NJ, and that they could select the location, we were able to hire six out of the seven key people. After this, on August 18, the RCA deal was completed and as a coincidence on the same day Solarex's merger with AMOCO was also signed. Solarex's a-Si division opened its doors in Newtown, PA, which was close to Princeton, NJ, on August 29, 1983.

A few weeks later on September 16, 1983, Solarex merged with AMOCO. Joseph Lindmayer and I did not wanted to stay in the management but had to sign a non-compete agreement and stay on as consultants. The program of Solarex's a-Si division did not go in the consumer product direction as I imagined when it became totally functional. As Solarex's supervisors at AMOCO were changed, so the direction of the a-Si division was also changed to chase the "big picture." Ultimately, AMOCO brought in Enron Corporation to finance the a-Si division's expansion to produce large PV modules for utility applications. This expansion was in 1996, when the entire operation was moved to Toano, VA. This was followed by the acquisition of AMOCO by BP on December 31, 1998, and the a-Si thin film operation was shut down in 2002 when BP Solar exited thin-film technology, which by that time included also a CdTe thin-film operation in California, to better focus on "core markets and current technologies."

When in 1983, Solarex acquired all of RCA's a-Si assets, including patents, some of the Japanese companies which received license previously were able to keep their license. These companies, such as Sanyo and Sharp, were the ones which started manufacturing a-Si for calculators. They continued manufacturing a-Si for calculators and even today practically all handheld calculators utilize a-Si as its power source.

It has to be mentioned that the thin film a-Si system had a handicap that its efficiency was in the range of 4–6%. This was obviously acceptable for calculators, but was not very attractive for terrestrial use. This limitation was overcome by using multilayer a-Si or more recently in addition to the a-Si layer an additional microcrystalline layer is added. This increases the produced thin film efficiency to 8–10%.

To get back for a minute to the first paragraph of this chapter, to the origin of the presently, widely used "flat screen monitors or TVs," the work after the discovery of the a-Si solar cells[149] continued by LeComber and Spear and in 1979 resulted in the first report on the fabrication of a TFT.[150] This was the basis of a liquid crystal (LCD) imaging system and the Sharp Corporation in 1988 developed the world's first 14" color TFT LCD TV, which was called the "Crystaltron,"[151] and now 25 years later even 70 inch (1 m 75 cm) TFT LCD TV can be purchased at a reasonable price, on which pictures can be seen even in 3D format.

Returning to the a-Si story, besides Solarex, in the USA three companies entered the a-Si field in the 1980s. One of them was Energy Conversion Devices (ECD) founded in 1964 by Dr. Stanford R. Ovshinsky. In the 1980s, ECD's fully owned company, United Solar Ovonic Corporation, was formed to develop the "roll-to-roll" production of a-Si solar cells. The roll-to-roll technology meant that the a-Si thin films instead of being on a glass sheet were deposited on a very thin steel material, which was in a roll. The steel sheet went through the machine, in which the a-Si layers were deposited and was winding up on the other end of the machine into a roll again. In the chamber where the a-Si layers were deposited, in order to achieve higher efficiency more than one set of layers were formed, what was called "triple junction solar cell" and achieved an efficiency of 6–8%, which was sufficient for many applications. The produced a-Si solar cells are good quality and United

---

[149]Fuhs W (August 2005) *J Optoelectron Adv Mat*, **7**(4), 1889–1897.

[150]LeComber PG, Spear WE, Gaith A (1979) *Electron Lett*, **14**, 179–181.

[151]http://www.ait-pro.com/aitpro-blog/tag/lcd-television-history-and-facts/.

Solar[152] was able to produce in excess of 100 MWp/year. ECD declared bankruptcy in early 2012 and its assets were auctioned off in June, 2012.

PowerFilm Inc.[153] (in Ames, IA) was founded in 1988 and developed a roll-to-roll production of a-Si solar cells. PowerFilm's carrying material was the flexible Kapton® film. The a-Si product is being used for military, consumer as well as building-integrated PV (BIPV) applications.

Another company was Chronar, which was started in the USA (NJ) by Dr. Zoltan Kiss around 1980. Chronar developed an interesting new technique to produce a-Si solar cells. They used a so-called "batch" deposition system. Only one vacuum chamber was used into which they loaded several glass sheets (batch), which were heated. The chamber was evacuated, the needed gas was admitted, and the gas after the proper thickness of Si layer deposited was pumped out. That was repeated until all of the needed layers were completed.

As the required machinery was relatively inexpensive, Chronar sold equipment to several companies which started to produce a-Si. Most of the a-Si solar modules were sold for consumer products.[154] Some of these a-Si manufacturing companies are still manufacturing products.

In Japan, several companies entered the a-Si field in the 1980s.

Sanyo was the first in 1980 to start manufacturing a-Si power sources in the amount of several 10 kWp/year for watches and calculators.[155]

Sharp started to manufacture in 1983 a-Si power sources for watches and calculators.

Mitsubishi, Fuji Electric, and Canon also started manufacting a-Si solar cells in the 1980s.

In Europe, several companies entered the a-Si field in the 1980s:

---

[152]http://www.uni-solar.com/wp-content/uploads/pdf/FINAL_TR_Presentation_6_1_10.pdf.

[153]www.powerfilmsolar.com/.

[154]http://kammen.berkeley.edu/ESMAP_Kenya_aSiPV.pdf.

[155]Private communications from Dr. Y. Kuwano of Sanyo.

Chronar established about three small a-Si manufacturing facilities in Europe.

Development of the thin-film solar cells at MBB (ASI®)[156]— its successor, Schott Solar, produced tandem a-Si PV modules in Jena, Germany.

Presently, there are manufacturers offering a-Si production equipment capable of producing very large sheets (1 × 1.5 = 1.5 m^2) of relatively high efficiency (7–11%) tandem junction (micromorph) (a-Si and μc-Si[157]) solar modules. These companies (Oerlikon Solar now: Tokyo Electron[158] and Ulvac[159]) also offer 50–100 MWp turn-key installations of such factories.

There are at least 30 a-Si and tandem μc-Si manufacturers worldwide. The yearly manufactured a-Si solar cell modules are being used for many applications and the quantity became substantial since 2005.

After 2009, the price for crystalline Si (c-Si) solar cells decreased very much, which resulted in a decline in the sales of the a-Si solar modules. a-Si solar modules are manufactured in a smaller quantity than they were before 2009.

Applications

(1) Grid and Off-Grid Power Production

a-Si modules were frequently used on the roof of buildings as well as for large PV fields. In these applications a-Si is competing with other thin film as well as with c-Si modules. With the decreasing prices of the CdTe the c-Si and lately the CIGS modules a-Si modules face serious competition.

In off-grid systems in developing countries a-Si modules were used since a very long time and they are being used as of today.

---

[156]http://www.schottsolar.com/global/about-us/history/.

[157]Fischer D et al. (May 1996) *The Micromorph Solar Cell*, 25th PVSC (IEEE), p. 1053, Washington DC.

[158]www.solar.tel.com.

[159]www.ulvac-solar.com.

(2) Consumer Products

From the beginning, a-Si thin film PV was used for consumer products. In previous paragraphs (and also in Chapter 17) the utilization of a-Si was described for watches and calculators. As of today, several watch manufacturers are still manufacturing solar watches and clocks with a-Si as their power source and practically all of the handheld calculators are powered by a-Si solar cells.

The first consumer application in which a-Si modules were able to replace c-Si modules were electric fence chargers. Electric fences powered by fence chargers are used to replace wood or metal fences to keep animals, for example cows, in or keep unwanted animals, such as deer, out of a certain area. The electric fence charger is connected to a wire which is surrounding the area and produces a tiny high voltage pulse, when an animal touches the wire. In many places where animals are kept, electric power is not available or would be expensive to install. Since PV modules were available, they were used to power electric fence chargers. The fence charger need very little electricity to keep its battery charged and therefore only small 5–10 Wp PV modules are needed. To manufacture these small modules from a-Si was less expensive than to assemble them from small c-Si pieces to obtain the voltage needed to charge a battery (for example 6 or 12 V). Therefore, a-Si was able to take this market over. There are many fence charger manufacturers all over the world. It was mentioned that Europe (mentioned also in Chapter 16) was a big market for PV modules for fence chargers, but large amount of PV modules were and are sold for fence chargers also in the USA, New Zealand, and Australia.

Since the beginning, a-Si modules were used also for lights around the houses, for example garden lights, decorating lights, and house numbers. As of today, especially by using LED lights, more and more of these products use a-Si PV modules.

(3) Building-Integrated PV (BIPV)

a-Si modules are being used for BIPV (Please read more in Chapter 25 about BIPV). a-Si PV modules have been and are being utilized for a variety of applications in buildings.

On the façade of the building, they are used as decorative elements and also integrated into the wall of the building.

a-Si is also used on the roof of buildings. a-Si roofing shingles are available in the USA in stores selling building materials. They are described as solar a-Si shingles and are structurally and aesthetically integrated roofing elements. They can be directly nailed on wood decks with fire resistant underlay and result in a wind and water tight roof.

(4) Windows

The utilization of a-Si for windows is an interesting application. If a glass sheet is covered by an a-Si PV coating obviously, the glass sheet will not be transparent, because in order to produce electricity the a-Si layer has to absorb the light. Scientists discovered around 1985 that if they produce holes in the a-Si PV module, its electric output will only decrease by the size of the removed a-Si material. An experimental equipment was designed in which a laser was moved over the a-Si PV module and produced a multiplicity of holes in the a-Si layer. If these holes were evenly distributed, the light transmission of the PV module increased by the percentage of the hole area. If the hole area was 25% of the total area the eye would see a perfectly good picture of the world outside of the PV glass therefore, it could be used as a window. This transparent a-Si glass sheet was named "PowerView" and was used for windows as well as for shadings. Today several companies are manufacturing this a-Si window.

ACT 2
SUNRISE
1985–1999

Chapter 23

Changing Landscape

Act 1 of this book describes as W. Palz writes, *The Pioneering Role of the US*[160] in the development of the terrestrial PV. Following were the results:

- The basis of the manufacturing technology of terrestrial PV was developed, which with modifications is still in use.
- The first large-scale terrestrial PV production facilities were established, financed mostly by the American oil companies.
- Research and demonstration projects were financed by the US Government. Its 1980 PV RD budget was $900 million.[161]
- Several consumer markets for terrestrial PV were opened; they still exist and are growing.
- The PV industry growth in the years 1973–1984 (11 years) was average 90%/year.
- The great majority of the terrestrial PV cells and modules in the world were manufactured in the USA.

---

[160]Palz W (2011) *Power for the World*, Pan Stanford Publishing, Singapore.

[161]Williams N (2005) *Chasing the Sun*, New Society Publishers, Gabriola Island, BC, Canada.

Sun above the Horizon: Meteoric Rise of the Solar Industry
Peter F. Varadi
Copyright © 2014 Peter F. Varadi
ISBN 978-981-4463-80-5 (Hardback), 978-981-4613-29-3 (Paperback), 978-981-4463-81-2 (eBook)
www.panstanford.com

All this came to culmination during the Carter administration when on June 20, 1979, President Carter along with his wife, Rosalynn, went up onto the roof of the White House followed by some members of his government and the press, in order to dedicate a solar hot water system. "No one can ever embargo the sun" Carter declared. During President Carter's time the Department of Energy (DOE) was created and the budget commitment for PV was about $1.5 billion.[162]

The transit into Act 2 started when President Ronald Reagan (January 20, 1981, to January 20, 1989), who followed President Carter, ordered the dismantling of the solar water heater on the roof of the White House and also slashed the government's PV budget to a minimum. W. Palz writes, "For all US Presidents since, it has been a case of 'business as usual.' Even the Clinton–Gore administration fared no better."

This was the beginning of the end of the pioneering role of the US PV industry. The US PV industry was dominant before 1985 by producing at least 80% of the world's terrestrial PV products. The demise of Exxon's Solar Power Corporation obviously benefited both Solarex and Arco Solar. The decline did not go fast. The US PV companies' market share dropped to about 40% by 1997 and in 2011 the amount was probably only 6%. But in spite of the US administration's torpedo, the PV business proved that it was already "unsinkable." The baton was picked up by Japan, Germany, and several other countries.

We entered Act 2, which lasted from 1985 to about 1999. The PV industry average growth rate during this time "sank" to about 15%/year. One should not belittle a yearly average of 15% growth. This means that the business would double every 5 years. Very few industries experienced such growth. This means that the total production of PV modules in 1985 was 20 MWp then by the end of Act 2 in 1999 would be about 100 MWp.

---

[162]Maycock P (1981) *Sunlight to Electricity in One Step*, Brick House.

Chapter 24

The Big Oils' Involvement in PV

The big oil companies looked at the situation from their point of view. Why should they get involved in photovoltaic (PV) power generation? Their incentive was that if the sun's nuclear energy was going to replace the oil or gas, which they were selling, then they should consider the sun's nuclear power also as fuel, and that was their business, they had to be in it.

By the end of 1983, the three major PV manufacturers in the world were owned by oil companies. It is, however, interesting to look at the entire picture. Which oil company got involved in PV and what were they doing?

Exxon started in 1973 with Solar Power Corporation (SPC). After Elliot Berman left in 1975, they discontinued SPC's direct European presence. They continued to be in commercial PV, but they mostly expanded their government project business. After President Reagan slashed the PV budget, SPC had excellent technical people and they tried to sell more to the commercial market, but commercial sales have to be built up and that cannot be done very fast. Losing the government demo business and not able to replace it with commercial, Exxon discontinued SPC in 1984 and sold its assets to Solarex.

Sun above the Horizon: Meteoric Rise of the Solar Industry
Peter F. Varadi
Copyright © 2014 Peter F. Varadi
ISBN 978-981-4463-80-5 (Hardback), 978-981-4613-29-3 (Paperback), 978-981-4463-81-2 (eBook)
www.panstanford.com

Arco acquired STI in 1977 (renamed it Arco Solar). Arco Solar had good technical staff as well as good management under Bill Yerkes and later under Dr. Charles Gay. Arco Solar extended its business in international areas, mostly where Solarex was not present such as India and Japan. But Arco Solar never diversified into consumer business. ARCO sold Arco Solar to Siemens in 1989.

I very thoroughly research the stories I am writing, but I would like to say that I am not guaranteeing the accuracy of the description of Shell's zigzag in PV. Shell first invested in "Solar Energy Systems" (SES) manufacturing CdS/Cu_2S solar cells in 1973 and more thereafter, approximately a total of $80 million.[163] SES was shut down because of the degradation of the CdS/Cu_2S solar cells in the terrestrial environment. Shell subsequently went in and out of the PV business. Shell acquired Holecsol in the Netherlands; the Holecsol machinery was originally built by Solarex. Shell obtained a large share of Photowatt in France and after a while got out of Photowatt. Shell built a PV cell factory in Gelsenkirchen, Germany, which was transferred to the Dutch Scheuten Solar, which went bankrupt in 2012. Shell bought Arco Solar from Siemens in 2002 and sold it to SolarWorld in 2006. In Japan, Showa Shell Solar, which recently was renamed Solar Frontier, is producing copper indium gallium (di)selenide (CIGS) thin-film modules.

Mobil was involved with Tyco in the development of the edge-defined film-fed growth (EFG). Silicon sheet development (described in Annex. 2), but Mobil-Tyco Solar Energy Company did not continue after the EFG process was developed. Schott Solar website[164] claims that it belonged to Schott by 1973. Schott discontinued crystalline Si solar cell and module production in 2012.

The interesting stories of two "Big Oil" companies have to be mentioned in more detail. British Petroleum (BP), which merged with AMOCO on December 31, 1998; by that time both AMOCO and BP had a solar PV division of which AMOCO's Solarex

---

[163]Böer KW (2010) *The Life of the Solar Pioneer*, iUniverse, Inc., Bloomington, NY, p. 221.

[164]http://www.schottsolar.com/us/about-us/history/.

was much larger than BP's BP Solar. The other was the French TOTAL oil company.

Solarex's big advantage over its competitors owned by the other oil giants Exxon and ARCO was its flexibility to adjust to the PV market when needed. Why did Solarex have this flexibility to deviate from only "chasing the core market," the "big picture" and go even into solar consumer products, sold through catalogs and also stores in shopping malls, making solar arrays ("micro-generators") for the solar watch, and solar calculator markets?

Solarex was practically the only solar company able to operate with this split identity, which means chasing the "big picture" but also able to enter the consumer and commercial market, was just a lucky coincidence. One of Solarex's investors was the French company Leroy-Somer, a large manufacturer of pumps. They invested in Solarex as described in Chapter 20, because they envisioned they would need solar electricity to power their pumps especially in the western part of Africa, where the underground water table is close to the surface which means that to pump up the water does not require much electricity but in most of those places no electricity except solar would be available. For these reasons, their interest was not in the futuristic view of central power stations, but immediate applications, and therefore, on the Board level they supported Solarex's thinking of not seriously believing in the central power station idea.

The other investor in Solarex was AMOCO, the Big Oil giant. AMOCO was investing in Solarex because they believed in the "big picture." Luckily, the people at AMOCO who were responsible for investing in Solarex and giving guidance to the company, such as Larry Fuller who was VP of AMOCO and the corporate development came into his area (later—in 1983—he became president of AMOCO), Gordon McKeague, manager of AMOCO corporate development, and John Johnson. They did believe that the future of PV was the centralized or decentralized utilities, as it was also believed by the other "Big Oil" companies at that time, but these people were wise enough to realize that until that time one had to go after all of the PV market, including consumer products, even stores in shopping malls, where people could see and touch PV applications and learn about them. It was a pleasure to work with them as they provided a great support for Solarex's split personality.

AMOCO invested in Solarex and in 1983 when Solarex was the world's largest manufacturer of solar electric products, acquired it, because Solarex was a rapidly expanding company, needed lots of cash for its operation and at that time to get listed on the stock market and raise money from the public was impossible. I know it, because we tried.

Solarex was purchased by AMOCO on September 16, 1983. Joseph and I decided not to continue our employment but had to sign a non-competing consulting agreement. A new president was appointed by AMOCO. It was very unfortunate for the new president and for Solarex that after AMOCO purchased Solarex, Larry Fuller was promoted and became President (and in 1991 CEO and Chairman) of AMOCO, and got very far away from what was happening with Solarex, which was a very small part of the giant AMOCO, and Gordon McKeague was reassigned. The new people in AMOCO Technology management, overseeing Solarex, had tunnel vision and saw only the "big picture." The new president was firmly advised to concentrate on what the "big picture" people called "core business" and that every effort of the company should be concentrated on chasing the utility-oriented "big picture" and he should bring in new money, possibly from utilities.

The result was that the corrosion protection, navigational aid, aerospace, and consumer product (which by that time had six "Energy Science" stores in the Washington—Baltimore area shopping malls, had also mail order business and the products were in many catalogs) divisions were shut down. Internationally only the Australian factory was left. Hong Kong and the Swiss operations were closed. The benefit for Europe was that Solarex's European joint ventures became independent PV companies: the Italian joint venture Solarex's Pragma including the Intersemix casting equipment and technology became Solare spa, Solarex's French joint venture with Leroy-Somer in Angouleme, France Photon, was taken over by Leroy-Somer and later it was merged into Photowatt, the Dutch joint venture. Holecsol was taken over by Shell and Solapac in Newcastle, UK, managed by Philip Wolfe became independent.

The new AMOCO management overseeing Solarex wanted to invest no more money in PV and therefore, when the requirement came that a new and extended a-Si plant had to be built, they took ENRON as a partner in 1996.

BP's PV division was started in 1979 and subsequently in 1981 acquired the part of the British Lucas Energy Systems which sold solar modules manufactured by SPC and also Arco Solar in the USA and was designing and assembling Solar Electric systems. After BP took over Lucas Solar, it was renamed BP Solar International. Philip Wolfe, who started the Lucas solar business in 1975, managed BP Solar International. With BP's name and money behind it very aggressively marketed PV systems in South America, Africa, and the Middle East. BP Solar established a solar cell manufacturing plant in 1980 in Spain and later participated in a 50/50 joint venture manufacturing solar cells and modules in India—BP-Tata—where Tata Power Company Limited, India's largest private sector power utility, was their partner.[165]

John Browne became Chief Executive of BP in June 1995. He made BP one of the world's biggest and most profitable oil giants. He managed to bring BP to America by engineering the merger with AMOCO and by the subsequent takeover of ARCO and Castrol. For his business accomplishments he was knighted in 1998 by Queen Elizabeth II and in 2001 became Baron Browne of Madingley and a crossbencher in the House of Lords.

BP's merger with AMOCO resulted in AMOCO's Solarex and BP's BP Solar being merged under the name of BP Solarex (it was renamed several years later to BP Solar) with its headquarters in Frederick MD, where the Solarex headquarters were (Enron's interest in the a-Si division of Solarex was bought out). Lord Browne initiated the phrase "Beyond Petroleum" and was a big supporter of BP's role in the expansion of BP's PV business. In November 2005, while Lord Browne was in

---

[165]BP sold its entire interest to Tata Power Company Limited.

Washington DC, he announced a major expansion of BP Solar "of $70 million to double output at the facility and initiated to erect a building to house the production lines"[166] and by adding large capacity to the casting and wafering production in BP Solar's Frederick MD plant. He visited BP Solar (formerly Solarex) in Frederick, MD on November 29, 2005, to announce this to the workers at that facility.[167] BP Solar also added another factory to its Spanish solar cell and module production, the Tata-BP factory in India was enlarged. BP Solar was then probably the tenth largest PV manufacturer in the world.

In 2009/2010, BP Solar was disassembled and became suddenly an insignificant PV company. It started with Lord Browne's resignation from BP on May 1, 2007. He was succeeded by Mr. Tony Hayward. Under the Hayward regime the first closing of BP's PV operation was the announcement "that it (BP) will cease the production of solar photovoltaic (PV) power cells and panels from its manufacturing plant in Sydney Olympic Park (Australia) at the end of March 2009." I opened Solarex's manufacturing operation in Sydney, NSW, Australia, in 1977—first to produce PV modules from solar cells manufactured in the USA— and later a solar cell line was added.

BP closed the PV production in Spain and also announced that it would stop manufacturing of Si wafers, PV cells and modules in its Frederick MD facility, and the plant was finally closed at the end of 2011 except for a few people to handle returns of under guarantee modules. BP also sold its shares in the Indian BP-Tata Solar to Tata and the result was that BP was no longer in the PV business.

The French TOTAL unlike the other oil giants has been steadily in the PV business since 1983 and is expanding its presence there. Now it became among the other oil companies the biggest in this field by starting or investing in solar companies. Arnaud Chaperon, President of Total SA, stated,[168] "For us it is very clear, we want to be the top players in the world

---

[166]*Washington Post*, March 27, 2010: Steven Mufson: BP closing Maryland solar manufacturing plant.

[167]A large plaque commemorating this event was removed from the wall in the lobby of that facility in December 2010.

[168]Gelsi S (29 June, 2011) *MarketWatch Website*.

of solar. Within a few years the costs will be close to equal—a term known as grid parity." Total Energy (near Lyon, France) was started in 1983. The company was mainly involved in designing and installing PV systems, primarily in France and the French-speaking part of Africa. The company started to manufacture PV modules in 1999. Électricité de France S.A. (EDF) joined in ownership and now each, Total and EDF, owns 50%. The company in 2005 was renamed and it is now Tenesol. Tenesol produced PV modules in a quantity of 85 MWp in 2009.

Total also owns 50% of Photovoltech which started to produce multicrystalline solar cells in 2003. Total had also investment in Konarka (USA), a company developing organic solar cell technology. Konarka went bankrupt in 2012. On June 21, 2011, Total purchased 60% of the American SunPower Corporation for $1.3 billion. SunPower (listed on the New York stock market: SPWR) is one of the largest companies[169] manufacturing primarily high-efficiency PV modules. SunPower has a very large division for designing and installing large PV systems, which is the second biggest in the world.

Interesting Conclusions

The oil industry is generally considered the Big Bad Wolf for PV. I disagree with this, because as it was shown, the oil companies provided money, which no other investor did, for the establishment of the first terrestrial solar cell and module factories. The examples are Shell (SES), Arco Solar, Mobil Solar, Exxon (SPC), AMOCO (Solarex), BP (BP Solar and Tata-BP), and TOTAL in France. Without big oil money, these companies would not have been able to establish their PV production, which served as the basis of the PV industry. Without those manufacturing facilities built on the big oil money the terrestrial PV manufacturing process and market would not have been established. An exception is Japan Sharp Corporation, as its expansion was financed by its parent company.

---

[169]SunPower revenue in 2011 was $2.2 billion, with a net income of $139 million.

Chapter 25

Building-Integrated Photovoltaics and PV Roofs

Today, BIPV is so popular that in Germany alone over 1 million homes and farms totaling 20 GW have "PV roofs." In the USA, there are also a large number of homes equipped with solar roofs and building supply stores are carrying large number of PV products in their catalog and/or on their shelves. As discussed before, the big advantage of PV systems over the utilities is that they are decentralized electric power sources. They can be set up anywhere to supply electricity and do not need long wires and distribution networks. It can be mounted on the roof or on the side of a building and the electricity is produced right there where it is being used. The person in such a building will immediately understand the advantage when a tree happened to fall on the electric power line and causes the neighborhood to lose its electric power, perhaps for days, sometimes for a week.

Another important matter is that the large numbers of buildings provide an incredible amount of surface on which PV systems can be deployed. It is obvious that PV can be deployed in many locations on a building and the "building-integrated

Sun above the Horizon: Meteoric Rise of the Solar Industry
Peter F. Varadi
Copyright © 2014 Peter F. Varadi
ISBN 978-981-4463-80-5 (Hardback), 978-981-4613-29-3 (Paperback), 978-981-4463-81-2 (eBook)
www.panstanford.com

photovoltaic" (BIPV) systems can be used to connect the PV system to

- The electrical cable supplying the building and selling the electricity to the utility.
- A "Home Power Storage" system (see Chapter 42). This is a concept offered now by several companies, including the German utility Rheinish-Westfälisches Elektricitätswerk AG (RWE) started to offer this system from the spring of 2013. In this in-house system, the electricity generated by PV is stored in storage media (batteries). The "smart" electronic system directs the electricity for usage in the house (building), or if electricity is needed buys it from the grid and if the batteries are fully charged sells the excess electricity to the grid.
- The electrical cable supplying the building and selling the electricity to the utility, but in case of power failure to be automatically disconnected from the power line and supply electric power to the most important equipment and/or machinery in the dwelling, for example, refrigerator, electric lights, oil, or gas heating system. In multistory buildings, there is usually an electric generator to supply electricity in case of a power failure for some lights and for some of the elevators, but as it was experienced, for example, in Florida, where after some of the hurricanes there was no electricity for days, gas stations could not operate to provide diesel to run the in-house generator. Private homes based on their experiences with the uncertainty of the utility power especially when it is most needed, for example in a snow storm, are now more and more installing a diesel or a gas-powered generator for such occasions.
- Certain equipment (for example, a solar-powered roof exhaust fan) used in the building saving money on sometimes difficult electrical connections to the power line and saving money by not using the electrical power sold by the utility.

For over 30 years, PV systems have been successfully integrated in buildings for many applications, which are as follows:

- Integrated in the building:
  - walls (façade: walls, spandrels)
  - windows
  - decorative purposes (for example on façades)
  - awnings or external shadings
- Integrated in roofing:
  - used as roofing materials (for example shingles)
  - mounted on the roof, but not as a roofing material
- Other applications:
  - skylights, ventilation, emergency, warning, or garden lights

PV-Integrated in Building Walls

Among the BIPV segments listed above, the first two, that is, the integration of PV using the building walls and building windows are mainly related to office structures and public buildings. These are the most impressive, most fascinating terrestrial utilization of PV.

The building walls are made sometimes from non-transparent materials, but many times the entire building walls are constructed from glass. The 1973 oil crisis focused the people who commissioned to build the building and the architect's attention to design and construct these large structures to be very attractive but at the same time to save energy. Before 1973, only the attractiveness and the image the building was projecting were the primary importance.

After 1973, energy saving became important too. This was evident in buildings constructed of glass. Before that date, in the USA even for apartment buildings single glass windows were used. Because of that, the buildings' heating and air-conditioning required a lot of energy. From there on, double windows were used. For all large glass buildings glass material was used, which had a sun reflection coating. Windows were started to be utilized, where the window had a utility even if its cost was double or triple or in the case of insulating windows even at a cost which was 20 times compared to clear window glass.

The reason for this was two-fold. The cost of the glass even if it was double that of the clear window was negligible compared to the assembled window cost, and besides the appearance, the utility of the window also became a serious consideration.

The utilization of PV for architectural purposes started after the energy crisis in 1975, when Foster + Partners and also other architects had the idea to include PV in buildings, but the technology was in its infancy, and not having prior experience, the architects had no assurance of success and could not convince the customers for the need.

As mentioned before, in the pre-oil crisis, to radiate the mage, the might of the owner was the primary importance. After to oil crisis, the utilization of PV on buildings would have contributed to the image of the owner, but because its infancy and the lack of experience on buildings the owner did not risk to do it.

An excellent example is the Aon building, which was originally built and was the headquarters of AMOCO oil company on 200 E. Randolph St. in Chicago, IL, near the shores of Lake Michigan. This 83-storey, 2.3 million sq. ft. building when it was completed was the tallest skyscraper in Chicago–1136' (346 m) high and the tower's base was 194 × 194' (59 m^2). It was at that time the fourth tallest building in the world. It was completed in 1972 and the steel structure was sheeted with 43,000 marble cladding panels from the famous marble quarry of Carrara, Italy. The quarry's white or blue-gray marble was used by Michelangelo to create his famous statue "David," which was completed in 1504 AD in Florence, Italy.

When the AMOCO building was designed and completed with the Carrara marble sheeting, it was unfortunate that the designers did not take into consideration that the weather near Lake Michigan in Chicago is different from weather in Florence. Marble is very sensitive to acidic rain and especially rain and moisture freezing and thawing; while Michelangelo's 6 ton heavy David sculptured from the marble of Carrara was standing already more than 450 years outdoors in Florence, Italy, the first marble slab of the AMOCO building detached from the façade as little as 1 year after completion of the building and fell down and penetrated the roof of the nearby Prudential Center. Unfortunately, by 1990 the deteriorating marble became so dangerous that all of the building's 43,000 marble cladding panels had to be removed and replaced with something else. Two thirds of the discarded marble slabs were in such a bad condition that they were crushed and used as landscaping decoration at AMOCO's refinery in Whiting, IN.[170] Solarex was by that time one of AMOCO's subsidiaries and proposed to replace the marble with a PV façade. In 1990, there was no BIPV system in existence which was a true façade of a building. Obviously, AMOCO did not dare to experiment on an 1136' (346 m)-high structure in the middle of a large city. Therefore, from 1990 to 1992 the Carrara marble slabs were replaced with 2-inch-thick light gray Mt. Airy (USA) granite panels at a cost of about $80 million.

PV was already used at that time as BIPV but not integrated in building walls. The first time it was used was in 1973 when Dr. Karl Böer installed PV modules on the roof of his house.[171] After the 1973 oil crisis, the US Government supported several demonstration building projects in which PV was used for one reason or another. In 1980, a 7.5 kWp residential application was designed and completed by Solar Design Associates. The Carlisle House as it became known was sponsored by the Massachusetts Institute of Technology and the US Department of Energy. The house was all electric with no fossil fuel burned on-site.[172]

The culmination of this era was when Solarex in 1982 built a new factory in Frederick, MD, and installed a 200 kW PV roof utilizing multicrystalline Si wafers produced by the casting technology developed by Solarex (Fig. 25.1).

The 200 kWp roof was not integrated into the building, but the way it was built reduced the cost of the building's underlying roof structure and provided shelter against the

---

[170]http://www.emporis.com/application/?nav=building&lng=3&id= aoncenter-chicago-il-usa.

[171]Böer KW (2010) *The Life of a Solar Pioneer*, iUniverse, p. 201, Bloomington, NY.

[172]Strong, SJ *A New Generation of Solar Architecture* (PDF), Solar Design Associates.

environment for equipment traditionally located on the roofs of commercial buildings, for example air-conditioning systems and cooling towers.

Figure 25.1 Solarex Building, 200 kW multicrystalline roof Frederick, MD, 1982. (BP closed the facility in 2012 and the building was demolished in 2013.)

Figure 25.2 Intercultural Center, Georgetown University (Washington, DC) 300 kW multicrystalline solar system, 1984.[173]

The first large-scale BIPV roof was a 300 kWp multicrystalline PV system completed by Solarex in 1984 on the Bunn Intercultural Center of Georgetown University in Washington, DC (Fig. 25.2). It was at that time the largest PV installation in the world 3318 m^2 (35,710 sq. ft.) of solar panels on the

[173]http://maps.georgetown.edu/interculturalcenter/.

roof of the center, facing south. The solar modules installed used rough glass that reduced efficiency, but prevented glare from the roof, which would have affected airplanes flying over the building when landing or taking off from the nearby Reagan National Airport. This system needed only very little maintenance and after almost 30 years was still in operation.

A list of the early (1982–1990) European PV-integrated solar buildings can be found in Annex. 4. All of those were standard production PV modules. These systems ranged from a few kWp to 53 kWp, which was installed on the roof of Aerni Fenster AG (window manufacturer) in Arisdorf, Switzerland.

It was a total departure from the "utility"-oriented PV modules when the true BIPV building-integrated façade became a reality.[174] On May 8, 1991, the very first of such a structure was inaugurated on the southern façade of the electric utility's headquarter (STAWAG) building, in Germany, in the city of Aachen (Fig. 25.3). The BIPV façade was designed by the architect Georg Feinhals. Flachglas Solar GmbH (later renamed to "Flagsol"), with the cooperation of Joseph Gartner & Co., a German aluminum window frame maker, presented the idea to install PV double glass façade elements, replacing the existing office building's south-facing glass wall which was planned to be refurbished.

The STAWAG PV façade had a chessboard pattern. A combination was used of double glass elements, some with blue Solarex multicrystalline PV cells and others just plane semitransparent glazing. The combination of these led to a surprising pattern both inside and outside. The wiring was installed in the aluminum frames so as not to interfere with the view of the façade from the inside. The PV façade electrical output was 4.2 kWp. The inverter which was used to connect the PV system to the grid was the first PV inverter produced by the SMA Company. The PV system after more than 20 years is still in operation. The project was a big success, a great number of architects and interested people visited it within a month to learn about this new technology.

---

[174]A part of the story described below is from the notes made by Mr. Joachim Benemann (at that time President of Flagsol GmbH) and provided to me.

Figure 25.3 STAWAG building in Aachen, Germany.

The development of a true PV façade was the result of Flagsol departing from the solar module technology used at that time. Joachim Benemann joined Flachglas AG in 1985 and had the responsibility to manage the company's solar business. Flagsol's solar business at that time consisted of the parabolic mirrors sold for solar thermal power plants.

At some point Mr. Benemann planned to expand the business into the PV area, but had seen no possibility of entry into this field to manufacture standard PV modules, because at that time established large manufacturers Arco Solar and Solarex, in Europe Photowatt and in Japan Sanyo and Sharp, dominated this market. Therefore, he decided to enter in a field where there was no competition and where a glass company had excellent access to the market, the building sector. He realized that architects did not like the small standard-size modules made in those days, so he installed a PV production line which was able to manufacture any sizes up to 6 m² (65 sq. ft.).

Mr. Benemann also realized that he could not use the conventional PV module lamination technology for architectural purposes, where the back of the module was a non-transparent material, but a double glass module had to be used in which the PV cells were sandwiched between two glass panels. He found that such a system was developed by Flachglass AG, the mother company of Flagsol for "noise protection windows," where the two glass panels were connected with a plastic layer, in which layer the PV cells could be encapsulated.

The STAWAG PV façade was made with laboratory equipment, but because it was extremely successful and an instant demand developed, Mr. Benemann was able to assemble a production line which could produce 15 MWp all-glass PV modules/year.

Flagsol was now able to buy PV cells, multi- or single-crystal types, based on the project requirements and to fabricate PV modules in any size or shape for the façades according to the architect's design. They could make PV modules with different glass types or thicknesses. One example was the PV modules needed for the roof over the platform of the Berlin main railroad station. Very special safety PV glass elements were required to protect against breakage to avoid a crash down on the platform or on the trains. The front glass of the module, which admitted light, was made of a 4 mm thick low-iron glass, which had the highest light transparency, while the back glass over the platform was a 10 mm thick float glass. Because of the design requested by the architect, each of the 3800 PV modules used for the roof had a different size, which as one can imagine, needed a very difficult operation for logistic, storage, and installation. But in spite of these difficulties, the project was successfully completed in 2002.

Not the most spectacular, but one of the most interesting and remarkable building-integrated PV system was the "Academy for continuing education" located near the city of Herne-Sodingen, close to the large city of Gelsenkirchen, where the Flagsol factory which produced the PV modules for the Academy was located.

The Herne "Academy for continuing education" of the Nordrhein-Westfalen state of Germany was built on the top of a coal mine which was started in 1871 and was shut down in 1978. During that time the depth of the mine reached 1100 m (3600 ft.) and in several mine accidents more than 100 people were killed.

The Academy was designed by the German architects Hegger Hegger Schleiff and by the French architects Jourda and Perraudin.[175] The building was constructed only of wood, glass, and concrete. The glass envelope of the building was 176 m (574.4 ft.) long, 72 m (236.2 ft.) wide, and 16 m (52.5) high. It is actually a glass box inside of which a climate is maintained similar to that in Nice, the city on the Mediterranean. The building's roof and also part of the south side glass wall has 3185 PV glass modules totaling 10,000 m^2 which produced 1 MWp electricity. This was the largest BIPV system in the world at that time when the building was completed in 1999.

Figure 25.4 Inside the Academy.

Mr. Benemann, who was the president of Flagsol, the company which manufactured the PV modules installed as part of this glass envelope, drove me to the Academy on a partly cloudy cool late-fall day and parked on the south side of the building. When we got out of the car, he pointed out the large number of PV modules which were a part of the building's south wall. When we entered the building, I was taken aback by what I saw. From the cold partly, cloudy German fall day we

---

[175]All the information about the Academy Mont-Cernis Herne was confirmed by Ms. U. Martin, Building Management, Herne.

The pictures of the Academy Mont-Cernis Herne are from its Web site: www.academie-mont-cernis.de.

entered an immense space which looked as if we were in Nice, on the Mediterranean coast of France. I saw a garden with palm trees. On the two sides of this garden were buildings, housing offices for the city management, a library, a restaurant, a meeting hall, the training academy of the State of Nordrhein-Westfalen and a hotel. All this in a glass box, where the ceiling much above these buildings was glass.

When I looked up to the flat glass roof of the "box," I saw that it was not everywhere transparent, because it was covered by the PV modules, but there were areas which had clear glass and one could see the sky. This can be seen in Fig. 25.5, which shows the building from above. Mr. Benemann pointed out to me that the reason for this was that the area where the PV modules were above the buildings, they absorbed the sunshine to produce electricity and not letting the sunshine through to warm up the building's roof. On the other hand, the open area was over the garden, so the plants got a sufficient amount of light.

Figure 25.5 Academy for continuing education, Herne, near Gelsenkirchen.

The interior of the glass "box" is shown in Figs. 25.4 and 25.6. The buildings can be seen, as well as the garden and the wooden structure of the building.

Figure 25.6 Inside the Academy.

Another interesting feature is that the top of the roof is designed so that the rain is channeled to flow down in a pipe and accumulate in a cistern. The water from the cistern is used to water the vegetation, for the fountains, and for the toilets in the building, saving a lot of money, because water is very expensive in Germany. The energy needed for the building is generated by the PV system and also from the methane gas from the shutdown coal mine. The entire PV system was working fine since its installation and required no repair. Only a few of the inverters needed repair.

Flagsol completed 122 BIPV façade projects in 12 countries around the globe during the time between the inauguration of the first one in Aachen in 1991 until the company was sold to Scheuten Company in 2004.

I always believed that building integration would be a sizable market segment of PV, but thinking about it and bringing it to reality in the style in which Mr. Benemann did is a big difference. Unquestionably, his contribution to the creation of a new PV market segment and the popularization of PV systems for decentralized urban application was a milestone for PV. Everybody could see it and realize that it was not a scientific dream but a reality.

The true BIPV building-integrated façade continued after Flagsol was sold to Scheuten Company in 2004. Several other companies are offering this true BIPV system,[176] but the BIPV segment slowed down since. This has several reasons. One reason is that the price difference between standard PV modules and the specially fabricated modules needed for this BIPV application became very large and PV cell and module manufacturers went after the easy feed-in tariff (FiT)-oriented PV market, where investors were able to get good returns. Another reason is that the PV manufacturers actually dropped the connection to and education of architects.[177] The upturn in this BIPV field may happen, because when a slowdown in the FiT type market occurs PV manufacturers will turn to the not so easy markets, which they dropped, such as the market in developing countries and also this market, the true BIPV building-integrated façade.

Decorative Façades

Solar modules were attached to many old buildings to generate electricity, and also to attract interest in the buildings. One example was the office building in Berlin (1994) which was named "ÖcoTec[178] House" (Fig. 25.7) because of the PV modules attached to the building for decoration as well as for supplying electricity for the building. Because of the PV façade, it was easy to find tenants in spite of the overcapacity of office space in Berlin.

PV for Windows

As described in Chapter 22, in the 1980s it was found that with a laser, millions of small holes could be made in a-Si sheet.

---

[176]www.abakussolar.com, www.nanocanada.biz/Solar-Panel%28BIPV%29.php.

[177]Private communication from Mr. J. Benemann.

[178]"ÖcoTec" translated to English "EcoTec."

Depending on the number of holes, the non-transparent a-Si PV film became transparent. Today, a-Si PV modules can be made this way that are 1–25% transparent. The electrical output of the a-Si PV cell decreases by the percent of the area from where the a-Si was removed. These semi-transparent PV modules can be used in windows or roofs.

Figure 25.7 EcoTec House in Berlin.

Figure 25.8 Train station in Wuppertal, Germany.

Semi-transparent windows can also be fabricated if the c-Si wafers are not mounted closely to each other. Such an interesting application of a c-Si PV system can be seen in the roof of the platform of the Wuppertal (Germany) train station (Fig. 25.8).

Awnings or External Shadings

PV modules have often been used as awnings on buildings. They have two functions: To prevent the sun's heat in heating the building by shining through the windows, and generate electricity thereby reducing the cost of electricity to power the building. A good example of this is the Samsung office building in South Korea (Fig. 25.9). The PV system was built and installed by Flagsol in 1999.

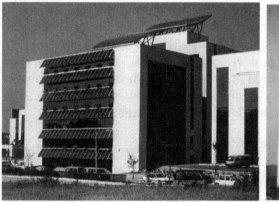

Figure 25.9 Samsung office building in South Korea.

PV Utilized on the Roofs of Buildings[179]

One of the largest utilization of PV is on the roofs of the buildings. There are two different application of PV on roofs:

- Used as roofing materials (for example, shingles).
- Mounted on the roof, but not as a roofing material.

---

[179]Additional information available: James T, Goodrich A, Woodhouse M, Margolis R, Ong A. (November 2011) *An Analysis of Installed Rooftop System Prices*, NREL Technical Report: NREL/TP-6A20-53103.

Used as Roofing Materials (for example shingles)

BIPV roofing, tiles, and shingles are at present the largest segment of the BIPV business. There are now several manufacturers in the world of PV roof shingles (solar tiles), which can be obtained from the manufacturers; their representatives, or even from building supply stores. These PV shingles or tiles can be used in roofs covered with ceramic tiles or can be installed like the asphalt shingle roofs.

The world's largest BIPV roofing system (as of 2011) was installed on a fruit and vegetable distribution center in Perpignan in the southern part of France (Fig. 25.10). After 2 years of construction, the roofs of the site of Saint Charles International have received a new building skin. The 68,000 m² of roof sheets were replaced by 97,000 PV roof tiles. With an installed power of 8.8 MWp, the system will annually produce 9800 MWh of solar electricity, which corresponds to about 10% of the electricity consumption of the city of Perpignan. The system was realized with the innovative PV roofing product SUNSTYLE, developed by Solaire France. The PV roof tiles are manufactured in partnership with the French company Saint Gobain Solar. There are several PV roofing tile manufacturers. One can find them by searching the Internet.

Figure 25.10 Perpignan, France. Solaire France "Sunstyle" PV roofing tiles (8.8 MWp).

Mounted on the Roof, but Not as a Roofing Material

Extremely large amounts of PV's are utilized attached to existing building roofs. PV is being used on either on sloped roofs as well as on flat roofs. Much more than a million solar homes were built in Germany (also many in other countries) and many companies which built them and are offering to build them and pictures of the houses they built can be seen on the Internet.

An interesting application of PV is on the German Bundestag[180] in Berlin. PV modules on the roof of the building have a peak capacity of 40 kWp. They provide energy for the ventilating plant and the operation of the sun-screening to the dome of the building.

Figure 25.11 is the Bundestag's view from the air where the PV modules can be seen on the roof. Figure 25.12 shows the Dome of the building and its reflection from the PV modules. The modules were fabricated by the German company Flagsol in 1998.

Figure 25.11 Bundestag in Berlin.

There are many farmhouses equipped with PV modules. Most are in countries where the PV system is connected to the grid, and because of the FiT, the electricity produces income

---

[180]The German "Bundestag" (German parliament building in Berlin).

for the owner of the farm. An interesting example is shown in Fig. 25.13. There are 424 solar panels on the roof at The Prince of Wales' Home Farm,[181] UK, capable of generating 100 kW electricity in total. They are expected to generate about 80,000 kWh electricity per year, saving over 40 tons of CO_2 emissions. The panels have been designed to keep the visual impact to a minimum, and cover an area of 690 m²—about the same size as a tennis court.

Figure 25.12 Dome of the Bundestag.

Figure 25.13 Prince of Wales' Home Farm, UK.

[181]http://www.flickr.com/photos/britishmonarchy/sets/7215762706-1005759/.

There are several buildings with large areas of flat roofs, where large PV systems can be installed and connected to the electric grid with the purpose of taking advantage of the FiT and recovering the investment and making profit for the investor(s).

In the future, the BIPV segment is expected to grow rapidly. Leading architect Norman Foster and others are integration solar cells into the skin of buildings and also to use solar modules on buildings such as the Masdar Institute building in Abu Dhabi or the stadium hosting the 2014 FIFA World Cup meeting in Brazil.

Other Applications

PV has been used for skylights, emergency lights, warning lights, and ventilation, since the time terrestrial PV modules became available. The utilization of these products increased even in the early days of terrestrial PV, for example between 1982 and 1989 by 127%.[182]

As of today, many manufacturers worldwide are producing these types of products. One of the largest US building supply chain offers in its catalog more than 40 different solar-powered roof and attic exhaust (see Chapter 18).

---

[182]Varadi PF (1992) *Architectural Utilization of Photovoltaics*, Report.

Chapter 26

Space Applications Used in Everyday Life: Communication and GPS

This chapter on certain space applications of solar cells had to be included in this book, which is devoted to the terrestrial use of PV, because during the 1990s for the everyday life, two types of satellites became very important for terrestrial use. They provide needed services to the population of the entire world and most people using them but they do not know that neither of these extremely important services would not operate without solar cells. These two solar cell-powered satellite systems are

(1) communication satellites
(2) satellites for the global positioning system (GPS)

The development of solar cells for space and terrestrial purposes were running on two different tracks, but the basic material, Si, for both types of solar cells was the same at the beginning. From 1988, the situation changed. The work

---

I would like to thank Dr. Joseph Pelton former Dean, International Space University (Strasbourg, France), for the information I received from him, which helped me to write this chapter.

Sun above the Horizon: Meteoric Rise of the Solar Industry
Peter F. Varadi
Copyright © 2014 Peter F. Varadi
ISBN 978-981-4463-80-5 (Hardback), 978-981-4613-29-3 (Paperback), 978-981-4463-81-2 (eBook)
www.panstanford.com

of Jerry Olson and Sarah Kurtz at the National Renewable Energy Laboratory (NREL) in Colorado[183] opened up the use of new materials achieving much higher conversion efficiency of light to electricity than those made on Si. As mentioned several times, for solar cells developed for space application high conversion efficiency and the endurance of the space environment were important and the price was not. These non-Si type solar cells had much higher efficiency and were more withstanding of the space environment, but they were also much more expensive, than the ones made on Si.

The importance of these new solar cell materials and systems for space solar cells can be best appreciated from their conversion efficiency of light to electricity (see Annex. 6). The Si space solar cells reached its best efficiency of 15% with Lindmayer's "violet cell" in 1972. By introducing the first non-Si solar cell: "GaAs" (Gallium Arsenide) in 1990 the efficiency reached immediately 20% and as the chart indicates (Annex. 6) with multilayer cells it reached 35% in 2012 and it expected to become higher due to research.

Communication Satellites

In October 1945, Arthur C. Clarke published a paper in the journal *Wireless World* precisely describing how and why the "space station" used for communication should be placed in a circular orbit 35,786 km (22,236 miles) above the Earth's equator (geostationary Earth orbit) and following the direction of the Earth's rotation. To ground observers, an object in such an orbit (which sometimes is called Clarke orbit) appears to be motionless at a fixed position in the sky. Obviously, he believed in 1945 that the station would require a crew to change radio tubes which frequently burn out and it was unclear how the batteries would be recharged. But he described the concept.

The beauty of his idea was that antennas communicating with the satellite can be pointed permanently in the direction of the stationary satellite and sending microwave signals after

---

[183]http://www.nrel.gov/news/press/2007/502.html.

being received by the antenna on the satellite can be transmitted back to the earth and received by another permanent or moving antenna located anywhere (except the northern- or southern-most part) of the Earth, thereby establishing communication with any distant location even across oceans.

In order to implement Clarke's vision only three *"minor"* problems had to be solved:

(a) Substitute the electron tubes with some other system, which does not have to be periodically replaced.

(b) Find a way to replace the batteries which are operating the system or find a way to charge them while the station is floating in the space.

(c) Find a way to lift the system from the earth to a circular orbit over the equator, which orbit is one-tenth of the distance from the Earth's surface to the Moon.

With incredible luck and coincidences, mankind was able to solve these "minor" technical problems in 17 years.

What was missing from Clarke's paper was making suggestions how to solve the global political complexity of a satellite communication system sitting at one point over the equator. One of the question is how far above a country is its domain? The other question is who can use the radio-frequencies used by these satellites? On top of that when all of these had to be decided it was during the so-called "cold war" in which the USA and the Russians had the capabilities to establish their own satellite communication, but the European countries which were in the process to rebuild themselves from WW2 believed they should also be included in the decisions and the last complication was the US legislation about what to do with the technology developed on US taxpayers' money.

One could write an interesting book about the twists and turns of US and global politics related to satellite communication but as this book is not about politics only the most important events will be mentioned. One can quote the unknown author: "Everything is okay in the end, if it's not ok, then it's not the end."

Slowly every needed ingredient became available.

The solution to the substitution of the electron tubes, which were needed to be periodically replaced, was found in December 1947 at Bell Laboratories, when William Shockley, John Bardeen, and Walter Brattain invented the substitute for the electron tubes—the transistors—making possible a breakthrough in the miniaturization of complex circuitry and needed only a fraction of the electricity used by radio tubes. The periodic replacement of the electron tubes was also solved by the practically unlimited life of the transistors.

The German rocket scientists who were split between the USA and Russia helped to develop the proper launch vehicle and satellites were launched. The first one was the Russian "Sputnik 1" in 1957. The Sputnik's electric power came from a battery and the signals continued for only 22 days until the transmitter batteries ran out. In 1958, the USA Signal Corps launched the first "Broadcasting satellite" named "SCORE." It was launched on December 18, 1958 and its batteries were exhausted in about 10 days. From these experiences, it became clear that to shoot up satellites to rely only on battery power is waste of money. The batteries must be recharged to have a useful satellite.

Daryl Chapin, who discovered the Si solar cells in 1953, achieved a 6% conversion efficiency, which was shortly improved by Morton Prince and Ed Stansbury also at the Bell Telephone Laboratory[184] to a conversion efficiency of 8–9%. It became obvious, that these "solar cells" when mounted on a satellite could be used to recharge their batteries. The first and very convincing application of solar panels for satellites was on March 17, 1958, when the Vanguard 1 satellite was launched.[185] This, by today's standard, "miniature" satellite weighed "only" 1.47 kg (3.25 lb.) and its diameter was 16.5 cm (6.6 inch), had two radio transmitters. One was powered by batteries, while the second was powered by 48 small solar cells[186] placed on the outside walls of the satellite to charge a rechargeable battery. The transmitter that operated only by

---

[184]Palz W (2010) *Power for the World*, Pan Stanford Publishing Pte. Ltd., Singapore.

[185]Easton RL, Votaw JM (1958) Vanguard 1 IGY satellite, *Rev Sci Instrum*, **39**(2), 70–75.

[186]Solar cells were manufactured by Hoffman Electronics of California.

batteries lasted 20 days, while the transmitter powered by the solar cells charged batteries operated 6.5 years, and worked until 1964, when the assumption was that it was not the solar panels that had become defective but the radio transmitter.

Four years after the launch of Vanguard 1, in 1962, the US telephone company AT&T launched a satellite "Telstar" for telecommunication purposes, to establish connection between continents. Telstar was launched only into a low Earth Orbit (not in a Geostationary Orbit) but it demonstrated that a live TV signal can be transmitted across the ocean. The required 14 W electricity was supplied by a battery charged by solar panels. Thus, for the satellites for communications and for other purposes, the successfully used solar cells (and panels) proved to be the solution to keep the satellite batteries charged.

The United States National Aeronautics and Space Administration (NASA), however, still did not trust completely the solar panels, and in April 1964 it launched a satellite for which the electrical energy was supplied by a nuclear reactor. During the launch of the satellite it broke up and the entire satellite, including the nuclear reactor and radioactive waste material, was scattered around the globe and could be found on every continent.[187]

Thereafter, only solar panels were used to provide electricity for satellites.[188]

The last milestone to be achieved was to launch a satellite into a Clarke—(Geostationary also called Geosynchronous)—orbit and to find out if it could be operated successfully from such a great distance.

This was demonstrated when on December 14, 1963, NASA launched "Syncom 2" and after that "Syncom 3" into a Geosynchronous orbit and successfully transmitted communications. The incredible advance of the satellite communication which was achieved in 8 years can be demonstrated with the fact that the events of the 1956 Summer Olympics which were held in Melbourne, VIC, Australia, could be seen in Europe or

---

[187]Johnstone B (2011) *Switching to Solar*, Prometheus Books, Amherst NY.

[188]The radioisotope thermoelectric generator (RTG) is a source of energy used only for space objects, which are sent so far away from the Sun that the solar energy produced by solar cells would be insufficient to operate it.

in America only a day later, as it had to be filmed and the film transported to Europe and to America by airplanes, but 8 years later the 1964 Summer Olympics which were held in Tokyo, Japan, could be seen in the USA and in Europe live via the TV signals transmitted by communication satellites.

The speed of the technological development was formidable, but the commercialization of satellite communication ran into governmental and political difficulties. The issue was how to transfer technological achievements financed by public funds and commercialize it, furthermore, what should be the participation of other countries. The result was in the United States the "Communication Satellite Act of 1962," which was a camel. A camel is a horse designed by a committee. The Communication Satellite Corporation (COMSAT) was created; it was a company registered on the New York Stock exchange, where 50% of the stocks were owned by a number of large corporations (AT&T, IT&T, RCA, Western Union, and Western Union International) and the other half of the stocks were sold to the public. COMSAT required also having a certain number of directors appointed by the US Government and had to get technical advice from NASA.

The other non-technical problem was that while to establish a submarine telephone cable only two entities had to agree (for example, on one end let's say the USA and on the other end the UK), in comparison, establishing a communication satellite over the equator, for example at the middle of the Atlantic Ocean, one can connect a large number of countries and to establish an agreement between that many countries is not easy. It took 2 years to arrive at an "Interim Intelsat Agreement," in which COMSAT became the manager of the Intelsat system in which many countries participated.

In spite of this extremely complicated operating structure, COMSAT managed to assemble an excellent technical staff and was able to successfully launch and put into operation the first commercial communication satellite (Figure 26.1) by April 1965, in an extremely short period of time.[189]

---

[189]The Soviet Union deployed their Molniya satellite network in a highly eccentric orbit also in 1965. This was the world's first domestic satellite system. To get continuous high elevation coverage of the northern hemisphere at least three Molniya spacecraft were needed.

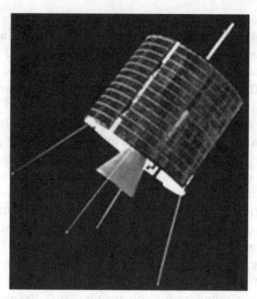

Figure 26.1 The Early Bird Satellite, world's first commercial communications satellite (Figure courtesy of the COMSAT Legacy
Project).

COMSAT's technical staff success can be seen from the
advances made during the ensuing 10 years.[190] It also demonstrates the need of more and more electric power delivered
obviously by solar cells:

Table 26.1 Increasing capacity and power requirements and decreasing
cost/circuits

	INTELSAT I	INTELSAT II	INTELSAT III	INTELSAT IV	INTELSAT IV A
Year	1965	1967	1968	1971	1975
No. Tel. Circuits	240	240	1200	5000	7500
Cost/ Circuit ($k)	20	10	2	1	0.7
Power (W)	40	75	120	400	500

[190]Edelson BI, Strauss R, Bargellini PL (1974) *Intelsat System Reliability*,
International Astronautical Federation, 15th Congress, Amsterdam.

By 1969, "Intelsat III" series were deployed providing a true global communication, telephone, and/or TV across the Atlantic, Pacific as well as across the Indian Ocean.

The problem with the politically negotiated solution creating COMSAT and Intelsat was that they had a monopolistic position. Starting with the Reagan era in the USA in 1984, the telephone communication monopoly of AT&T was broken up. It took years until COMSAT and Intelsat were truly privatized and were competing in a field where private equities or countries could set up their own satellite communication networks for domestic use, too. The first was Canada, where its population lived in a huge territory which was impossible to wire up, but with their domestic communication satellite was easy to provide both telephone and TV services for the people living there.

The international, regional, and domestic needs for telephone, TV, and even radio connection resulted that today (2012) at least 300 communication satellites are in Geosynchronous orbit. In addition of these there are hundreds of communication satellites in low and "Medium Earth Orbit" (MEO) constellation for a variety of services. Today 103 countries have their own domestic Satellite Systems or Satellite Leases. Twenty-eight countries have more than one. The total constitutes over half of the countries and territories of the world relying on long distance overseas and/or domestic communication (telephone, TV, and Internet).[191]

The United Nations has a specialized agency, the International Telecommunication Union, which provides the technical coordination.[192]

It must be mentioned that none of these satellites would be in operation without solar cells and without these communication satellites the global telephone, TV, and Internet service would not be possible.

---

[191]Pelton JN (2005) *Future Trends is Satellite Communications: Markets and Services*, International Engineering Consortium, Chicago, IL.

[192]Pelton JN, Madry S, Camacho-Lara, S (eds) (2013) *Handbook of Satellite Applications*, Springer, New York; Heidelberg; Dordrecht; London.

GPS (Global Positioning System)

Global positioning system (GPS) is called GPS in every language. The "word" GPS is being used even in France or Hungary, the countries where it is the national pride to use their own language. The public knows what the GPS system is used for. It tells them where they are located on the Earth and if they indicate where they want to go it tells which direction one should go including the streets in a city. Millions of people use it every day, but I think that very few people in the world know that without the electricity supplied by the PV modules on the satellites there would be no GPS system.

The GPS system was developed for military purposes but since it became operational in 1994 the United States released it to the public. Today GPS receivers are used practically everywhere, people have them mounted in their automobiles, they have them in the form of a portable gadget and they are also included in people's cellular telephones or watches on ocean going ships, airplanes, and for many other uses. One can pick up the cell phone, push a button and in seconds the screen of the cell phone produces a map and a point on the map indicating where the person is located on the Earth. It also provides the exact latitude, longitude, and elevation of that point. A person going by foot, bicycle, car, boat, or an airplane can be directed by the GPS receiver to reach the destination.

A very accurate solution to terrestrial navigation is this recently established satellite GPS. Today people know how to operate the GPS receiver, but extremely few people know

- how it works
- that *without solar cells the GPS system would not have existed*
- that at present two systems are operational and they are of the "United States" and the "Russian federation".[193] Taxpayer's free gifts to all of the people in the world (Annex. 7).

---

[193]GLONASS, Russia's global navigation system was made fully available to civilians in 2007.

In order to appreciate this quantum jump in navigation on the surface of the Earth or in the air (for example, for airplanes—but it cannot be used under the surface of the Earth, for example, under water or in tunnels or in space), one should look back to appreciate what a miracle is the GPS that mankind developed in the last 40 years compared to what was used before.

It is not well understood how people in ancient times on land or on water left their base and arrived at their destination. On water it was obviously more dangerous to leave your home and become lost. There are many theories how in ancient times people were able to navigate to arrive to their destination. It is believed they followed the coast line, but even in the Mediterranean the history of which is well known, it is not clear how they were able to cross open waters and arrive to the right port? So, how did people navigate before? It is assumed they used celestial navigation, based on stars, moon, and the sun. But what about cloudy days?

It is believed that about a 1000 years ago, the Chinese discovered a magnetic iron ore (loadstone) which would point to the North and they used it as a compass. The compass helped on cloudy days. The idea of a compass slowly moved to the west and it is believed that an Italian mariner and inventor Flavio Gioja of Amalfi, Italy (c. 1300) perfected the sailor's compass by suspending its needle over a fleur-de-lis (lily) design, which pointed north. He also enclosed the needle in a little box with a glass cover. Although modern scholars dispute that he ever, in fact, existed, nevertheless his statue as the inventor of the magnetic compass was erected in Amalfi in 1900. The compass only gave the direction but was not useful to determine where the ship actually was.

Eratosthenes of Alexandria, Egypt, in about 250 BC assumed that the Earth was round and calculated the circumference of the Earth. It is believed that he was the first who proposed the latitude and longitude for a map of the world, where the lines of latitude would be parallel to the equator, while the lines of longitude would connect the poles of the Earth. To determine the ships location one has to establish the exact position in latitude and longitude.

Determining latitude on clear days was relatively easy. One has to measure the angle where the Polaris (North Star) is. If the angle of Polaris is zero, which means it is exactly on the horizon then the ship is on the equator. If it is just above the person's head, then the person is on the North Pole of the Earth. The angle in between Polaris and the horizon tells which latitude the person exactly is. Several angle measuring instruments were used in the centuries. The last one was the sextant.[194]

On the other hand, to determine the longitude eluded mankind until recently. European seafaring countries offered prizes for the development of a system to establish the longitude: Spain in 1567, Holland in 1636, France in 1666, and Great Britain in 1714.[195] The best solution would have been to use the time difference between a known point and the location of the ship as the Earth is rotating 360° a day, 1 hour difference would be 15° of the Earth's circumference. The problem was to develop a marine chronometer, which would be accurate in sea environment on a moving ship. The British Parliament rewarded John Harrison for his marine chronometer in 1773 and the 0 (zero) line of longitude was established to be at the Royal Observatory, Greenwich, England and the difference between the time at Greenwich and the time where the ship was would indicate the longitude. From the beginning of the twentieth century radio signals were used for synchronization of chronometers.

The last attempt to solve navigational issues was the "gyroscope" which was invented in the nineteenth century but was made functional only in the early twentieth century. The gyroscope is a spinning wheel or disk in which the axle is free to assume any orientation. It does not solve the problem of showing where the person or ship is located on the Earth's surface, it only can tell where the true north is and is a device for measuring or maintaining orientation. It is extremely useful under water and in space and it is excellent as backup system to the GPS system for ships and airplanes.

---

[194]Ifland P, *The History of the Sextant*, http://www.mat.uc.pt/~helios/ Mestre/Novemb00/H61iflan.htm.
[195]http://en.wikipedia.org/wiki/History_of_longitude.

So that brings us back to the ultimate solution of navigation on the surface of the Earth, namely the recently established satellite-based GPS (see Annex. 7 for the listing of the functional and planned GPS systems), which enables land, sea, and airborne users to determine their three-dimensional position, velocity, and time 24 hours a day, in all weather, anywhere in the world, with a precision and accuracy for the precise positioning on the surface of the Earth. The development of the GPS was initiated in 1973 by the USA.

The Operation of the GPS System

The principle of the operation of the GPS system is very simple. As an example, we can take a thunderstorm. A lightning hits somewhere. The observer sees the lightning and somewhat later hears the thunder. A lightning can be seen instantly (the speed of light is 299,792 km/s), while the sound travels only 340 m/s. If the observer hears the sound 3 seconds later than the lightning is observed, this means that the lightning hit 3 × 340 m = about 1020 m away. He can only tell how far away the lightning struck but cannot tell the exact location.

One would need three observers and their exact location to find where the lightning struck (Fig. 26.2). This is called in science trilateration.[196] Obviously, we have to know the exact locations where the observers were standing and drawing circles around the observers positions with the corresponding diameter will have one point where they cross and that will show the exact position where the lightning has struck.

The GPS system operates exactly the same way. Where the lightning struck is where the GPS receiver is and each of the three observers represent a satellite. Users figure their position on the earth by measuring their distance of the receiver from the satellites. The satellites act as precise reference points. Each GPS satellite transmits an accurate position and time signal. The user's receiver measures the time delay for the signal to reach the receiver, which is the direct measure of the distance to the satellite. This will assure extremely accurate three-dimensional (longitude, latitude, and altitude) location. This means that in order that the receiver should function it

---

[196]http://en.wikipedia.org/wiki/Trilateration.

has to see a minimum of three satellites, to determine the altitude require four satellites at the same time or better if there were more visible.

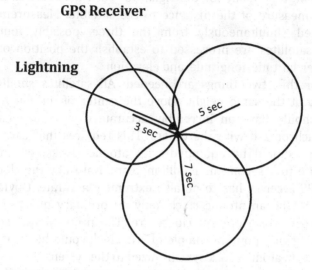

Figure 26.2 "Trilateration" to find the exact location.

The difference between the example given considering the lightning and the thunder is that the sound of the thunder is traveling 0.340 km/s, while the GPS system emits a radio signal which is traveling at the speed of light (299,792 km/s). The user's receiver has to measure the time delay for the signal to reach the receiver at the speed of light. The satellites are positioned about 20,200 km above the earth.[197] If the satellite happens to be exactly above the receiver when it emits the radio signal, traveling the distance to the receiver the delay will be 0.0740513 seconds. To put this time in perspective, at the London Olympics in 2012 Usain Bolt won the gold medal for the 100 m running which he did in 9.63 seconds. The silver medalist Yohan Blake finished at 9.75 seconds. The difference was 0.12 seconds. Obviously to measure this small difference required a very accurate stop watch. To measure 0.0740513 seconds will need an incredibly accurate time piece.

---

[197]The mean distance from the middle of the earth is 26,560 km. With a mean earth radius of 6360 km.

In the case of the GPS satellite signal measurements, the one mentioned above will be the first to arrive, the others which are not positioned exactly above the receiver will arrive later. The time delay for the signal to reach the receiver is the direct measure of the distance to the satellite. Measurements collected simultaneously from the three, possibly four, or more satellites are processed to establish the position of the receiver, latitude, longitude, and elevation.

For this, two things are needed: All satellites should be exactly at the same height above the center of the Earth and they should have an incredibly accurate clock which has to be synchronized with the others. This requires that each GPS satellite should be equipped with atomic clocks, which are accurate to 1 second in 1 million years. Naturally the clock of the GPS receiver has to be also extremely accurate. Obviously it cannot be an atomic clock, only an ordinary quartz clock, like most of the present clocks are. The problem was solved very elegantly how the simple quartz clock could be tricked to become an atomic clock synchronized to the system.[198]

The GPS system has three segments: space, control, and user (receiver).

The space segment

The first GPS satellite was launched on February 22, 1978. The full operation of the system was announced on July 17, 1995. The Space Segment consists of a minimum of 24 operational satellites (Fig. 26.3) in the constellation of six circular orbits in the "Medium Earth Orbit" (MEO) a mean distance from the center of the Earth is 26,560 km and from the surface of the earth 20,200 km (10,900 NM) above the earth at an inclination angle of 55 degrees with a 12 hour period.

At present the satellites are spaced in orbit so that at any time a minimum of six satellites will be in view to users anywhere in the world (Fig. 26.4). The satellites continuously broadcast position and time data to users throughout the world. The system is operational 7 days a week for 24 hours.[199]

---

[198]http://www.kowoma.de/en/gps/positioning.htm.
[199]http://www.navcen.uscg.gov/?pageName=GPSmain.

Solar array

Figure 26.3 GPS IIIA satellite[200] The Navstar GPS Constellation.

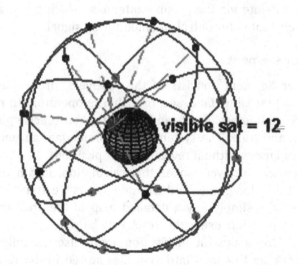

visible sat = 12

Figure 26.4 A simulation of the original design of the GPS space segment, with 24 GPS satellites.[201]

---

[200]Images have been released by NASA into the public domain and may be reused or published (http://www.gps.gov/multimedia/images/).

[201]http://en.wikipedia.org/wiki/File:ConstellationGPS.gif (By copying this on the Internet will provide *animated information* about the system).

Each of the GPS satellites are operated by *the electricity provided by the 750 Wp solar arrays.* The weight of the GPS satellite is 2 tons. *The 31 satellites in operation in December 2011 had a total 23.25 kWp PV systems.*

The control segment

The control segment consists of a master control station in Colorado Springs, with five monitor stations and three ground antennas located throughout the world. The monitor stations track all GPS satellites in view and collect ranging information from the satellite broadcasts. The control segment is also responsible to synchronize all of the atomic clocks on board of the satellites. The monitor stations send the information they collect from each of the satellites back to the master control station, which computes extremely precise satellite orbits. The information is then formatted into updated navigation messages for each satellite. The updated information is transmitted to each satellite via the ground antennas, which also transmit and receive satellite control and monitoring signals.

The user segment

The User Segment consists of the receivers, processors, and antennas that allow land, sea, or airborne operators to receive the GPS satellite broadcasts and compute their precise position, velocity, and time. The system can accommodate an unlimited number of users without revealing their position.

Once the receiver makes this calculation, it can tell you the latitude, longitude, and altitude of its current position. The user can simply buy a detailed map of the area and find the exact position using the receiver's latitude and longitude readouts. To make the navigation more user-friendly, most receivers plug this data into map files stored in the receiver's memory.

A standard GPS receiver will not only place you on a map at any particular location, but will also trace your path across a map as you move. If you leave your receiver on, it can stay in constant communication with GPS satellites to see how your location is changing. With this information and its built-in clock, the receiver can give you several pieces of valuable information:

- how far you have traveled
- how long you have been traveling
- your current speed
- your average speed
- showing you exactly where you have traveled on the map
- The estimated time of arrival at your destination if you maintain your current speed.

Just to remind the reader I am repeating two sentences from the beginning of this chapter:

Neither of these extremely important systems would operate without solar cells. These two solar cell powered satellite systems are

- communication satellites
- global positioning system (GPS)

These two satellite systems consisting of about 500 satellites are mentioned, because everybody uses them. But there are about over 3000 (some sources puts the number up to 8000) operational satellites in orbit in use for various applications, all powered by PV.

Chapter 27

Development of the Global PV Quality Program

In general, wild animals and birds enjoy more protection than human consumers. However, if somebody is locked up for 8 hours in an airplane, which was standing somewhere at the end of an airport tarmac, without food and water and with an overflowing toilet, realizes that the only protection they have is God. Governments have many ministries and agencies for all kinds of purposes but none for "consumer protection," except in Germany, which as of today (March 27, 2014) according to my knowledge is the only country, with a minister of the federal cabinet responsible for consumer rights and protection (*Verbraucherschutzminister*). The USA recently established the Federal Trade Commission's Bureau of Consumer Protection.[202]

Manufacturers in general have no interest to establish standards for their products or services. Governments establish regulations or standards usually only after there is a strong public demand. (Recently after strong public demand a USA ruling imposed a 4-hour limit on tarmac delays, after which the airline has to return to the terminal and let passengers out

---

[202]http://www.ftc.gov/bcp/about.shtm.

Sun above the Horizon: Meteoric Rise of the Solar Industry
Peter F. Varadi
Copyright © 2014 Peter F. Varadi
ISBN 978-981-4463-80-5 (Hardback), 978-981-4613-29-3 (Paperback), 978-981-4463-81-2 (eBook)
www.panstanford.com

of the airplane.) In general the consumer can only depend on the reputation of the manufacturer of a product, or the service provider.

At Solarex, we obviously manufactured PV modules with quality second to none, but perhaps I am prejudiced. I must admit that our major competitors also made excellent products. By 1995, I was out of the trenches, the daily grind of worrying about production, quality, sales, and marketing. I was in an ivory tower mainly involved with new PV markets, and the future of PVs. So I had a rude awakening from the information I received after I met Richard Spencer, whom I knew because of my involvement in the EU PV financing project when he worked in London. By this time he was working at the World Bank (WB) in Washington, DC. During our discussion he told me that the WB has projects to use PV to provide electricity as there were about 2 billion people having no access to electricity and that most of the WB Renewable Energy programs were in the Bank's Asia Alternative Energy [Asia Alternative Energy Program (ASTAE)] program, headed by Ms. Loretta Schaeffer. He suggested that I meet Dr. Anil Cabraal, a leader in ASTAE's PV projects. I immediately contacted Dr. Cabraal and met him on August 21, 1995 in his office. He told me about ASTAE's PV projects. He mentioned the failures they were encountering. When utilization of PV in developing countries was initiated by the WB to provide electricity for people who could not be reached by the grid, the WB had no way to differentiate among manufacturers in terms of good or bad reputations, nor as to which produced good PV products and could advise customers who could not distinguish good products from inferior ones. Obviously such customers bought the cheapest products, which in the most cases failed. ASTAE was in the process of drafting PV specifications and was about to publish a report entitled *Best Practices for Photovoltaic Household Electrification Programs: Lessons from Experiences in Selected Countries.* One of the crucial requirements listed in this report was to "provide quality (PV) products and services."

This news was very disturbing to me. That not the PV industry, but an important customer such as the WB in its own interest and desperation due to failures encountered, was resorting to draft PV specifications and write directives to its clients. So I believed that quality had become a big problem not only for the WB, but also for the entire PV industry, since defective products could ruin the entire industry's reputation. I believed the PV industry needed to take action immediately to improve the situation. This problem had to be solved.

After this introduction, the reader may be surprised since Chapter 9 has already noted that PV had an excellent start with the JPL program to protect consumers. The JPL Block program correctly specified that in order to achieve good quality PV modules, not only product testing was needed, but also that manufacturing required a Quality Management System (QM) to assure that all products would have the same quality. The JPL Block Program also requested that an outside inspector be given the opportunity to inspect the facility and verify the adherence to the QM.

Once I became aware of the quality problem affecting PV modules, I found out why this was so. After the JPL "Block buy" program ended in 1985, despite the fact that the QM's idea as standard can be traced back to World War II, to the US Department of Defense MIL-Q-9858 in 1959 and the British BS 5750, it was not until 1987 that a QM standard was adopted by the International Standards Organization (ISO) as the ISO 9000 standard. This meant that until that time PV manufacturers had no standard to establish their QM and that there was no commercial entity to certify a manufacturer's QM.

That also meant that until the 1990s PV module manu-facturers could only voluntarily evaluate their products by complying with the International Electrotechnical Commission (IEC) PV module standard, testing them in their own testing laboratory and verifying the results by an independent testing laboratory. To add to the problem, only a very few independent PV testing laboratories existed.

Those companies which took part in the JPL program knew that they benefited from testing every type of modules they manufactured and using independent testing laboratories to confirm their results. But in the late 1980s, globally more and more companies started to manufacture PV cells and modules and as the last JPL "Block buy" was completed by 1985 most of them were not familiar with JPL quality requirements.

Interestingly, after the very successful JPL program was completed, not even the US Government procurement requested PV products complying with the JPL requirements. The only thing which survived from the JPL program was the type testing standard (see Chapter 9). But for smaller manufacturers, even to comply with testing of PV modules in-house and verification of the results in an independent testing laboratory was too expensive compared to their sales volume. Therefore, several manufacturers sold untested and many times unreliable products which obviously failed. It became evident that if this continued, PV would develop a very bad reputation and would not be considered as a serious source of reliable renewable electrical energy.

I discovered that many reputable manufacturers were seriously worried about this problem. I discussed the matter with John Wohlgemuth of Solarex Corporation who was very active in PV standardization. He said that the US had already started an organization: "PowerMark," which was supported by Dick DeBlasio of NREL (see Chapter 9). The purpose of "PowerMark" was also to establish a requirement for QM programs and to develop a label for PV product which qualified by conforming to specifications developed by PowerMark, which was basically similar to the JPL approval system.

I was very pleased that something was started, but my opinion was that this had to be a global organization and not just an American one, because if an American system existed, a similar but separate system would be established in Europe and also in Japan. For manufacturers this would become a serious trade barrier, since it would mean their product had to undergo retesting at every market. The PV quality program had to be global and somehow it had to be part of IEC which was accepted globally. This was suggested to me at a meeting by Dr. Heinz Ossenbrink who was the head of the EU's Joint Research Laboratory's PV section in Ispra, Italy. I told Wohlgemuth that perhaps we should start a global organization in which

PowerMark would be a participant. I also talked about this with Scott Sklar, who was then the Executive Director of the US Solar Energy Industry Association (SEIA). Sklar agreed that most SEIA members believed it was in their interest to start a global PV quality program.

Dr. Cabraal also mentioned that I should discuss this matter with the United Nations Development Program (UNDP) people, because they too were very much involved in utilizing PV in developing countries. I subsequently visited UNDP in New York and met Prof. Thomas B. Johansson who was at that time the Director, Energy, and Atmosphere Program—Environment Division of UNDP and Susan McDade who worked in his division. Both agreed on the need for a global PV approval program (PV GAP).

I attended the workshop on "Financing solar energy in the developing world" on October 11–13, 1995 at the Pocantico Conference Center of the Rockefeller Brothers Fund (RBF). The meeting was organized by Neville Williams who founded the Solar Electric Light Fund (SELF), devoted to bringing PV to rural areas of developing countries. During the dinner of that meeting I had a long discussion with Dr. Charles Gay— who at that time was the Director of the National Renewable Energy Laboratory (*NREL*)—about the quality problems of PV modules and systems caused by some of the manufacturers. We concluded that the evident quality issues of PV modules and systems would frighten away investors, banks, and customers and could make the credibility of PV questionable. Charlie thought it urgent that something to be done. He noted that shortly after the Pocantico meeting, the European Solar meeting was set to be held in Nice, France. Since I had many friends there, why not ask their opinion, and, if they agreed and also believed it was an important issue, then we could start the ball rolling. Charlie assured me that he backed this idea and that NREL would strongly support it. Neville joined us during our talk and drawing on his experiences in developing countries fully agreed with our ideas. Neville later wrote in his book:[203]

---

[203]Williams N (2005) *Chasing the Sun*, New Society Publishers, Gabriola Island, BC, Canada.

"The Conference was a huge success, for it motivated a dozen people and the outcome was the idea of a PV quality program."

On October 26, 1995, during the 13th European PV Solar Energy Conference in Nice, I had lunch with Wolfgang Palz. We discussed PV quality problems which would make it probably impossible to finance PV manufacturers and systems. Palz concurred that a system had to be set up, so customers could easily distinguish reliable PV products from PV products of unknown quality. A global PV quality certification system was necessary and for consumer protection a visible recognition was needed to identify a quality product by placing a PV quality label on it. This is a customary practice in many other industries. Examples are a recognized quality label on wool products, the "wool quality mark" or the *Underwriters Laboratories* (UL) safety label. The PV industry should establish a "PV GAP" and develop a PV quality mark.

Being a good organizer, Wolfgang immediately outlined what had to be done. He decided that I was the only one who could start something like this, since at that time while I was not affiliated with any PV manufacturing organization, I was well known in the PV industry and could discuss the matter with the many people needed to support such a project.

I told Wolfgang that while I knew something about standards and standardization, I knew nothing about accreditation, certification, and quality labeling. Wolfgang next provided me with a list of people and organizations with whom I should discuss the matter. I started to contact them to create a consensus of how such a PV GAP should be started and structured.

My first question was, is the PV industry interested in a global quality certification and labeling program? I received great support from the PV industry: John Bonda, the General Secretary of the European Photovoltaic Industry Association (EPIA), Scott Sklar the Executive Director of the US SEIA as mentioned before, Allen Barnett, President of SEIA and also Steve Chalmers, president of PowerMark Corporation, all agreed that a worldwide approval program for PV product should be initiated. Dr. Heinz Ossenbrink as mentioned above gave his support for the project. Heinz was involved in PV standardization and reiterated his advice that ultimately this PV certification program should be merged into the IEC program.

The consensus of the PV industry associations was that a PV quality mark was needed since the majority of their members at that time were established PV companies and the interest of these companies was to have a means to differentiate their products from low quality PV products. Dr. Cabraal also supported the idea of establishing a PV quality certification program as well as a Global PV quality mark to be displayed on a PV product, thereby providing a visible recognition to distinguish a quality product from inferior or falsely represented ones.

Now that a consensus was reached to start PV GAP and establish a quality mark for PV products and the next question was how to implement this? As I started to develop PV product certification and quality labeling, I realized that when I told Wolfgang in Nice, that I knew something about standards and standardization, but nothing about certification and quality labeling, I had made a huge understatement. Product certification and labeling is an entirely different world that uses a different language. Experts in this field were people and organizations I never interacted with before, but now I had to.

I first got acquainted with the American National Standards Institute (ANSI) which was established in 1918. I found out that ANSI is also actively engaged in accrediting programs that assess conformance to standards. I learned the words "accreditation," and "conformity assessment." Luckily, I was advised to meet John Donaldson who I was told was the best known and biggest name in the country in this field. At that time he was the vice president for conformity assessment at ANSI, as well as the secretary of IAF.[204]

---

[204]International Accreditation Forum (IAF) is the world association of Conformity Assessment Accreditation Bodies. (I would wager a bet, that 99.99% of the readers of this book never heard anything about this organization. But when I got acquainted what this organization is doing, I realized how important their function is.)

Fortunately, John lived in the Washington, DC area, not far from me. He was nice and joined me for lunch. John was fascinated with, (a) the idea of producing electricity with PV to replace fossil fuel, and, (b) the fact that I, an ignoramus on the subject of certification, accreditation, conformity assessment, quality labeling, had the audacity to say that I was involved in developing and establishing a Quality Label for PV.

Nevertheless, John said he would help us. The first step would be to give me a brief education on the subject, so that when I would talk about it I would not discredit PV. After about our fifth lunch he declared that I was getting so dangerously close to the level, that if I should talk about the subject, experts in his field might possibly understand me.

Donaldson next connected me to the UL people. They educated me concerning the UL label which indicates a product has passed a safety test, and the "Wool Mark," a global label to indicate that "real" wool was utilized in the manufactured product. I learned from these organizations how their labeling system is structured and operated. It allows customers to identify the quality products easily. It would mean in our case that a PV quality label would allow PV manufacturers to distinguish their quality products from those of unknown quality not having the quality label on their product.

Establishing a PV approval (now I knew from John, that it is an approval and not an accreditation) program was also supported by a number of NGOs. The RBF offered their Pocantico Conference Center where PV GAP could organize a meeting where all interested parties would be able to express their opinion: Are a PV quality approval system and a PV quality label needed, and, if so, how do we proceed? The meeting was organized by Mark Fitzgerald of Sustainable Power and by Deborah A. McGlauflin of insights in action. It was scheduled to take place on June 24–25, 1996.

When the meeting date was established, I thought some more about WB's participation. Cabraal had already agreed to attend the meeting and give a presentation. However, when I phoned Cabraal at the end of May, I found out that he was on a trip for several weeks. I decided to visit his boss Ms. Loretta

Schaeffer, who headed the ASTAE[205] program, to ask that she too participate in our meeting because I believed that WB support and the story of their problems with the quality of PV modules would help us establish a quality assurance program for PV.

The WB's huge building is located in Washington, DC, 2 blocks from the White House. As I lived in the Washington area it was easy for me to pay a visit to Schaeffer on May 21, 1996. I was escorted to her office. It was a large, but not rectangular-shaped office. At one end was her desk and at the other end was a large circular table with chairs. As she asked me to sit down she indicated that she had very little time for our meeting.

I immediately got to the subject. I explained briefly who I was, that I and several of my colleagues (I mentioned some names) were planning to establish a system to differentiate between good and bad PV products, that I had discussions with Dr. Cabraal about our plans to establish a PV GAP, that he had indicated it was a good idea and that he planned to attend our meeting on June 24–25 at Pocantico to give a presentation.

Schaeffer said that she had been informed of this. Somehow she apparently assumed from my introduction that I was looking for money, because she very abruptly informed me:

"The Bank has no money to support this sort of quality-development program."

It looked like our meeting was about to end. Therefore, I tried to make the purpose of my visit very clear and said:

"As The WB has quality problems with PV and is going to issue its 'Best Practices Report' I feel very badly about the fact that the PV industry has been developing such a bad reputation. It is the PV industry's duty to bring law and order to the quality issue. As the Bank has experience in this quality issue, what we need is not money, but WB moral support and guidance. What I actually came here for was to invite you personally to attend the upcoming meeting on June 24–25, in

---

[205]http://web.worldbank.org/WBSITE/EXTERNAL/COUNTRIES/EAST-ASIAPACIFICEXT/EXTEAPASTAE/contentMDK: 21122177~menuPK: 3144 322~pagePK:64168445~piPK:64168309~theSitePK:2822888,00.html.

the Pocantico Conference Center in Tarrytown, NY. The purpose of this meeting is to get the input of relevant organizations and people as to how we should proceed. The meeting is sponsored by the RBF. **I came here today recognizing that it would be very helpful if, besides Dr. Cabraal, you** yourself would also participate in that meeting and offer guidance with your experiences."

It seemed I may have hit the right tone. I sensed that our meeting's end was no longer so imminent. Her response surprised and heartened me.

"I think what you propose is the right thing to do. I think Dr. Anil Cabraal is the right person to join your meeting, but if for some reason he will be unable to participate, I would be willing to attend the meeting."

It was now clear to me that PV quality issues were important to the WB. I thanked Ms. Schaeffer, promised I would send all information to her as well, and then I left.

The June 23–24, Pocantico meeting was chaired by Wolfgang Palz. All invited people came. From Europe as well as from the USA, the European and US PV industry Associations were represented, also from The WB (A. Cabraal), the UNDP, RBF, and several delegates from NGOs. It was equally important, that John Donaldson and representatives from UL attended and spoke about a possible PV approval system.

One conclusion of the meeting was that the PV GAP should be instituted as soon as possible. Ossenbrink offered ISPRA to host the next meeting to formally start PV GAP. He suggested it take place in Belgirate, Italy near ISPRA, so that those attending the meeting would be able to visit the JRC PV test facilities. He proposed the dates: October 27–30, 1996.

John Bonda of the European PV Industry Association (EPIA) and Scott Sklar of the US PV industry association (SEIA) expressed their support and Heinz Ossenbrink and Wolfgang Palz said that the Japanese PV association would be invited to come to the Belgirate meeting.

The other conclusion was that PV GAP should be affiliated with IEC with the idea that when it would be developed, IEC could take it over. In this respect Donaldson advised, that the most important person to be invited would be Mr. Richard Kay

(Fig. 27.1), the General Secretary of "The IEC Quality Assessment System for Electronic Components" (IECQ) located in the IEC headquarter in Geneva, Switzerland. PV GAP ultimately would become a part of that department of IEC and Richard was the most knowledgeable person, who could help establish and guide the development of PV GAP and PV quality labeling. Donaldson advised us, that it would be very important to get Kay interested in the project and to have him participate in the work.

At the end of August, I was in Switzerland. I called Richard Kay who agreed to meet me on Saturday, August 24, at 9 a.m. at the Boccard coffee house in Rolle, a village on the shores of Lake Geneva, not far from where he was living. The meeting went very well. Kay was very much acquainted with PV. He felt that an approval program and a PV label were needed, and that IECQ could be interested in adding to its program. I asked him to attend the Belgirate meeting and to give a talk for those of us who know very little about the subject of "quality assessment" and "quality labeling."

Figure 27.1 Richard Kay.

By the end of August, Heinz Ossenbrink sent out the Belgirate meeting invitations. The indications were that everybody important for any major decisions would be there. The Belgirate meeting (October 27–29, 1996) was attended by representatives of the European, Japanese, and USA PV industry associations and PowerMark, people from NREL, ISPRA, The WB (Loretta Schaeffer), UL, IECQ, etc.

Based on the decisions taken during the Belgirate meeting and on the follow up meetings of the Executive committee, PV GAP was incorporated by Dr. Markus Real and by Dr. Peter F. Varadi in Geneva, Switzerland as a not for profit organization for which IEC agreed to provide secretarial services. It also established the PV quality marks (Fig. 27.2) following the advice by UL, one mark for PV components and one for PV systems.

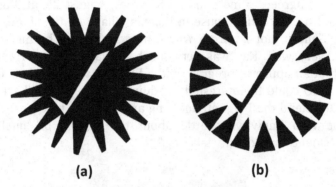

(a)   (b)

Figure 27.2   (a) PV GAP Mark for PV components. (b) PV GAP Seal for PV systems.

Under Richard Kay's direction, conditions were established concerning how a manufacturer would have the right to display the PV GAP PV quality mark on product(s). Also developed were the standards the IECQ member testing laboratories should use to assess the manufacturer and product to inform PV GAP in awarding a certificate for the manufacturer of a product approved by the IECQ organization.

The requirements for a manufacturer to receive IECQ approval were very similar to what the JPL requirement was:

- Product manufactured and tested according to International Standards (IEC) or provisional international specifications (PVRS) by an accredited (ISO/IEC 17025) Testing Laboratory.
- The manufacturer should have a Quality Manufacturing

System (ISO 9001) certified by an accredited ISO Registrar.

- Periodic unannounced inspection of the facility and the possibility that the inspector may select products for testing.
- Based on IECQ approval of a product, PV GAP will issue to the manufacturer a free of charge license to authorize the display the PV GAP quality mark and/or Seal on the approved product(s).

In the following years, the modules of several European, American, and Indian PV manufacturers received IECQ approval and licenses from PV GAP to display the PV quality mark on their products.

In order to promote the PV module and components approval and certification system as well as to help (Fig. 27.3) PV manufacturers, the WB sponsored the creation of a PV GAP "Quality Management in Photovoltaics" training manual which was completed in 2000[206] and published in French, English, German, Spanish, and Chinese. The training program was sponsored by the WB and also by the EU. The manuals on CD were available from the WB free of charge. Cabraal also proposed training courses for employees of PV companies and PV practitioners in Quality Management and Quality Control. These training courses were also sponsored by the WB and the EU. They were held in Geneva, Berlin, Barcelona, South Africa, India, and also in China.

In 2005, IEC transferred the administration of the PV quality mark and seal from IECQ to IEC's Worldwide System for Conformity Testing and Certification of Electrotechnical Equipment and Components (IECEE).[207] IECEE's global performance certification scheme is used for a very large number of products throughout the entire world. It is also available for PV products based on the GAP as described above. The Global PV quality labels can be obtained from IECEE free of charge if the product was certified by an IECEE member testing

---

[206]Varadi PF, Dominguez R, Schaeffer L: Quality Management in Photovoltaics, PV GAP QM 2.0.

[207]www.iecee.com.

laboratory. All information about this can be obtained from Mr. Pierre de Ruvo, Executive Secretary of IECEE.[208] The IECEE PV program is available on the Internet.[209]

Figure 27.3 PV GAP Executive Board, Dresden, Germany, 2006.[210]

As of today the WB recommends that PV modules (or other PV components or systems) purchased for projects supported by the WB, which have the IECEE certification and the PV GAP quality mark or Seal[211] should be preferred.

Australia decided that solar modules must be tested and certified against IEC standards according to IECEECB-FCS (including factory inspections) by testing laboratories and

---

[208]http://www.iecee.org/pv/html/index.html.

[209]IECEE PV Program (OD-2051-Ed.1.0), *Procedure for Certification of photovoltaic (PV) Products and the Use of the IECEE PV Quality Mark and PV Quality Seal.*

[210]From left to right: Richard Kay, Wolfgang Palz, Qin Haiyan, Markus Real, Bernard McNelis, Stephan Novak, Minoru Takada, S. Vasanthi, Eric Daniels, Peter Varadi, Frank Wouters, Pierre de Ruvo, WG van Sark, Anil Cabraal.

[211]http://web.worldbank.org/WBSITE/EXTERNAL/TOPICS/EXTENERGY2/ EXTRENENERGYTK/0,,contentMDK:21820033~pagePK: 210058~piPK: 210062~theSitePK:5138247,00.html. Also, http://www.worldbank.org/ retoolkit/.

certification bodies (and thereby can obtain the PV GAP quality mark) in order to enter the country.[212]

Interestingly, the WB-supported *PV Manufacturers Quality Training Manual* and Seminars had a great effect on the events related to PV in China. In China, the "Golden Sun Mark"[213] the quality requirement which is similar to PV GAP's, was adopted by the "China General Certification Center" (CGC) for PV products. The details of this interesting story are presented in Chapter 37, "The Chinese Miracle."

---

[212]Information from Mr. de Ruvo P, *Executive Secretary IECEE.*

[213]http://www.cqc.com.cn/english/Newscopeofcertification/webinfo/2009 /12/1260927918520352.htm.

Chapter 28

World Bank's Trailblazing Role for PV in Developing Countries

Back in the 1970s, the leaders of the terrestrial PV industry were aware that a very large population in Africa, Asia, and South America lacked access to electricity and that these people had little chance of ever having grid connections to centralized power generation. What they needed was distributed electric power generation which would then offer a tremendous potential for the PV market.

Statistics clearly show this. World population has been growing at a very rapid pace. An estimated 275 million people inhabited this planet in 1000 AD. By the year 2000, the world's population had increased to 6.1 billion. In 2012, the number was 7.1 billion and an estimated one-quarter of the world's population (1.5–1.75 billion people) has no access to electricity. Population growth in recent years amounts to well over 80 million people/year.

South Africa is a good model for how PV might be used in the rest of Africa when distribution and money problems are solved. Writing about South Africa, Neville Williams

Sun above the Horizon: Meteoric Rise of the Solar Industry
Peter F. Varadi
Copyright © 2014 Peter F. Varadi
ISBN 978-981-4463-80-5 (Hardback), 978-981-4613-29-3 (Paperback), 978-981-4463-81-2 (eBook)
www.panstanford.com

observed:[214] "As a consultant to Solarex, then (1989) America's largest solar PV manufacturer, I learned that their biggest market for PV modules was South Africa. PV was widely used for telecommunication, wireless telephone, radio, and TV broadcast translators, railroad signaling, and lighting."

In the rest of Africa, a very large population lives under different conditions from South Africa. In sub-Saharan Africa, a large portion of the population has no electricity. In several countries less than 5% has electricity. In "better developed countries" 10% of the people have electricity. The power demand of those without electricity would be very low. They use kerosene lamps for lighting when it is dark but kerosene is both expensive and dangerous. Kerosene lamps can cause fire, destruction of property and injury, or even death. The PV industry was long aware of this situation but also recognized that missing distribution infrastructure of PV products and very low family incomes hampered their ability to sell products in these areas. We should not ignore that the world has changed during recent decades for certain technologies. Wireless communication, radio, TV, and telephone now reach even these remote areas of the Earth. Yes, telephone too. One can have good cell phone reception in remote places, such as the Darjeeling area in India. Wireless services rely on electricity. PV is excellent for this purpose. It is modular and thus available to small users. In the last 30 years, PV prices have decreased so much that, as described in Chapter 20 (Water Pumping), PV modules are not even worth stealing anymore.

In 30 years, millions of people have gained electricity provided by PV, a distributor infrastructure has developed (for example, PV-powered water pumping systems as was shown in Chapter 20) and some of the developing countries have even became leaders in the PV business? How did this happen?

Starting in the 1980s and the beginning of the 1990s, many efforts were made to help people in developing countries who lacked electricity to utilize PV. A broad range of organizations were involved. These included a fair number of NGOs such

---

[214]Williams N (2005) *Chasing the Sun*, New Society Publishers, Gabriola Island, BC, Canada.

as the organization that wanted to establish a birth control program in Nepal (see Chapter 12), as well like the Solar Electric Light Fund (SELF) that Neville Williams started in June 1990. SELF is still in operation and has succeeded with many programs. SELF has also set up PV manufacturing operations in India, Sri Lanka, and Vietnam.

Various government aid organizations also supported PV projects in developing countries. Help came in particular from the EU, as well as certain European countries and the USA, although there was little or no coordination of programs among donors.

A list of European Commission (EU) projects in this field is given in Annex. 4.

France primarily helped its former colonies. Some examples are given in Chapter 20.

Germany contributed through its bilateral aid program under the German Agency for Technical Cooperation (GTZ) (http://www.giz.de).

In the US both the US Department of Energy (DOE) and the US Agency for International Development (USAID) initiated several projects.

The UN operated through its United Nations Development Program (UNDP).

However, the ultimate effectiveness of NGOs, bilateral government aid agencies, and the UNDP owes much to the multiplier effect of the World Bank Group and the work and imagination of a small number of its staff. It was they who first guided renewable energy (RE) and energy efficiency (EE) projects through the bureaucratic maze of the WB (which had previously dealt only with huge conventional power projects) and through at least 30 governments of miscellaneous orientation.

The World Bank's effective and well-organized program deserves a special look. One key advantage of the WB approach was that it encouraged partnerships between the private sector and the public sector in developing countries to create and implement large programs. The WB also provided technical help and capacity building so that local agencies could implement PV and other programs. In addition, WB funds helped finance projects whether in conventional IBRD[215] loans for middle income countries or in low interest International Development Association (IDA) credits which are available to poorer countries with low per capita incomes. (The World Bank often combined its IBRD loans or IDA credits with subsidies in the form of grants from the Global Environmental Facility (GEF),[216] established at the UN in 1991.[217]) Finally, and of great importance, the World Bank realized early on that uniform quality PV standards were essential and helped to ensure that these were provided.

Historically, the WB had helped financing in very large energy projects, such as big hydropower or fossil fuel (oil or gas) development, as well as large power transmission projects. The interesting story of how it began to fund wind farms, village hydro or solar PV development is described for the first time below.

In 1989, the World Bank's Energy Sector Management Assistance Program (ESMAP) together with bilateral donors, including the USDOE, the Netherlands Ministry of Foreign Affairs Directorate General for Development Corporation (DGIS), and the UNDP initiated a study on the financing of "micro energy projects." The study was called "Financing Energy Services for Small Scale Energy Users" or FINESSE. Its outcome was presented by ESMAP's study team at an October 1991 meeting in Kuala Lumpur, Malaysia. The meeting was attended by representatives of the donor countries and various Asian countries. The study's attractive project proposals, its strong arguments for developing RE and EE, plus the interest shown by the Asian country representatives made for a very positive meeting.

---

[215]The WB comprises the International Bank for Reconstruction and Redevelopment (IBRD) and the International Development Association (IDA), both of whom lend only to governments. The World Bank Group also includes the International Finance Corporation (IFC) which lends to the private sector and has its own separate staff.

[216]The Global Environment Facility (GEF) unites 182 member governments— in partnership with international institutions, civil society organizations (CSOs), and the private sector—to address the global environmental issues.

[217]http://www.thegef.org/gef/whatisgef.

Paul Hassing, who then headed the Netherlands DGIS, consulted with Robert (Bud) Annan who led the USDOE Solar Energy Programs on the follow up to the FINESSE study. The two Dutch and US aid representatives decided to each offer grants to the WB of $1 million annually for 3 years to help cover costs for a WB pilot program to develop RE and EE projects in the power sector portfolio of Bank-assisted loans and credits in Asia.

This generous donor grant offer was a substantial incentive for WB action. In 1991, Dan Ritchie headed the Bank's Asia Technical Department (AST) responsible for operations in the Asia region. Ritchie readily accepted the Netherlands DGIS/ USDOE offer. He agreed that something should be done to develop the pipeline of potential projects identified by the FINESSE study. Ritchie was willing to house the pilot unit in AST, his department. He also pledged to identify and fund a full time staff member ASAP to manage this FINESSE follow-up unit. An ongoing World Bank reorganization at this time made available Ms. Loretta Schaeffer, a World Bank professional with more than a decade of experience working with Vice-Presidents and Department Directors on various critical issues. The Asia region Personnel Officer alerted Dan Ritchie that Schaeffer was strongly recommended by very high-level bank managers. Ritchie had little choice. He agreed to consider Schaeffer as the FINESSE Unit Manager. Next, Personnel phoned Schaeffer to offer her the job of managing this pilot unit in Asia region which would develop the World Bank's first RE and EE program.

The offer surprised Schaeffer. Her professional back-ground prior to joining the World Bank was a Masters in City & Regional Planning from MIT plus 10 years of city, regional, and national physical planning in Turkey and Libya. In the World Bank, she had subsequently worked closely with Department Directors and Vice-Presidents on a score of issues. However, none of Schaeffer's assignments concerned the power sector and Schaeffer was no power engineer. She was also unfamiliar with either RE or EE. Schaeffer promptly voiced her concerns about leading FINESSE to Dan Ritchie when she met with him. Ritchie wisely decided not to say that he too shared her objections. He glossed over her concerns, noting that the staff of the unit would have the requisite technical expertise in these areas. Instead, Ritchie stressed that the pilot FINESSE unit needed a strong WB manager to maneuver the program through the ins and outs of bank operations. Schaeffer's first task would be to "sell" RE and EE projects to bank staff and managers in the Bank's Asia region. Their support was essential to introduce RE and EE into the bank's program of assistance to the power sector in Asian countries. Reluctantly, Schaeffer said OK, but then asked for time to think it over.

As she thought about managing a pilot unit in a field she knew practically nothing about, Schaeffer identified yet another problem. Suppose she could overcome the double handicaps of being the lone woman (the bank's power engineers were all male) and of not being knowledgeable about the power sector. Another cause for worry was the pilot unit's name. The bank used acronyms (such as AST for the Asia Technical Department) to name the elements in its organizational structure. In fact, FINESSE itself was an acronym derived from *FIN*ancing *E*nergy *S*ervices for *S*mall-scale *E*nergy users. Well and good. But what if this FINESSE pilot unit were to be managed by a woman? To do her job well, Schaeffer would need to convert crusty World Bank power engineers to the cause of RE and EE development. Instead everyone could easily associate FINESSE with the name of a well-known shampoo! She could already imagine a line of macho jokes behind her back about Schaeffer and her "FINESSE" unit.... No way!

After much thought, Schaeffer decided to accept the job, only if she were allowed to change the name of the program. But change FINESSE to what? Well, the unit would be housed in AST. Why not add an AE (for Alternative Energy) and call the pilot program ASTAE, meaning the "Asia Alternative Energy" Unit? As an acronym, "ASTAE" would fit perfectly well with divisions in AST such as ASTEN (for the Environmental division) and ASTEG (for the energy division).

It worked.[218] She and Dan Ritchie made a deal. The "ASTAE" acronym was approved and the ASTAE pilot unit was created

---

[218]In 1991, the term "alternative" and not "sustainable" was used for Renewable and Energy Efficient technologies in the energy sector. Eventually, the name ASTAE was modified so that it currently stands for the Asia Sustainable and Alternative Energy Program.

in early 1992. Those associated with ESMAP's FINESSE study were understandably disappointed by the name change but they did recognize that ASTAE was orthodox bank nomenclature. One comical problem did crop up. Paul Hassing, the Dutch donor, pronounced ASTAE as though it were a drink like coffee! His pronunciation glitch was cleared away and Schaeffer settled down to recruit staff and manage ASTAE's budget of donor funds.

The Netherlands, through Paul Hassing, soon delivered the promised $1 million for ASTAE's budget. Next appeared what seemed to be a serious dilemma. It turned out that Bud Annan had no USDOE money he could actually earmark and transfer to ASTAE. The initial disappointment soon turned into a decided advantage. Instead of cash, Annan offered to provide ASTAE with USDOE-funded RE consultants. With this and other help he would fulfill his $1 million pledge.

What a piece of luck for Schaeffer! She did not have to search for or vet any RE experts. She got them immediately. Judy Siegel, who had worked on the FINESSE study as Annan's RE consultant, was delegated by Annan to take charge of recruiting RE experts to be paid by USDOE. One of the first experts recruited by Judy was Dr. Anil Cabraal, the most competent RE expert (especially in PV applications) whom Loretta could have ever found. Other RE experts soon joined the ASTAE team full time including Mac Cosgrove-Davies (bringing knowledge of biomass energy options) (Fig. 28.1) and Scott Piscitello (wind). The team developed good working relations with other parts of the World Bank Group such as the Environment Department's Global Environment unit, where Charles Feinstein worked on RE. A close ASTAE connection was built to the UN's GEF which would supplement ASTAE's RE programs with grants for project development and project implementation.

The Bank staff visited wind farms outside Oakland, biomass cogeneration (almond husks) in the Diamond Nut plant, solar thermal installations in the Mojave Desert, and the Siemens Solar PV factory (formerly Arco Solar) in California. The Bank's power engineers were understandably impressed by the potential of these unfamiliar technologies.[219] By the summer of 1992, Schaeffer even led a team of World Bank power engineers on an educational tour of US RE installations out West that Annan helped fund.

Figure 28.1 Mac Cosgrove-Davies, Anil Cabraal, and Loretta Schaeffer.

The goal of ASTAE's PV program was to provide electricity to people not able to get electrical energy. Initial solutions included, for example, a "small home PV system" to provide families with light, connect them to the world by radio, and maybe power a black and white TV for a few hours. PV could also power vaccine refrigerators for rural health clinics or it could connect school computers to the Internet. A minimum target was to eliminate the costly and dangerous kerosene lamps villagers were using by replacing them with a solar PV light

---

[219]After a few years, both Cabraal and Cosgrove-Davies were hired by the WB as regular staff. So were other experts such as Susan Bogach (energy economics) who first joined the ASTAE team as a consultant. World Bank recruitment of these experts as full time regular bank staff is another indication of the mainstreaming of RE and energy efficiency within the institution. In fact, Mac Cosgrove-Davies is currently Manager, Energy Sector in the Sustainable Development Department of the Bank's Latin America region.

source. ASTAE's PV program began in Indonesia, soon expanded to Sri Lanka and China, and has eventually spread to at least 25 more countries. A good description of the World Bank PV program is given in a paper by A. Cabraal.[220]

The ASTAE program was highly successful. From 1992 until Schaeffer's retirement in 1998, the total cost of projects or project components in the pipeline of Bank-supported RE (including PV) and EE projects in Asia grew from zero to $2 billion! ASTAE had successfully mainstreamed sustainable energy into WB financing. The program sparked initiatives not just in the WB Asia region but in other bank regional offices as well.[221]

But the road for PV was not easy. Cabraal's paper, mentioned above, describes many problems which had to be solved so projects could succeed. WB and IFC loans and supplementing GEF grants were given to the governments of recipient countries. Funds then had to pass through Finance Ministries, down through the public sector, private sector, and NGOs before reaching the final user. Eventually, this meant dealing with about 30 governments in Africa, Asia, and Latin America. Each posed a different challenge for project approval. Multiple experiments were needed in each country to identify the best financing system for customers who wanted to acquire the PV systems. The same holds true for the establishment of the infrastructure for PV distribution as well as the deployment and maintenance of PV systems.

The one key objective which was identical for every project and in every country is described by Cabraal: "Quality (*of PV systems*) must not be compromised."

The World Bank's program was very much affected by the quality problem of PV products especially the quality problem of the PV module, since at that time this was the most

---

[220]A summary of the World Bank Group's program in Photovoltaics is described: Cabraal A (2011) Photovoltaics in the World Bank Group Portfolio, in *Power for the World*, Pan Stanford Publishing Pte. Ltd., Singapore.

[221]ASTAE is still an important WB program. Any search engine, such as Bing, Google, and Yahoo will identify a large number of references related to ASTAE. However, no reader will find on the Internet the story you just read about the beginnings of ASTAE, nor the history of this strange acronym.

expensive part of the PV system. The reason was that once the Bank helped establish a PV program in a country, every PV manufacturer had an incentive to market its products there. Those manufacturers of PV modules who spent no money for testing and quality control obviously could sell modules more cheaply than those of a manufacturer whose modules were properly made and tested.

Customers, believing that all PV modules are equally good, would obviously buy the less expensive ones, the quality of which was often inferior, and would not last. The WB was unable to advise people as to which products were good and which were of bad quality. To sell these untested and maybe unreliable products in a poor developing country is morally and financially a horrible act. Selling a bad product to people in an affluent country is an aggravation for the customer, especially if they cannot return it. But selling a substandard solar-powered lantern to people who cannot return the product, and for whom the cash they paid amounts to a major household cost item, should really be considered a criminal act.

This was the subject of a 2002 Study of Photovoltaic Module Quality used in the Solar Home Systems (SHS) Market in Kenya. The study made confirmed that in that country and probably in other countries too, SHS systems were sold where the PV modules were "sub-standard." They proposed that a quality assurance program such as PV GAP should be required for all PV systems[222].

The only organization which realized this, and was trying to introduce tested and reliable PV products, for example modules, was the World Bank. In his 2011 paper,[223] Cabraal notes that the WB had already started to include quality require-ment enforcements in 1997 and that the Bank was a great supporter of PV GAP when it was created in 1996.

PV GAP's (see Chapter 27) purpose was to establish a system to enable customers to visibly distinguish a quality PV product from a product of unknown quality by creating a PV Quality

[222]Duke RD, Jacobson A, Kammen DM (2002) *Photovoltaic Module Quality in the Kenyan Solar Home Systems Market*, Elsevier: Energy Policy.

[223]A. Cabraal (2011) Photovoltaics in the World Bank Group Portfolio, in *Power for the World*, Pan Stanford Publishing Pte. Ltd., Singapore.

Mark to be displayed on the product and also to define the requirements for how a manufacturer could obtain the right to use the PV Quality Mark.

As mentioned in the previous Chapter 27: "The World Bank recommends that photovoltaic modules (or other PV components or systems) purchased for projects supported by the World Bank, which have the IECEE certification and the PV GAP quality Mark or Seal[224] should be preferred."

The ultimate question is: What was the result of all of these efforts of the World Bank and did they help the one-quarter of the world's population which had no electricity?

The answer is: Yes, they were successful. I can say this as I have an impartial opinion. I was neither the originator nor the executor of these projects. I did some work on some of them and am familiar with the twists and turns on how these projects were implemented.

In conclusion, the WB's PV projects have yielded two important results:

(1) They opened a new and important PV market in developing countries. The importance is that with rapidly declining module prices, developing countries represent a huge market. Until now this market was not developed by existing PV manufacturers since it was easier to sell PV modules in countries where the FiT system was introduced rather than to build a market in developing countries. But now some large PV module manufacturers realize that, while establishing a PV market in developing countries is slower and needs more effort, it is a stable and not subsidized market, compared to the FiT market, where local governments can change ground rules instantaneously.

(2) WB projects were very successful in Africa, in Latin America, and also in Asia. As summarized below, some of these WB Asian projects became leaders in the production and utilization of PV and contributed to its world-wide success.

---

[224]http://web.worldbank.org/WBSITE/EXTERNAL/TOPICS/EXTENERGY2/ EXTRENENERGYTK/0,contentMDK:21820033~pagePK:210058~piPK: 210062~theSitePK:5138247,00.html Also, http://www.worldbank.org/ retoolkit/.

India: In the early 1980s, India was the world's first country to create a Ministry of Non-Conventional Energy Resources. India was also one of the first countries where the WB (ASTAE) started its PV program in 1993. Although many difficulties were encountered, after a slow start it was somewhat successful. As time went by, more and more PV systems were installed for a variety of applications. A 2010 ESMAP[225] report states that: "India today finds itself on the path of becoming one of the leading nations in solar energy by taking steps toward implementing large MW scale solar power projects and is poised to position itself as one of the world's major solar producers as well as a manufacturing hub for solar power plants."

Good quality PV products are now made in India. Two manufacturers of PV modules, TATA BP Solar (renamed recently as Tata Power Solar Systems Ltd.) and Websol Energy Systems Ltd.,[226] have been qualified to display the PV GAP Quality Mark on their products. Also among the 21 suppliers of PV Lanterns tested according to PVRS 11A[227] for the WB, 10 PV Lanterns were found to be the best. Six among these 10 best were made in India by Indian lantern manufacturing companies.[228]

China: In 1995,[229] an estimated population of about 70 million people living in Tibet, Mongolia, and western China lacked access to electricity. The World Bank Group and the GEF offered financial help so these people could utilize electricity produced by wind and/or PV. After year-long negotiations

---

[225]Report on Barriers for Solar Power Development in India, South Asia Energy Unit, Sustainable Development Department, The World Bank, 2010, http://www.esmap.org/sites/esmap.org/files/The%20World%20Bank_ Barriers%20for%20Solar%20Power%20Development%20in%20India %20Report_FINAL.pdf.

[226]http://www.webelsolar.com/.

[227]PV GAP Recommended Specification, "Portable solar PV lanterns for everyday use, Design qualification and type approval."

[228]Testing Solar Lanterns against Global Technical Performance Specification (PVRS 11A). The World Bank Group Contract #7144845.

[229]Junfeng L (1995) *Commercialization of Solar PV Systems in China*, China Environmental Science Press, Beijing.

this project was approved with some modification. Preparation of the Chinese Rural Energy Development Project (REDP) was started in 1998 with a goal to boost the use of PV solar-home systems in off-grid areas in western China. This was achieved through (1) support to the Chinese PV industry to improve the quality of its PV modules and other components; (2) technical and management assistance to local installation companies; and (3) GEF subsidies to lower sales prices.

After 2008, Chinese PV module manufacturers produced about 50% of the entire world's PV modules. The detailed description of the success-story of the WB program in China is given in Chapter 37.

Chapter 29

Japan and Then Europe Take Over the Baton

The research for space application of solar cells began in the USA, but by 1973 when the terrestrial industry was started, the level of space-oriented solar cell research was the same in the USA and in Europe. Europe had excellent research teams working on projects for space solar cells but did not specifically work on how to adapt solar cells for terrestrial use and put this into mass production until about 1980.[230]

The need for terrestrial solar systems in the USA and in Europe was very different. I considered this issue many times and finally came to the conclusion that while in Europe the countries were well served with power lines and only small areas had no electricity, in the USA where the distance between the Atlantic and the Pacific coast, for example between New York and San Francisco is 4126 km (2564 miles), and also adding to it the vast area of Alaska and a coastline

---

[230]Palz W (May, 11–15, 1981) *Photovoltaic Outlook from European Community's Viewpoint*, IEEE Photovoltaic Conference, Houston, TX.

Sun above the Horizon: Meteoric Rise of the Solar Industry
Peter F. Varadi
Copyright © 2014 Peter F. Varadi
ISBN 978-981-4463-80-5 (Hardback), 978-981-4613-29-3 (Paperback), 978-981-4463-81-2 (eBook)
www.panstanford.com

of 142,640 km (88,633 miles),[231] there was a great need for electricity for communication, meteorology, navigation, etc., in areas where the electric grid was not even near. This provided great business opportunities for the terrestrial applications of solar cells. In addition, the Carter administration in the 1970s established a large budget to support the utilization of solar energy for terrestrial applications.

In Europe on the contrary, in the 1970s there was no natural need for solar electricity. The distance in Europe from Dublin to Volgograd (at that time Stalingrad) is "only" 3512 km (2182 miles). In the area between Dublin and Vienna (distance 1680 km) the infrastructure was built up in 2000 years and not much area was uninhabited and without electricity. In the area between Vienna and Stalingrad (distance 1832 km) there would have been a need to use solar energy to produce electricity, but at that time there was no interest to establish communication, meteorology, or any other conveniences for the people, which means the demand for solar electricity was much less in Europe than in America where huge territories were without the electric grid and electricity was needed. The USA was a ready market and PV was needed for it. This could be compared to the development of wind turbines, in which Europe was the engine of development, because of the application on its north-western coast, while in the USA there was no such clearly defined market and therefore, was trailing behind the European results.

With so much important research conducted in Europe on space-oriented solar cells it is hard to understand why only very little terrestrial PV manufacturing was established[232] specifically

---

[231]Figures were obtained in 1939 to 1940 with recording instrument on the largest-scale maps and charts then available. Shoreline of outer coast, offshore islands, sounds, bays, rivers, and creeks is included to head of tidewater or to point where tidal waters narrow to width of 100 ft. *Source:* Department of Commerce, National Oceanic and Atmospheric Administration, National Ocean Service.

[232]In the 1960s, Henry Durand of Philips in Paris provided PV for terrestrial application in Chile (information from W. Palz).

in France and Germany before 1977 or 1978.[233] The need to use PV to generate electricity for terrestrial purposes was fulfilled by import from the USA, Solar Power Corporation, Solarex, and later by Arco Solar. The need for solar electricity grew in Europe for weekend houses, second homes, navigation, and for exporting PV systems, for example to Africa and South America. The first to manufacture terrestrial PV products in Europe was when Solarex established a factory in Switzerland (1975) and set up joint ventures in France (France Photon in 1978), Italy, and the Netherlands.

The need to supply PV electricity to areas with close connections to Europe, in which areas there was no electricity, became the first urge to start producing PV cells and modules in Europe by European manufacturers. The other driving force was that people in Europe realized the need to curtail the use of oil, gas, and nuclear power for environmental reasons. The French company Photowatt was established in 1979.[234] In Germany, NUKEM's solar division was also started in 1979 (later it became ASE—Applied Solar Energy GmbH) but it really took off after Dr. Winfried Hoffmann took over its leadership in 1985. BP Solar established a solar business by exporting PV systems which were utilizing modules made in USA but in 1980 set up a cell manufacturing plant in Spain.

In the USA, the government's terrestrial solar program started under the National Science Foundation's RANN program in about 1972. It continued in 1975 under the Energy Research and Development Administration (ERDA) which was transferred to the DOE which began its operation on October 1, 1977 under the Carter administration. As mentioned before, DOE received a very large budget for renewable energy including solar. On the other hand this did not last very long, because the Reagan administration which started in 1981 drastically slashed the budget for PV.

---

[233]Starr MR and Palz W (1983) *Photovoltaic Power in Europe: An Assessment Study*, D. Reidel Publishing, Holland.

[234]http://www.photowatt.com/en/, As of March 2012, Photowatt is wholly owned by EDF ENR, a subsidiary of EDF Energies Nouvelles. Its corporate name is now EDF ENR PWT.

The EU was a late starter. In Europe the European Commission's support of renewable energy, including solar, started only in 1975, but it really became an important program in 1977 when Wolfgang Palz became the manager of the Commission's Development Program for Renewable Energies. His work included policy development and management of the Commission's budget (about $1 billion over the period of 1977–1997).

Palz's PV program obviously supported research to lower the cost of PV cells and modules, but many demonstration projects (Annex. 4) were also supported to open up new market areas, such as architecture (see Chapter 25), water pumping especially in Africa (see Chapter 20), a variety of systems around the Mediterranean, and in developing countries in Africa as well as in Asia around the Pacific Ocean. Wolfgang Palz initiated the "Power for the World" program in 1993[235] to address the huge market in the less developed countries (LDC).

Important outcome of projects initiated by Palz[236] indicated that without waiting for a breakthrough in research, present technology could yield very low prices for PV if it could be produced in mass production. The importance of these studies was that it became the fundamental basis a few years later to the introduction of the FiT system, which proved it right.

To achieve high quality terrestrial PV product was also a goal. In this respect, the European Commission's Joint Research Center (ISPRA) did a very important job. This project established connection and continuation with the US quality programs started with the JPL work. Palz was responsible for establishing the European Solar Energy Conference, the first of which was held in Luxemburg in 1977. In the USA, the IEEE Photovoltaic Specialist Conferences were started several years before but were mostly concerned with space-oriented research. The first IEEE conference where terrestrial applications were featured was held in Washington, DC, in 1978. The subjects of the USA IEEE conferences stayed mostly in the research area, the European PV solar conference

---

[235]Palz W (20 August 1993) *Power for the World*, ISES Conference, Budapest.
[236]EU APAS contract CT 94-0008 "MUSIC FM."

became mostly oriented for terrestrial applications. Especially, when Palz in 1983 asked Peter Helm to take over the organization of the European PV conferences. Helm enlarged the exhibit area giving opportunity for the PV industry to show its progress.

It was also notable that the fledgling European PV industry encouraged by Palz and organized by Joachim Benemann started the European Photovoltaic Industry Association (EPIA), which became much enlarged and politically very important, when in 1986 John Bonda became its Secretary General. He was able to promote PV solar energy as a well-known technology to decision-makers globally. Bonda's untimely death in 1999 was a big setback for EPIA in which he was able to raise the membership considerably.

In the decade of 1980–1990, PV was not anymore a rarity. It was already used in many countries in the world. At the beginning, manufacturers such as Solarex, Solar Power, and Arco Solar established their sales offices or production facilities at various parts of the world or established appointed sales organizations. The maturing of the PV industry resulted that PV manufacturers as before had sales offices and authorized distributors to sell the company's products, but the change was that independent companies were started. They did not manufacture PV; they were buying PV products from many sources and were selling them or even installing systems. This took place in the USA as well as in Europe and also in other parts of the world. This happened because people realized that PV was a business, but also that in countries where large installations were started, for example by the telephone companies, they insisted that a local company should be responsible for the systems and not a manufacturer in a distant land that if the system fails to function there should be a local company responsible for it. Therefore, the bidding had to be done by the local company but that company could select the vendor with whom it wanted to team up.

The appearance of an independent distribution network was an important step in the development of PV manufacturing, marketing, and sales infrastructure, which exist for other developed industries. Obviously the result was that the distribution network expanded and the selling of PV became more competitive. By that time not only American, but also European and even Japanese companies could be selected. This happened as by 2007 Europe became the largest PV cell and module producer in the World (see Annex. 8).[237] Some of the major European PV cell and module manufacturers are as follows:

Isofoton[238] of Spain started in 1981. In 2011 produced solar modules totaling of 230 MW. At present the company is in financial difficulties.

SolarWorld AG of Germany was founded in 1988 by Frank Asbeck and went public on August 11, 1999. SolarWorld built some of their PV factories but also acquired some existing ones. It acquired Shell's c-Si PV manufacturing facilities in Europe as well as in America which included also Si wafer manufacturing. By 2011, the total sales for SolarWorld were well over $1 billion. However, SolarWorld recently announced financial difficulties.

Another company Q Cells AG of Germany started in 1999 to produce solar cells in a little village of Thalheim in the eastern part of Germany and was founded by Reiner Lamoine, Holger Feist, Paul Grunow, and Anton Milner who became the CEO of the company. Already the first fiscal year of Q Cells was profitable. In October 2005, Q Cells went public. In 2007, 8 years after it was founded it became the largest solar cell manufacturer in the world employing 1700 people. The total sales were over $1 billion and had close to $200 million profits. The downturn started in 2008. The reason for this is discussed in many publications and could be the subject of many Business School home works. It winded up in bankruptcy and the Korean business conglomerate Hanwha Group bought it. The appropriately named Hanwha Q Cells with the purchase of the bankrupt solar cell manufacturer Q Cells now claims ownership of a total of 2.3 GW of manufacturing capacity across Germany, Malaysia, and China.

---

[237]Mints P (5 October 2011) *Reality Check: The Changing World of PV Manufacturing*, Renewable Energy World.
[238]http://www.isofoton.com/en/index-en.

Renewable Energy Corporation (REC)[239] Norway was a poly-silicon manufacturer but diversified to manufacture solar cells and modules. In 2012, it produced 780 MW solar modules.

In the USA and in Europe the solar cell research and its utilization started in 1953 and for 20 years it was only for space use. The research, development, and utilization of solar cells for terrestrial purposes started only in 1973. In Japan the R&D and utilization of PV started for terrestrial purposes and not for space use. Japan became the first to use solar cells for terrestrial applications when in 1961 NEC installed the first PV-powered lighthouse.

Why the development of PV in Japan was different from the USA and from Europe can be explained considering the market requirements. The Japanese story started that, until the San Francisco peace treaty of September 8, 1951 Japan was prohibited developing aircrafts. From January, 1953, Professor Hideo Itokawa stayed in the United States for 6 months and found many books on space exploration. Returning to Japan in 1954 Hideo Itokawa, formed a research group called Avionics and Supersonic Aerodynamics (AVSA) aiming at rocket development. The development of rockets and space-oriented programs started in Japan only in October 1969 when the National Space Development Agency of Japan (NASDA) was founded, 11 years after NASA was formed in 1958. So there was no market for solar cells used for space research.

Japan developed a strong semiconductor industry making transistors and integrated circuits, and could easily make solar cells. Japan being an island nation with naval connections between its many islands had a great need of PV providing electricity for navigational aids, such as lighthouses and that is for which the terrestrial PV business started. NEC started to provide solar cells for this purpose but after installing a few, the manufacturing of solar cells for this market was taken over by Sharp Corporation.

The next incentive of going into the solar cell market was for consumer products. Japan's semiconductor industry became the leader to fabricate integrated circuits for digital watches

---

[239]http://www.recgroup.com/.

and calculators. The Japanese watch and handheld calculator manufacturing took over the world leadership for these products. As described in Chapter 17, digital watches needed PV recharging in the beginning, but when the digital watch circuits and displays used only minimum amount of electricity, this business faded but was replaced by the need of solar electricity for the handheld calculators. At the beginning Solarex and Sharp supplied c-Si type "micro generators" to this business, but when Sanyo in Japan discovered that a-Si is better suited for this application Sharp also started to manufacture a-Si "micro generators" and Sanyo and Sharp took over the handheld calculator PV market.

This was the point when Sharp and Sanyo started to export PV products. Kyocera joined these two companies and started to get into the PV business in 1975 and shipped the first products in 1979.

In the 1980s and in the 1990s the world PV market increased a substantial 15%/year but the US PV manufacturers suffered primarily that the oil companies which provided the money for adding production capacity pulled out of the PV business, and they could only slightly increase their production. In Europe the production increased but was also little supported and the Japanese PV manufacturers, Sharp, Sanyo (recently sold under Panasonic label), Kyocera, Mitsubishi Electric, and Showa Shell solar (now renamed Solar Frontier—manufacturing the thin film CIS solar modules (see Chapter 34), all supported by their parent companies were able to produce substantially more for domestic use as well as for export.

In spite that in the USA the PV production increased Japan took over the lead in 1999 but a few years later in 2007 Europe took over the baton from Japan.

Chapter 30

Three Axioms That Will Shape the Future of PV

When Wolfgang Palz in 1977 became the head of the EU Renewable Energy program, the European terrestrial PV industry practically did not exist. About $1 billion EU money spent under the direction of Palz between 1977 and 1997 for research, development projects, and infrastructure resulted that in 1997 a European terrestrial PV industry existed, which manufactured 29.1 MWp (Annex. 8), close to 60% of what the dominant USA PV industry produced in the same year.

It was unfortunate that Palz's extremely well-structured program was basically dismantled, when Ms. Edith Cresson, who so far had spent the shortest time in office (May 1991 until April 1992) for any French prime minister of the Fifth Republic, because in a speech she compared the Japanese to "yellow ants trying to take over the world,"[240] had to resign and subsequently in January 1995 became the EU Commissioner for Science and Education to which area the Renewable Energy program also belonged. One of her first activities was to approve and authorize that one third of the PV proposals

[240]http://en.wikipedia.org/wiki/%C3%89dith_Cresson.

Sun above the Horizon: Meteoric Rise of the Solar Industry
Peter F. Varadi
Copyright © 2014 Peter F. Varadi
ISBN 978-981-4463-80-5 (Hardback), 978-981-4613-29-3 (Paperback), 978-981-4463-81-2 (eBook)
www.panstanford.com

which under strictly followed rules were just evaluated and rated for approval, should be arbitrarily downgraded to a level of not acceptable to be funded. These were proposals totaling more than 30 million euros. The "saved" money assumingly was transferred to nuclear research. This falsification of the records was done by a highly incompetent person, who neglected to remove from the computer the original ratings. So both the original and the altered ratings were still on it. The result was that a journalist cleverly obtained a copy of both ratings and the story was aired on a 20-minute segment of the ARD[241] TV station's "Monitor" news program in 1995. The EU Parliament investigated this issue. Ultimately Commissioner Cresson resigned from the Commission. *"Subsequent to a fraud inquiry the European Commission said that Cresson in her capacity as the Research Commissioner 'failed to act in response to known, serious and continuing irregularities over several years'. Cresson was found guilty of not reporting failures in a youth training programme from which vast sums went missing."* But none of her criticism said she had gained personally.[242]

The PV program was continued by the EU, but lacked Palz's vision and energy. The following three of Palz's axioms survived everything and wound up in the political arena which shaped the future of PV.

A study which was started under the Palz (Fig. 30.1) regime was carried out under the EU contract CT 94-0008 "MUSIC FM[243]" to find out what PV module prices could be achieved in large-scale mass production in a facility producing 500 MWp/year utilizing then existing technology. The interesting finding was that mass production will result in a drastic price reduction of PV modules. At that production level prices of $1/Wp or below would be feasible. The study in its "Conclusion" stated: "Indeed the study shows that it is market size which is one of

---

[241]ARD is one of the world's largest broadcast organizations. The public network is the market leader of news programs in Germany.

[242]BBC News, March 16, 1999 Published at 12:12 GMT, http://news.bbc.co.uk/2/hi/europe/255053.stm.

[243]Bruton TM, et al. (1997) *A Study of the Manufacture at 500 MWp p.a. of Crystalline Silicon Photovoltaic Modules*, 14th European PVSE Conference, Barcelona.

the most significant factors in achieving cost reduction." This indicated that not even considering a breakthrough in research, just utilizing the 1995 technology, extremely low PV module prices could be achieved if PV would be produced in a facility of 500 MWp/year or bigger. This study was presented in 1997 in a year in which the world's total PV production was only 114 MWp (Annex. 8). The prediction was proved to be correct in 2010 when more than one production facility was manufacturing 500 MWp/year.

Figure 30.1 Prof. Dr. Wolfgang Palz—2001.

At a time when the general belief was that research will bring prices down and mass production will be achieved by building central PV power stations, Palz in 1981 proposed[244] that because of the decentralized nature, PV systems at any location could be set up and could be and will be grid connected and what is needed is that a favorable "feed in" rate should be established with the utilities. From these predictions one can derive two more axioms:

(1) The first axiom can be established from this that, PV mass production utilizing even the then (1995) existing technology could achieve PV prices approaching utility acceptance.

---

[244]Palz W (11–15 May 1981) IEEE Photovoltaic Specialist Conference, Houston, TX.

(2) The second axiom is that because of the decentralized nature, PV systems at any location could be set up and could be and will be grid connected. (The "grid connected" PV system was described in Chapter 12.)

(3) The third axiom is that despite that utilities are and will be reluctant to do, a favorable "feed in" rate should be somehow established.

Obviously these axioms indicated that not technology, but rather political solution was needed to cut the Gordian knot of inexpensive PV. The political solution was interestingly achieved not at the sunniest part of the world but in Europe and Germany.

During the 1992 European Photovoltaic conference in Montreux, Switzerland, my friend Wolfgang Palz got hold of me and said:

"I want you to get acquainted with Hermann Scheer (Fig. 30.2). He is a member of the German Parliament and he is the greatest promoter of solar energy in the important political arena."

Figure 30.2 Hermann Scheer (1944–2010).[245]

I did not want to offend Wolfgang by telling him that I have very little interest to get acquainted with politicians. We sold Solarex in 1983, but in 1992 I was still consulting for

---

[245]Photograph from 2008.

the company, making studies to explore new markets for PV. When we developed Solarex, I had very little interest to get involved with government contracts, by 1992 my interest shifted from business to the more basic issues affecting the future success of PV in which I still did not believe that government subsidies were the solution. However, not to offend Wolfgang, I said:

"I would be delighted."

At some point, Wolfgang managed to get hold of both of us and we got introduced. Wolfgang must have already told Hermann about my background because the first thing that he told me was that some time ago he visited Budapest as a member of the German water polo team playing against the Hungarian water polo team. In Hungary water polo is like baseball in USA, a national sport. The Hungarian water polo team usually wins the Olympic gold medal, and in the year they do not win, when they arrive back to Budapest, they probably need police escort from the airport to a secret place. In my hometown in Hungary (Szeged), if the local water polo team lost, the referee certainly needed police escort from the pool to a place well outside the city. When I learned from Hermann that he played water polo on that level, I immediately forgave him that he was a politician and started to pay more attention to what he said.

He immediately started to describe the global connection for PV and politics. He opened a new world for me with the perspectives, connections, and possibilities. His logic was second to none and very convincing. As I learned later, that was the time when he wrote and published his book *Sonnen Strategie—Politik ohne Alternative* (Solar Strategy—Politics Without Alternative). That book and the many other books he wrote contain incredible amount of information, but more important, they contain outlines of strategy and connections between issues which are unique but extremely important and timely.

In Germany, ultimately a few years later, after a great amount of work in politics, Hermann Scheer and Hans Josef Fell, both members of the German parliament—Fell for the

"Green Party" and Scheer for the "Social Democrats"—opened the way for large decentralized grid-connected PV systems that could be financed privately, as a result of which the mass production of PV was achieved in a short period of time.

It is interesting to follow this winding road to the solution.

Chapter 31

Does PV Need Government Subsidy or Can It Be Financed Privately?

When the terrestrial PV business was started most experts believed that this research would yield a breakthrough which would result in inexpensive PV products. People starting the PV industry in 1973 believed that increased production volume would bring prices down and not research. In 1989, 16 years later, the belief of the people who started the terrestrial PV industry was supported by the fact that the yearly global PV production as reported by Paul Maycock[246] was 33.6 MWp and the price was down to $6 Wp. As described in T. M. Bruton's[247] study (see Chapter 30), when a PV factory's production level reaches 500 MWp, prices could go under $1/Wp. We considered this as the first axiom. Obviously this would require a considerably large market and proper financing for PV modules. The question was how this large market and financing to establish mass production level of solar cells and modules to become competitive with other means of electricity production could be achieved in a short period of time.

---

[246]Maycock P (ed) (February, 2002) *From PV News*, yearly editions.

[247]Bruton TM, et al. (1997) *A Study of the Manufacture at 500 MWp p.a. of Crystalline Silicon Photovoltaic Modules*, 14th European PVSE Conference, Barcelona, p. 11.

Sun above the Horizon: Meteoric Rise of the Solar Industry
Peter F. Varadi
Copyright © 2014 Peter F. Varadi
ISBN 978-981-4463-80-5 (Hardback), 978-981-4613-29-3 (Paperback), 978-981-4463-81-2 (eBook)
www.panstanford.com

It was obvious by that time that there was a big need for using PV, but that market growth based on those needs would result "only" in a yearly 12–20 percent steady increase in production (which would be a dream for any industry), but would not achieve a quantum jump, where mass production would happen in a span of only a few years.

The 1972 report, *Solar Energy as a National Energy Resource* (described in Chapter 3), envisioned that the utilization of PV should be for "buildings, central ground stations, and central space stations." By 1989, it was obvious that the idea of a "central space stations" was not feasible. The idea of a "central ground station" to achieve mass production of PV was, using an American expression, a "catch 22." It certainly would result in mass production, but to set up a central ground station one would already need mass production to yield inexpensive, close to grid parity PV systems. On the other hand, to utilize PV for buildings could be a possibility to wind up with mass production. This could be one of the reasons that the German, Japanese, and US Governments initiated solar roof programs. The US Government started with demonstration projects, and these demo houses were all successful. Interestingly the person who first picked up the idea in 1986, to try to mass produce PV to install the modules on the roofs of houses, was a young Swiss entrepreneur, Markus Real (Fig. 31.1).

Figure 31.1 Dr. Markus Real.[248]

---

[248]Photo: Ruedi Staub, Zürich.

He called his plan "Project Megawatt," because he planned that he would get 333 customers to install 3 kW on the roof of each of the houses which in total would amount to the 1 MW. In 1988, with clever promotion and support from the most important Swiss newspapers, his company got the 333 customers to realize the project without any governmental support or subsidies. All of his customers were wealthy and able to buy the expensive PV systems. And with no difficulty at all, the 333 PV systems were installed and connected to the Swiss grid within a year, which at that time was a novelty. It was also Real's achievement that some of his customers were able to sell the solar-generated electricity at the same price the utility was charging them. At that time "Net metering" also reflected a breakthrough in the utility world in Europe. The 333 roof-mounted PV indicated that the generation of electricity in decentralized PV systems and connecting them to the utility grid was very much feasible, but to extend such promotion on a large scale would need financial support until the mass production brought the price of the PV systems down.

Governments tried to inject money to achieve large-scale production. In the USA and Europe and in Japan, solar roof programs were started. In Germany a 1000-roof program was initiated in September 1990, which was followed by a 100,000-roof program. The 100,000-roof program was budgeted with about $700 million. It was started in 1999 and ended successfully in 2003. During those 4 years, 346 MWp were installed.[249] For the 100,000 roof program a 0% 10 year loan was provided by the German state-owned bank, Kreditanstalt für Wiederaufbau (KfW).

In Japan as a result of the leadership of Professor Yoshihiro Hamakawa, the Ministry of Energy, Trade, and Industry (METI) started in 1974 the "Sunshine project" and in 1993 embarked on the "New Sunshine Project." In Japan the "Sunshine program" in 20 years spent $5 billion of the taxpayers' money.

---

[249]Stryi-Hipp G (19 October 2004) *Experience with the German Performance-based Incentive Program—Solarpower 2004*, San Francisco, October 19, 2004.

In June 1997 in a speech at the United Nations, President Bill Clinton proposed a Million Solar roof program.[250] This could have resulted in the mass production of PV, but ultimately the US Federal Government program invested only a total of $16 million in this program,[251] which would have been only a $16 subsidy per roof, but even that money was not spent on roofs, but was spent mostly on grants to some 90 cities.

In retrospect, one can find several reasons why these "roof programs" did not yield the mass production of PV. This can be best understood from an old Hungarian story. The Bishop came to visit a very small town and the priest waited for him at the train station. They start to walk toward the nearby small church and the Bishop remarked, "You know I expected to receive a bigger reception." The priest answered, "There are 101 reasons we could not do it." "What are these reasons?" the Bishop asked. The priest answered, "Well one of them is that we do not have money and..." The Bishop interrupted him. "You do not have to tell me the other reasons."

Actually these government-sponsored "roof programs" were insufficient to bring up PV production to a level that Bruton had predicted it was needed to establish mass production to bring down PV prices to the $1.00/watt W level. The other fault in these programs was lack of sustainability for PV production. For example, the German program assured 4 years of 100 MW/year production, but what would happen after 4 years? Who would invest money in a production facility with that future?

It became clear that governments are not going to be able to invest the amount of money needed to achieve mass production of terrestrial PV modules. If PV would have expected to be used by the military, like nuclear or the semiconductors, the governments would have not hesitated to invest money. But PV is a source of electricity in which, in spite of nuclear electricity is being on shaky grounds was hard to convince taxpayers to invest lots of money when oil and gas were abundant. The only argument for PV was that oil and gas will run

---

[250]The program was recommended after a meeting with the US PV industry by Dr. Allan Hoffman who was in 1997 the Acting Deputy Assistant Secretary for the Office of Utility Technology of the US DOE.

[251]Johnstone B (2011) *Switching to Solar*, Prometheus Books, Armherst, NY.

out at some point, and the present danger is that the quantity of oil and gas being used is ruining the planet we are living on and therefore, they should be replaced with energy sources which will not harm us. This was a good argument, but one could wait to spend money to replace them. It is true that oil companies received large government subsidies, but in those days there was no consensus to provide large subsidies for the small PV industry. At that time, a program to raise taxes for RE was dead at arrival in any countries' legislation.

Obviously one needed to look for financing PV outside of government subsidies, in private capital. In his book published in 1993, Hermann Scheer emphasized the importance of investigating the non-government financing of renewable energy projects. Having had experience with the 1000- and the 100,000-PV roof programs, Scheer realized that private financing and not government subsidies was one of the most important issues to be solved for PV. The need of private financing for PV became evident in Europe as well as in the USA in the early 1990s. This is evident, for example, in that the EU's renewable energy program during the Palz regime issued contracts to study the possibilities to finance renewable energy. One of the workshops was held on June 7, 1995, in the UK. In the USA about the same time, on May 15–16, 1995, with the International Energy Agency (IEA) and the World Bank participation, Allan Hoffman of the Department of Energy (DOE) organized a workshop on "Financing the Development and Deployment of Renewable Energy Technologies."[252]

Scheer and supporters founded in 1988 EUROSOLAR in Bonn, which became the most important independently operating association for renewable energy in Europe. Scheer, as mentioned before, was already aware of the importance of private financing of PV and decided that EUROSOLAR should organize conferences to bring together leaders of the PV industry, financial institutions, and government representatives. Scheer asked the author of this book to help him organize this conference for EUROSOLAR.

---

[252]The Workshop Proceedings was produced by the Oak Ridge Institute for Science and Education (ORISE) for the US DOE under contract DE-AC05-760R00033.

The first conference[253] on "Financing Renewable Energy"[254] was held in Bonn September 1–3, 1997. For me, it is interesting to read again my opening remarks at that conference, which was held a quarter of a century after the terrestrial PV industry was born and only a little more than 15 years ago when this was written. I said:

"Many meetings are organized to discuss the technology, research, development, applications, cost effectiveness and the future of PV. But only lately it became evident that the widespread utilization of PV is mostly dependent on financing and much less on technology or cost. This meeting is finally a departure from the usual PV meetings. This will address exclusively the issues related to financing of PV and not to explain all the available technology and how it works. When somebody asks a banker to finance an automobile one does not have to start explaining how the machine works and what its efficiency and expected life is. After a quarter of a century PV is like an automobile: an established product. It proved its usefulness, it is needed and it has a long life. Financing anything means simply that one invests money and expects to recover it with profit."

The Feed-in Tariff program introduced a few years later proved me right.

The subject of the 3-day conference was all related to financing. The purpose was to actively involve banks, financial institutions, insurance companies, non-governmental organizations, and also private investors in the subject by giving them the opportunity to tell their views of investing in PV. One of the papers presented was Mike Eckhart's, who developed financing

---

[253]The presentations and the results of the two international conferences mentioned here on financing renewables held by EUROSOLAR in 1997 and 1998 were published. The book presents banking concepts and applications for financing renewables, public frameworks for renewable energy market introduction, financing renewables in developing countries (http://www.eurosolar.de/de/index.php?option=com_conten t&task=view&id=497&Itemid=20).

[254]On October 11–15, 1995 a conference on "Financing Solar Energy in the Developing Countries" was held at Pocantico, NY.

for solar energy under the Solar Bank Initiative in South Africa and India.[255]

The conference was so successful that a second conference was held a year later on November 16–18, 1998, in Bonn. The author and Loretta Schaeffer of the World Bank helped Hermann Scheer to organize the meeting. In the USA, there was also a follow-up to the 1995 meeting, organized by Dr. Hoffman. The US DOE National Renewable Energy Laboratory issued in September 1998 a booklet entitled *The Borrower's Guide to Financing Solar Energy Systems*.

The positive side, especially of the European meetings, was that a stage was given to people dealing with money to consider financing PV. They represented a variety of organizations and interestingly it became evident that because they understood that PV was a clean and renewable energy source, money would be easily available if there would be proper projects to be financed. Obviously the invested money was expected to be recovered with some profit.

The very positive result was that it became evident that private money without the government subsidy would be available for large number and/or size PV systems. What would be needed is long-term security that the invested money and some profit would be assured. This attitude proved to be very useful a few years later. Interestingly, the financial solution that PV without government support would achieve mass production was already in motion.

---

[255]http://www.citigroup.com/citi/about/leaders/mike-eckhart-bio.html.

Chapter 32

Grass Roots—Common Sense

According to the second axiom, because of the decentralized nature, PV systems could become at any location "grid connected."

It was difficult to make this a reality. It is clear that utilities were in the business to manufacture electricity and sell it. They were at that time not in the business to buy electricity especially not from Independent Power Producers (IPP) and resell it. As discussed in Chapter 10, the electric utilities' aversion against the distribution-level PV systems may be the result of their "corporate culture." The utilities seemed not to understand the changing times; this needed a political solution.

Interestingly, the string of events to create the political solution started in the USA during the Carter years when the "Public Utility Regulatory Policies Act" (PURPA) was passed in November 1978 by the United States Congress as part of the National Energy Act. This law contained a provision little noticed at that time that resulted in huge unintended consequences. This paragraph states that utilities were required to purchase the power generated by private companies that produced electricity.[256] According to this law, utilities must pay

---

[256]Johnstone B. *Switching to Solar*, Prometheus Books, Amherst, NY.

Sun above the Horizon: Meteoric Rise of the Solar Industry
Peter F. Varadi
Copyright © 2014 Peter F. Varadi
ISBN 978-981-4463-80-5 (Hardback), 978-981-4613-29-3 (Paperback), 978-981-4463-81-2 (eBook)
www.panstanford.com

an equitable rate, determined by each US state's Public Utility Commission. Generally, this could be the equivalent to the dollar amount the utility would have had to spend to generate (but not distribute) the amount of power it receives.

The first result of this law was the introduction of "net metering." Net metering requires only one meter which can turn in both directions. The meter turns in one direction when the customer uses electricity and turns in the other direction when the customer provides electricity to the utility. Net metering also originated in the United States, where small wind turbines and solar panels were connected to the electrical grid, and consumers wanted to be able to use the electricity generated at a different time or date than when it was generated. Minnesota is commonly cited as passing the first net metering law in 1983, and allowed anyone generating less than 40 kW (which is now extended to 1000 kW) to either roll over any kilowatt credit to the next month, or be paid by the utility for the excess.[257]

The utilities resisted in the USA to accept the "net metering," the same way when AT&T resisted the connection of telephone sets to the telephone line when it was not made by its subsidiary Western Electric. AT&T had to give up on this at some point and the electric utilities had ultimately to do the same, when it was defined what electrical characteristics the connection between the PV system and the grid, the so-called inverter, should have to be compatible with the grid.

In 1991, the "Feed-in Law" in Germany was established and was modeled according to PURPA. This law promoted primarily the utilization of wind power and not PV. The Feed-in Law required that the German electric utilities had to buy electricity generated from "renewable electricity" generators at a higher fixed rate than the utility was charging customers and letting the electric utility recover the expense by adding pro rata to the electric bill of the customer as they would do with other "fuel charges."

It is easy to understand the difference between "net metering" and Feed-in Tariff "FiT", which was the result of the

---

[257]http://en.wikipedia.org/wiki/Net_metering

Feed-in Law. As described above in the "net metering" system only one electric meter is used. Accordingly the customer depending on its utilization of the electricity either pays for the electricity or receives money from the electric utility. In the "FiT" system two meters are used. One meter measures the electricity used by the customer at the rate the utility is charging for the electricity. The other meter measures the electricity produced by the customer's renewable electricity generator and the utility is paying the customer at the rate prescribed by the law.

In Germany, four major electricity-generating companies existed and still exist (E-ON, RWE, EnBW, and Vattenfall); besides them a large number of local electric utilities exist. These are corporations in which these small utilities' stocks are owned by the municipalities to which they supply the electricity. One of these utilities was Stadtwerke Aachen AG (STAWAG) serving the city of Aachen. Aachen's history goes back 5000 years and its citizens made history again in 1994.

Aachen's history goes back to the "stone ages" to about 3000 BC. It was much later a Roman spa, and later King Charlemagne resided there and made Aachen the center of his empire in about 768 AD. Aachen remained the coronation place of the German kings for about 600 years. On May 8, 1991, as described in Chapter 25, when the BIPV building integrated façade became a reality, the very first of such structure was inaugurated on the southern façade of the electric utility's (STAWAG) headquarters' building in the city of Aachen.

In Aachen, in 1994, some of the "green activists" got together and developed a FiT scheme, which was known as the "Aachen model." This "Aachen model" showed the possibility to achieve the result that those financial conferences suggested, in other words that large amount of private money for PV systems can be attracted if the stability of a reasonable return would be guaranteed. Aachen, with the steady work of locals, Harry Lehmann (who was also an official of EUROSOLAR) and Wolf von Fabeck, were the first to develop a useful version of the FiT. von Fabeck calculated how much the utility should pay for electricity produced by PV providing a 20-year contract which would return the investment with a reasonable profit to those who financed to set up the PV system. (One should notice that the predicted 20-year life of the PV module was based that the module was manufactured to comply with the JPL quality system, which predicted to achieve it.) He also calculated that in the town of Aachen because of the FiT system the average increase of the electric bill for a family would be about $1.50/month. von Fabeck's calculation was reviewed by a commission set up by the city of Aachen and they found it to be exactly right. The citizens of Aachen all approved it, because after the April 26, 1986, atomic catastrophe in Chernobyl, they all wanted to replace nuclear with RE, similarly like in the USA the Sacramento Municipal Utility District (SMUD) utility's rate-payers voted in the 1980s to shut down the SMUD-owned nuclear power plant and build the USA's first utility-scale solar array (Chapter 10). The Aachen model was approved and published on November 29, 1994.

In Aachen, as a result of the decision of the people living in the area served by their utility and their willingness to pay the small surplus on their electric bills, a favorable "FiT" was established. The Aachen model showed that PV was a good investment. If somebody invests and establishes a PV facility, the cost would be recovered during the 20-year contract and the invested money would yield a decent yearly interest. On top of that, the people in the city were able to decide themselves as no government subsidy was needed.

The next years indicated that the Aachen model was successful. During the next 6 years, the "Aachen model" was widely adopted in many towns in Germany which had their own utility.

At the 1997 EUROSOLAR Conference, "Financing Renewable Energies," Andreas Wagner, who worked at EUROSOLAR, presented a paper: "Renewable Energy Feed-in Tariffs in Europe: An Overview." The paper concluded based on the 6 years' result since the 1991 German Feed-in Law was passed that favorable tariff schemes caused wind power to have a rapidly expanding market and proposed the introduction of a common European legislation granting minimum tariffs for renewable energies which were fed into the electric grid. This would have been the ideal situation, but in the complex quilt of the countries forming the EU, the reality was that at least one country had to start the action, and if it would be successful, the others would follow.

Anne Kreutzmann, an Aachen-based PV activist, predicted, "A federal law could be written to make rate based incentives sufficient to encourage the utilization of PV for all of Germany."[258] Her prediction was fulfilled when Hermann Scheer and Hans-Josef Fell (Fig. 32.1), both members of the German Parliament, were able to achieve that on February 25, 2000 the Renewable Energy Act [Erneuerbare-Energien-Gesetz (EEG)] became law. Based on what was learned from the 1991 law, changes were introduced to the 2000 law. This law made PV systems very attractive investments and resulted in mass production of PV cells and modules. With this Germany became the first country in the EU and in the world to set up a system to attract large amounts of private money to create a financially sustainable system without government subsidy to develop the mass production of PV.

Figure 32.1 Hans-Josef Fell.

The Sun for PV rose above the horizon. The stage was set for Act 3.

---

[258]Johnstone B (2011) *Switching to Solar*, Prometheus Publisher, Amherst, NY, p. 176.

ACT 3

TOWARDS HIGH NOON

2000–2013

Chapter 33

The Winding Road to the Feed-in Tariff

As the "Aachen model," which was adopted by many other German cities, indicates that Germany, which is not known to be the sunniest country in the world, was interestingly among the few countries where the population was trying to find a solution to introduce electricity generation utilizing solar energy. Perhaps it started in Germany because it was one of the most industrialized countries which had no connection to and was distant from the sources of oil and gas and contrary to the USA and some other countries; there is and was no large oil or gas lobby and government leaders had no connection to the oil and gas business. Perhaps also because by that time, all but eight of the black coal mines which provided its industrial power in the nineteenth and twentieth centuries were shut down and the country had to spend about $3.3 billion a year until the last mine was closed. An additional reason could be that while the United States became a country of throwaway things, such as batteries, Europe came out of the war very poor and thrifty. They started to use things that were meant not to be thrown away, for example, rechargeable batteries, which were reusable.

Sun above the Horizon: Meteoric Rise of the Solar Industry
Peter F. Varadi
Copyright © 2014 Peter F. Varadi
ISBN 978-981-4463-80-5 (Hardback), 978-981-4613-29-3 (Paperback), 978-981-4463-81-2 (eBook)
www.panstanford.com

Or perhaps because the war was a catastrophe for Germany and they did not want to experience another one like the Chernobyl catastrophic nuclear accident that occurred in 1986, which happened 1044 miles (1680 km) from Aachen but its fallout still caused problems. I am mentioning Aachen because as described above, a new, so-called "green" movement, emerged especially in Aachen, Germany, where the population decided to spend their own money for generating electricity utilizing renewable energy, which resulted in the first useful version of the so-called Feed-in Tariff, (FiT) which was introduced there and called the "Aachen model."

The Aachen model became successful for PV because it was very simple and very understandable, compared to many others, for example, in the USA, the Renewables Portfolio Standard (RPS) or the Renewable Energy Standard (RES). It needed no government money and therefore could not be called a government subsidy. It was subsidized only by the users of electricity. When it was implemented, it was added to the electric bill like a fuel surcharge but it was only a very small amount. It was the decision of the population to be willing to pay a little more for electricity produced by RE, but also satisfied the investors who were willing to invest their money to set up the PV system. After this system was adopted by many cities in Germany, which had municipally owned utility where the city's population could decide to introduce it and was willing to pay a little extra for "green electricity" generated by PV, the question was how to extend this system to all of Germany, where the electricity supply was dominated by four major utilities.

The solution became clear to Hermann Scheer and Hans-Josef Fell, both members of the Bundestag (German Parliament). A German Federal law was needed to achieve this. The first and important step was already taken in the 1991 law, which forced the utilities, in spite of their resistance, to accept the electricity produced by RE, for example, by wind turbines to be connected to their grid and pay for the electricity provided by the turbines. This law was good for wind energy, but was not suitable for PV. The 1991 law, the "Electric Feed Act" (Stromeinspeisungsgesetz) had to be updated and replaced by a new one, which would be tailored to achieve exactly what was needed for all RE sources, including PV, and would not require money from the government, similar to the Aachen model.

They planned to change the 1991 law by specifying a FiT for each type of RE system, for example for PV. Harry Lehman[259] worked out the formula for each of the RE systems based on the system cost and a reasonable profit guaranteed for 20 years. An innovation of the new law was the lowering of the FiT by a so-called "degression rate" of yearly 5%. The "degression rate" meant that Germany's Renewable Energy Sources Act (EEG) reduced the tariffs for new projects each year. This was an important element of the law. It encouraged investors not to wait for their investment until the next year as module prices may not keep up with the degression rate and the return on investment would be lower. On the other hand, it also forced the PV industry to continuously improve production to reduce prices, because if they would not be able to reduce module prices, the profit for investors may decrease to a level where they would not invest and modules would not be sellable.

The problem was how this law could be put through the Bundestag. The Bundestag had at that time 550 members. Scheer and Fell needed at least 276 votes to get the law passed. Fell in his Green party had 47 deputies. It was easy for him to line them up to vote yes. Scheer's Social Democratic Party (SPD) had 298 deputies. He needed at least 230 to vote for the new law.

Scheer was an orator second to none. I speak English and German with a very distinct Hungarian accent, but I understand both languages perfectly well. I heard Hermann give speeches in German as well as in English, in which language he had a small German accent. I listened to his speeches in German and in English and in either of these languages after his first sentence one was riveted to what he was saying and came under the spell of his speech. I talked with Hermann the first time in 1992 when, as I wrote before, he opened for me—who

---

[259]Lehmann H, Reetz T (1995) *Zukunftenergien*, Birkhäuser Verlag, Berlin, Germany.

by that time was an "old timer" in PV—a new world of PV. He was a magician with words. He gave speeches in the districts of the deputies needed to vote for the law he and Fell proposed, and the people got convinced that it was good for them, which was also the truth. An additional boost was that Hermann Sheer was awarded on December 9, 1999 the so-called "Alternative Nobel Prize."[260] Receiving this prestigious award also united his party to support him to vote for this law. He managed to assemble all the votes needed and the EEG was passed on February 25, 2000.

The result of the 2000 FiT law was unbelievable. In 1999, the world's total PV module production was about 200 MWp. By 2003, it approached 680 MWp, an increase of 240%. One should have expected it, based on the financial conferences which indicat8ed that large private investments would be available if the return was sufficient and secured for at least 20 years. On the other hand, the entire PV manufacturing community was caught by surprise, but capacities were added and new companies entered the field. There was no need any more to have any government subsidized PV program in Germany.

During the following years, it became evident that the addition to the electric bill for the charge resulting from the FiT was very little. It was less than 50 (US) cents a month for the average household. During the initial, years it also became evident that some changes needed to be made in the 2000 FiT law. The tariff should be different depending where the PV system was located and also according to the size of the system. It also required some changes in the amount of the tariff per kWh. The government also realized that the PV segment created a large number of jobs and therefore, it was in the interest of the country to make proper changes, so it should be more successful. For this reason, the 2000 law was amended. This Amended Renewable Energy Sources law was enacted on July

---

[260]The Right Livelihood Award, also referred to as the "Alternative Nobel Prize," is a prestigious international award to honor those "working on practical and exemplary solutions to the most urgent challenges facing the world today."

21, 2004. It was also decided to establish that the "degression rate" should not be a preset fixed amount, but to periodically fine tune it, change it yearly or even monthly higher or lower based on the market conditions. The "degression rate" adjustment became a well-established system.

The 2004 FiT law was more perfect and gave a bigger incentive to install PV systems. In 2003, the production of PV modules was 680 MWp. Two years later, the PV module production doubled to 1400 MWp and in 2007 it reached over 3000 MWp. This meant a phenomenal increase in production. The FiT program was a huge success, as it can be seen in Fig. 33.1.

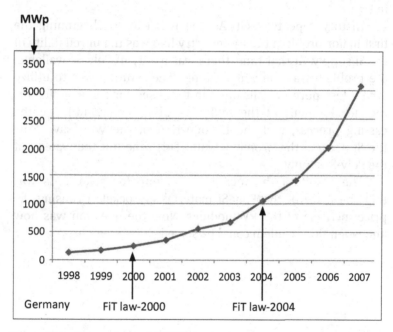

Figure 33.1 The success of the FiT laws.

The problem was, however that until 2003, the price of the solar modules decreased as it was expected because of the mass production. On the other hand, from 2003 the price of the modules unexpectedly started to increase. As a matter of fact, the 2004 new law caused a higher need of the PV modules but the prices instead of going down started to increase. In 2003, the price of 1 Wp was $2.65, and by 2007 it rose to $3.50, which was a 32% increase (see the chart in Annex. 8).

It turned out that the increase of the PV module production caused a great shortage of the Poly-Si material, from which the Si wafers used for PV were fabricated. The solar cells used basically the same Poly-Si as the semiconductor industry. The year 2006 was the first year when the need for Poly-Si for solar cells equaled the amount needed for the entire semiconductor industry in the world. Obviously the seven major Poly-Si manufacturers did not anticipated the tremendous success of the German FiT law and did not increase their production in time.

History repeats itself. As you recall at the beginning the first major problem the PV industry had was the unreliability of the Si supply. At that time, there was plenty of Poly-Si available; the problem was the wafer supply. The solution was to utilize for PV less pure polysilicon than that used for semiconductors, the development of the multicrystalline wafers utilizing the casting process, and the introduction of the wire saw. Now the Si became the problem again. This time not the wafer, but the Poly-Si supply.

The result of the increased demand for solar cells was that the price of the Poly-Si material skyrocketed causing the price increase of the PV modules. Now the question was how and when this problem could be solved.

Chapter 34

A New Silicon Enigma and Thin-Film PV Alternatives

As mentioned, because of the very successful FiT program the PV Module production from 2000 to 2003 increased by 240%. The problem arose that the manufacturers of polysilicon (poly) the basic material of the silicon wafers, did not anticipate this in time and did not add to their production capacity. Slowly there was not enough poly to sustain the growth of the PV industry. This problem became worst as from 2003 to 2006 the PV module production increased by another 200%. As mentioned in 2006 the polysilicon usage for PV became equivalent to the poly need for the entire global semiconductor industry. In the 1990s one could buy poly material for the production of the so called "solar grade" wafers for about $20/kilogram but starting in 2003 as the demand increased, the price of poly started to increase. By 2006 the poly manufacturers started to sell their product to whoever paid the most and the PV cell manufacturers to be able to fulfill the need were bidding the price of pol higher and higher bidding also against the semiconductor manufacturers. In 2007 the poly price reached $500/kg and on top of that the poly manufacturers requested a one to two year prepayment if a company wanted to have assured delivery.

Sun above the Horizon: Meteoric Rise of the Solar Industry
Peter F. Varadi
Copyright © 2014 Peter F. Varadi
ISBN 978-981-4463-80-5 (Hardback), 978-981-4613-29-3 (Paperback), 978-981-4463-81-2 (eBook)
www.panstanford.com

This resulted that manufacturers were forced to raise the price of the PV modules as mentioned by 32%. This was threatening the entire PV business, because at some point the yearly degression rate of the FiT and the module prices which increased by 2007 would make the investment in a PV system impossible.

The fallback position for PV was the thin-film PV systems. The thin-film amorphous silicon (a-Si) production was started in the 1980s with the idea, that silicon prices are high and that the a-Si thin-film system is using much less of that material. Now it was considered an alternative to the c-Si modules, which because of the shortage of the poly material may not be available in the needed quantity. The a-Si module was a choice as it had a long history and was already in production. It was also easy to make at that time large size a-Si PV modules, as the a-Si was also used to produce flat TV screens which by that time were made in large sizes and the PV module production could use the same machinery.

Besides of the a-Si there were several other "thin film" possibilities to be used for PV modules. The research and development work to find a suitable thin-film PV system to reduce the cost of making solar modules started not much after the 1953 discovery of the c-Si solar cells but until this time there was little incentive to use them. Now the situation changed. A large number of the various materials and compositions were discovered and tested, including solar cells made from plastics.

From the many thin-film systems, only three types are presently in production: a-Si, copper indium diselenide (CIS) or the presently used relative to CIS—copper indium gallium diselenide (CIGS), and cadmium telluride (CdTe).[261] The utilization of thin-film solar cells was feasible only if their price was lower than the price of c-Si solar cells. Obviously, the tremendous increase in the price of Poly-Si made the future of adequate supply of the c-Si solar cells questionable and thin-film solar cells gained more and more acceptance.

---

[261]Manufacturers of a-Si for the CIS family and CdTe thin-film solar modules are listed: http://www.pv-insider.com/thinfilmeu/content2.php.

Figure 34.1 shows that the great majority of the terrestrial solar cells used c-Si wafers as their base material. The only change was that more and more multicrystalline and less single c-Si wafers were used, and by 2013 the multicrystalline constituted more than 50% of all the PV cells made. EFG was used only in a small quantity, but it was discontinued.

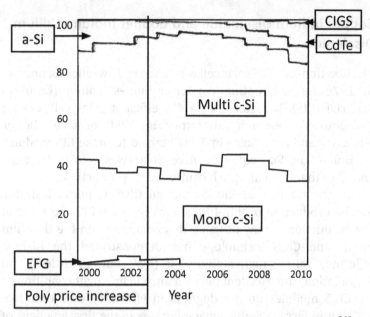

Figure 34.1 Utilization of the various technologies for solar cells.[262]

A-Si used mostly for consumer products was at the beginning the dominant thin-film solar cell type, but after the increase in the poly price, CdTe thin-film solar modules started to become the lead thin-film system, and recently the CIGS entered in small but rapidly increasing quantities.

Amorphous Silicon (a-Si)

The a-Si was the first thin-film solar module which was in industrial production. First by Sanyo in 1980 and later in 1983

[262]Kurtz S (May 13, 2010) *PV Technology for Today and Tomorrow*, NREL/ PR-520-49176, PHOTON International.

by Sharp and also by Solarex. Sanyo and Sharp manufactured and are still manufacturing a-Si for consumer products such as calculators. Solarex started to manufacture a-Si modules for commercial products. The story of a-Si is described in detail in Chapters 17 and 22. Lately it lost ground to the CIS and tthe CdTe thin-film systems.

Copper Indium Diselenide and Copper Indium Gallium Diselenide

The first thin-film CIS solar cell was made by Lawrence Kazmerski in 1976 by evaporating copper–indium–selenium ($CuInSe_2$) material (CIS). It resulted in a 4% efficient solar cell. Boeing Corporation's research laboratory in 1980 achieved better efficiency and Arco Solar in 1987 started to make CIS modules in production. Subsequently, Arco Solar was sold to Siemens and after that to Shell, which continued its production.

It was found that the compound CIGS is more desirable for the production of thin-film solar cells than CIS and most of the production is now utilizing this compound. In the thin-film group, the CIGS technology has demonstrated the highest efficiency rating, high stability in (kW) output, little or no degradation, and excellent performance in low-light conditions.

CIGS modules do not degrade in efficiency over time like a-Si, but in fact CIGS efficiency increases in the first few days of "sun soaking." They generate consistent (kW) output regardless of long periods of exposure to sunlight. Efficiency-wise, CIGS could achieve in mass production 18–20%. It is one of the best thin-film PV systems available.

At some point (around 2011), about 50 manufacturers were listed[263] to produce CIS or CIGS solar modules. Because of the big market erosion in 2012, many of these companies were either closed or sold. There are probably 10 manufacturers left. The largest of these is Japanese Solar Frontier, which is a subsidiary of Showa Shell Sekiyu. This manufacturer's production capacity in 2013 was over 500 MWp CIGS modules/year.

---

[263]Thin-film Manufacturers Directory 2011–2012, www.pv-insider.com/thinfilmeu/directory.shtml.

Cadmium Telluride Solar Modules

CdTe was identified in the 1950s as a good material for solar cells. Research to find the proper deposition and cell/module assembly was initiated. By 1983, the thin-film a-Si solar cell was already used for consumer products, but CdTe was not advanced to be used for anything despite that in small samples its efficiency to convert light to electricity was better than the a-Si solar modules. This was most likely because the CdTe solar cells/modules had some major handicaps.

The first and most important was that Cd is one of the most toxic materials; therefore, some countries may have restrictions to use it or there may be a requirement that the manufacturer must also have a safe recycling program.

The other is that tellurium (Te) is a rare material. It is a by-product of the copper manufacturing and only a small amount of it is at present available. Manufacturing of 1 GW of CdTe solar modules require (considering the presently used efficiency and thickness) 90 metric tons. The present world production is estimated to be 500–550 metric tons per year.[264] Te is not used for many applications; therefore, the price in 2011 was only about $184/kg. Calculating from these, the cost of the Te used to produce 1 W of solar cell would be less than $0.02, which is not much.

The third problem is that CdTe is moisture sensitive, and therefore, when the CdTe active layer is formed between two glass plates, the edges of the glass plates have to be sealed with a material which is impervious to moisture.

A large number of companies got in and out of CdTe research. Some even started a small production. At present, about six companies indicate that they are involved in CdTe module research and some production, but interestingly there is only one company—First Solar—which not only is involved in CdTe module production but also became one of the world's largest and lowest price manufacturer of PV modules in a very short time.

---

[264]USGS 2–11 Minerals Yearbook, Selenium, Tellurium (Advance Release), http://minerals.usgs.gov/minerals/pubs/commodity/selenium/myb1-2011-selen.pdf.

It all started when Harold McMaster, who invented the "tempered glass," which is used today in many applications, in 1984 decided that the main element of a solar module is glass, and the thin solar cells are merrily a different coating on the glass. He started Glasstech-Solar in Colorado to produce a-Si solar modules and built a very good production line. By that time Solarex acquired all the patents and key people from RCA and was already manufacturing a-Si modules. At some point, McMaster abandoned the idea of manufacturing a-Si modules and closed Glasstech-Solar. That was the time when Solarex owned by AMOCO asked me to look at the closed factory and if I felt it is OK to negotiate a deal to buy the machinery. I visited the premises and found that it was an extremely well-built automated system to produce a-Si solar modules. On my advice Solarex bought the machinery and transported it to Solarex's a-Si factory in Newtown, PA.

After closing the a-Si project, McMaster started a new company Solar Cells Incorporated (SCI) to develop the production of CdTe thin-film solar cells. By 1997, SCI had a prototype production machine. In February 1999, McMaster sold SCI to True North Partners, which was an investment arm of the Walton family, owners of WalMart. John Walton joined the board and Mike Ahearn of True North Partners became the CEO. The company was renamed "First Solar."

The story of First Solar is quite remarkable. By 2002, a fully automated production line was producing commercial quantities of the CdTe solar modules. In 2005, their manufacturing capacity was about 25 MW. Six years later in 2011, they were able to increase it to 2370 MW, and the company's sale for 2012 was $3.37 billion.

First Solar was very profitable and also expanded production because of the expected increase of business. First Solar was also caught by the large price reduction of the c-Si manufacturers and by the slower business because of the recession. But First Solar was able to decrease the manufacturing cost of their CdTe modules to well below $1/W and shifted its marketing and sales effort to developing utility scale projects some of it in excess of 500 MW in size.

A a result of the shortage of Poly-Si, thin-film PV modules were easier to sell. They were able to reach mass production quite fast, and therefore, price-wise they were even somewhat lower than the c-Si modules, the prices of which, as described above, were climbing.

Now that we have seen that thin-film PV systems were established as acceptable alternatives, let's review what happened to the Poly-Si problem.

You may recall that in the 1970s the Si industry did not take PV seriously. They considered the PV industry as scavengers. If they had leftovers from their semiconductor business, they would sell them to PV manufacturers. If they had no leftovers, they would not even return telephone calls. The idea to add capacity for the PV industry never crossed their mind. It is interesting to realize that 30 years later PV became an extremely important business for them. When they started to realize that the PV business suddenly is becoming very large, the seven large poly manufacturers started to expand their facilities. Seeing the shortage and the growing demand for Poly-Si, many companies and people dumped a large amount of money to establish new poly manufacturing facilities. There are records which indicate that at some point about 170 companies started to produce or planned to produce Poly-Si. With this very large amount of poly material flooding the market, the prices of poly collapsed very fast. This started in 2008 when by the end of that year poly prices were around $300/kg. From here on, it took only 6 months and prices were around $60/kg and by December 2011 it was all the way down at $30/kg. As of April 2012, the price of Poly-Si was down to $19/kg. The result was that the number of major Poly-Si manufacturers grew from 7 to 14 (Annex. 1). Some of the smaller ones which they call "Tier 2" or "Tier 3" manufacturers are still around, the rest of them are not in business anymore.

The continuation of what happened to the PV industry is described in Chapter 35.

Chapter 35

PV Becomes Unstoppable

The PV industry recovered from the shock caused by the Poly-Si shortage and the resulting astronomical $500/kg price. Encouragement came from the ensuing collapse of poly prices to $19, even below the price which was in the 1990s, which resulted in a drop in module prices, which in turn extremely increased the demand. As a result, the PV industry embarked on an unprecedented expansion. In 2007, many new PV manufacturing companies arrived. Several manufacturers in Europe as well as in China raised substantial amounts of money on the stock exchanges and added very large capacities to their production, producing Si and Si wafers, others solar cells and modules. The quantity of the globally manufactured PV modules, which was 3070 MW in 2007, increased 10-fold in 5 years and reached 31,500 MW in 2012 (a 220 times increase compared to 1998). This resulted in a further sharp decrease of the PV module prices, which was $3.50/Wp in 2007 but by 2012 dropped below the magic $1.00/Wp.

The main market was in Europe especially in Germany where new companies were able to raise money for PV production equipment and produced 32% of the entire world's

Sun above the Horizon: Meteoric Rise of the Solar Industry
Peter F. Varadi
Copyright © 2014 Peter F. Varadi
ISBN 978-981-4463-80-5 (Hardback), 978-981-4613-29-3 (Paperback), 978-981-4463-81-2 (eBook)
www.panstanford.com

production, exceeding Japan's 29%. The USA, which was dominant until 1998, produced in 2007 only 8%, but that was kW-wise five times more than what was produced in 1998. The year 2007 was also the year when more PV modules were produced in China than in the USA (details in Annex. 8).

The world's PV module production during the period of 1998–2012 is shown in Fig. 35.1. The results of the introduction of the German FiT law and its amendments on the PV module production can be seen in the figure.

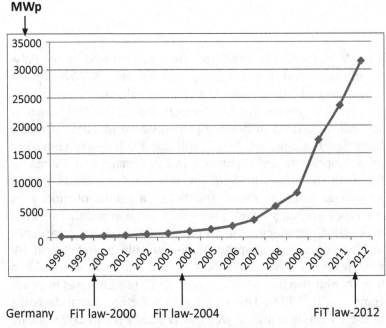

Figure 35.1 World's yearly PV module production 1998–2012.[265]

The last revision, the EEG 2012, was adopted by the Bundestag on June 30, 2011, and took effect on January 1, 2012. The purpose of this new amendment was to improve the FiT system utilizing what was learned during the period since the previous version enacted in 2004. With the FiT system,

---

[265]Private communication from Paula Mints, paulamspv@yahoo.com, SPV Market Research.

the mass production of PV was achieved and the genie was let out of the bottle.

The cumulative installed PV capacity in Germany[266] by the end of 2011 was about 27,700 MW. The Deutsche Bank Group report issued in September 2012 stated that 4400 MW of PV systems had already been installed in Germany by the first half of 2012 and the share of PV-generated electricity increased to 5.1%[267] of the German electricity production.

It is interesting to compare at the end of 2011 the cumulative installed Renewable Energy (RE) and nuclear capacity in Germany. Table 35.1 shows the results of FiT on the utilization of the various RE systems, biogas, PV, wind, and the comparison to the nuclear capacity, which interestingly became irrelevant and therefore, the German Government decided to order to close several nuclear plants and made 2022 the deadline for a full nuclear phase-out.

Table 35.1 Cumulative installed RE and nuclear electric capacity in Germany, 2011[268]

	MW
Biogas	19,200
PV	27,700
Wind	29,060
Nuclear	20,379

Seeing the success of Germany, several other European and also other countries enacted their own FiT laws. By now at least 50 countries in the world, including some parts of Canada and the USA, have introduced the FiT system. While the FiT system

---

[266]Deutsche Bank Group: http://www.dbresearch.com/PROD/DBR_INTER-NET_EN-PROD/PROD0000000000294376/The+German+Feed-in+Tariff%3A+Recent+Policy+Changes.pdf.

[267]*Sitzung der Arbeitsgemeinschaft Energiebilanzen,* July 26–27, 2012, Berlin.

[268]Biogas, PV, and Wind data from the Deutsche Bank Group report. Nuclear data from World Nuclear Generation Capacity, http://www.uxc.com/fuelcycle/utility/uxc_RxCapacityCountry.aspx.

was introduced in many countries, it is worth to follow what happened in Germany as they started this approach and their experience will probably be used in other countries.

The EEG 2012 law opened a new phase of the FiT. It established that when the cumulative PV capacity in Germany reaches 52,000 MW the FiT system will be phased out. By that time, electricity production from PV systems will be close to 10% of the German national electric supply. The FiT system was an excellent program to establish the mass production of PV and thereby drastically reduce its price. As this is now more or less achieved and obviously will be achieved by that time, one can say *The Moor (FiT system) has done its duty, the Moor can go.*[269] It is now realized that PV became an established part of the world's electricity-generating system, and 100,000 MW of PV systems were deployed worldwide by 2013. The future of PV became unstoppable.

The other interesting decision of the EEG 2012 was to create an incentive that 10% of the generated PV electricity should be used "*onsite*" and not fed into the grid. This was a very foresighted rule trying to turn people to the direction which is PV's big feature, that it is a decentralized source of electricity. This means that while the electricity generated at a power station has to be transported to the user by wires, a PV electric power system can be built anywhere where electricity is needed and used *onsite*. The EEG 2012 law was designed to nudge people in the direction to establish what can be called the "wireless environment."

It is a legitimate question to ask: Who *was* buying that large amount of PV and where was the money coming from? The other question is: Who *is* buying PV systems and who is going to buy them in the *future*? It is interesting to follow the events in Germany as it is the most advanced in the utilization of PV since the FiT system was started there. The total of the installed PV systems in Germany by the end of 2011 was 24,700 MW. Table 35.2 shows the total of the various-size PV systems installed.

The German experience shown in Table 35.2 indicates that the majority of installed PV systems were under 1 MW and

[269]Friedrich Schiller's play: "*Fiesco's Conspiracy at Genoa.*"

totaled 20 GW. This is 81% of the total capacity installed in Germany and it is quite unexpected that this large amount was installed mostly on private houses or farms. This has the equivalent capacity of several nuclear power plants. These PV systems the great majority of which were in the 10 kW range were installed by a large number of individual people, including farmers. According to the Climate Policy Initiative study, 85% were roof mounted. Where did the money come from to buy these PV systems? This means that the major investors in Germany were mostly individuals. Considering the average price in 2011 of a small residential PV installation in Germany to be $3.2/W, this added up to about $64 billion. This large amount of money apparently came from individuals.[270] It could have been from savings, loans, or from investors from whom the property owner leased the system. (In the USA, the majority of the PV units were also installed by individuals on their homes.) The money for the other PV systems totaling 4700 MW about $13 billion considering an installed price for larger systems of $2.80/W came from investment funds, investors, banks, and companies. The utilities did not buy any.[271]

Table 35.2 Germany (2011): MWp of installed PV according to the size of the systems[272]

Size of the PV system	MWp installed	Percentage of the total number of 24,700 MWp installed (2011)
1 M/MWp or smaller[273]	20,000	81
1–10 MWp	3050	12
10 MW and larger	1650	7

[270]2011 edition of Tracking the Sun, an annual PV cost-tracking report by the Lawrence National Laboratory (Berkeley Lab) as reported by PV magazine. http://www.pv-magazine.com/news/details/beitrag/us-installed-price-of-PV-systems-continue-to-fall_100009368/#axzz2JUQmlF9G.

[271]Wolfe P (2013) *Solar Photovoltaic Projects*, Routledge, London and New York.

[272]Wolfe P. Private communication, philip@Wiki-Solar.Org.

[273]Wynn G. *German Power Generation Moving Against Utilities*. http://www.reuters.com/article/2012/11/27/column-wynn-germany-renewable-idUSL5E8MQ6O120121127.

It is clear that whoever purchased the PV system did it because they expected to receive a decent return on their investment. The resulting mass production of PV modules as predicted caused the price to go down. The "degression rate" ensured that the return on investment stayed in an acceptable range and finally PV prices neared the rate which the utilities charged for the electricity. Not surprisingly the situation of who was buying PV systems and where the money was coming from started to change already in 2011 not only in Germany but worldwide. The reasons to invest in PV systems were realized by the users of electricity and by investors; the utilities could have but did not realize this except a few of them recently.

One of the reasons was that in the United States and also in Europe, peak demand hours for electricity consumption occur around noon and during afternoon hours, especially during the summer months when the air conditioning load is high. Electric utilities have to install "peaking power plants," which they call "peakers" that generally run only when there is a high demand for electricity, known as peak demand. Because "peakers" are inefficient investments as they are used only when needed, utilities are charging a higher rate for electricity users in the peak hours.

On the other hand, PV systems produce the most electricity during peak demand hours in the middle of the day and afternoon. The result was that by 2011, for example, probably all of the grid-connected residential PV systems totaling about 20,000 MW were feeding electricity into the grid and therefore, the "peakers" were actually not needed and the utilities had no income from them. As it will be shown, the loss of this income was a very unpleasant surprise for the utilities.

Companies and investors, however, realized that electricity produced by PV systems became cheaper than the peak electricity prices charged by utilities and as a result PV systems are now installed for "onsite" use, exactly what the German new FiT law was steering people to do. Example is Walmart, which installed many PV systems on the roof of their stores, for example, in California alone, on more than 100 of its stores,[274] generating up to 70 million kWh of electricity in a year. According to the US Solar Energy Industries Association (SEIA), more than 1.2 million solar PV modules were installed by the top 20 corporate solar users. Walmart and Costco combined have more solar PV on their store rooftops in the USA than all of the PV capacity deployed in the state of Florida. During peak power time, these stores use their own electricity, and if the store is not open during that time, they feed it into a grid and collect money.

Another reason why companies and investors are establishing PV systems is to lock in the electricity rate for 20 years. As the PV electric generator requires perhaps some maintenance, but during the life of the PV system, which is guaranteed for 20 years, the price of electricity will not change. It is not influenced by inflation or changes in the prices of the fuel. PV electricity became a fixed and not a variable expense.

Lately, a large number of large size PV fields were installed and are in the process of being installed or planned. These are developed and purchased by investors to become independent power producers. An interesting example is Warren Buffett's MidAmerican Energy, which is part of his holding company Berkshire Hathaway. The company has announced that it is buying SunPower Corporation's Antelope Valley Solar Projects, located in California. The field is a total of 579 MW.[275] This is an extremely interesting and important development with far-reaching consequences for the entire utility industry and will be discussed in detail in Chapter 42.

One issue has not been touched until now. Where were these large amounts of PV modules manufactured? In Chapter 29, it was described that until about 1999, the US PV module production was the largest. But reviewing the numbers (Annex. 8), one can see that the USA was overtaken by Japanese production. After that, as the result of the FiT

---

[274]http://www.pv-tech.org/news/20719.
[275]Richard MG, *Treehugger*, http://www.treehugger.com/renewable-energy/warren-buffett-buys-worlds-largest-solar-project-sunpower-25-billion.html.

program from 2007 for 2 years, Europe was the largest producer. Then suddenly in 2009, the Chinese PV module production soared above Europe, and by 2011 46% of the world PV module production was made in China, while only 3% was made in the USA, 7% in Europe, 12% in Japan, 17% in Taiwan, and 15% in the rest of the world. This does not mean that USA, Europe, and Japan produced less than before. No, their production increased from 2003 to 2011 by 7-12 times. The interesting question is that China in 2003 produced 0 (zero) percent of the world production and in 2011—only 8 years later—it produced 46%. How did this miracle happen? This interesting story is told in Chapters 36 and 37.

Chapter 36

The Chinese Miracle (中国奇迹)—Part 1

On October 6, 1983 our delegation was on an airplane flying from Hong Kong to Beijing. The background of this was that at the beginning of 1979, Lewis Rutherfurd, our partner and manager of Solarex's Hong Kong operation, called me and said that he had received enquiries from the People's Republic of China (PRC) that a delegation would like to visit Solarex in the USA. I told him we would be very glad to see them and asked when they were planning to come. He said I should send him an invitation for them and he was going to let me know when they would like to visit Solarex.

PRC's "New Energy Sources" delegation's visit took place in the summer of 1979. The delegation consisted of eight members,[276] including Lin Hanziong, chairman of the delegation (Figs. 36.1 and 36.2).

The Chinese visitors spent several days and toured Solarex's manufacturing operation and research departments. They must have had a good impression from their visit, because the PRC

---

[276]Lin Hanziong, chairman of the Delegation, Qian Hao, Wu Zengfei, Gong Bao, Ge Xinshi, Lin Hanxiong, Chen Ruchen, and Hu Chengchun.

Sun above the Horizon: Meteoric Rise of the Solar Industry
Peter F. Varadi
Copyright © 2014 Peter F. Varadi
ISBN 978-981-4463-80-5 (Hardback), 978-981-4613-29-3 (Paperback), 978-981-4463-81-2 (eBook)
www.panstanford.com

colleagues and we had continuous contact, primarily through our Hong Kong operation.

Figure 36.1 Lin Hanziong, Chairman of the PRC delegation with the author of this book.

Figure 36.2 The PRC delegation with Jay Conger, manager, and Maurene McHugh, member, Solarex International Sales.

Our trip started when a few days after Solarex was sold to AMOCO at the beginning of September 1983 and I was only a "consultant" with a 5-year "not to compete" agreement, I received a call from Solarex's president, who asked me to see him. I hoped that after running that business for 10 years, I was going to have a few weeks of vacation. However, because of my agreement,

I had no choice. I had to see him, but anyway if he asked me, I would have to see him, as he was a nice person, with whom I had a very good relationship. When I arrived in his office and was seated, he offered me coffee, which I turned down. I saw his face and knew whatever he was going to say, he was not comfortable to tell it to me. He started:

"I know that you probably wanted a few quiet weeks after having no days off in the past 10 years, but we just received an invitation from the PRC Government inviting Solarex to send some of our people to Beijing and discuss matters related to PV technology transfer. I talked to Gordon McKeague at AMOCO and he told me that it is important for them that we should go to Beijing to discuss the matter. I also talked to Lewis Rutherfurd and he told me also that we have to go to that meeting. They both told me and I agreed that as I have so many other things to attend, that you have to lead the delegation, because they met you before and they know your reputation. As a matter of fact, I want you for the time being to continue to oversee Solarex's Asian operation and therefore during this trip spend time in Hong Kong and also in Bangkok."

"John, I do not understand this. PRC used solar cells already in 1971 on the Dogfanghong telecommunication satellite, and for terrestrial application in 1973 they used solar cells on the Tianjin Harbor lighthouse. They visited us a few years ago. It is true that at this time (1983) their terrestrial PV production is nearly nothing and has very high selling price."[277]

"It looks like they want to talk about technology transfer for the mass-produced terrestrial PV. Look, today is September 22 and you are supposed to go and lead a Solarex delegation from Hong Kong to Beijing on October 6. We have only a few days to get your Chinese visa and prepare everything for your trip. Gordon said that he arranged that AMOCO should give you the title 'vice-chairman' to have the authority to properly represent Solarex. I just got your new business cards a few minutes ago. You know this business better than I do. Achieve what you would have wanted to achieve if you would run this joint. Good luck and get to it."

---

[277]Sicheng W (February 2–21, 2006) *Current Status and Future Expectation of PV in China*, China-EU Energy Collaboration Forum, Shanghai.

With this he pushed a little box toward me. I opened it and saw that I was the "vice-chairman" of Solarex.

Next afternoon I went to his office.

"Here is what I suggest. We are leaving on October 3, Monday. I said 'we' as I would like to take with me from here Ramon Dominguez[278] and from AMOCO John Johnson.[279] Ramon, because he would be able to answer all the technical questions, and John, so that he should give me advice what AMOCO would say to whatever I am doing."

"I agree. What else do you want?"

"I talked to Lewis and we suggest that from Hong Kong he should come with me, as you know he speaks Mandarin and also two of his Chinese people David and Joseph from our Hong Kong office. They both are very competent and we would need them too. We would have a delegation of six people, which according to Lewis would give them the impression that we are very serious."

"Agreed."

So there we were, the six of us flying on October 6 afternoon from Hong Kong to Beijing. All the arrangements were made between our Hong Kong office and the PRC Ministry, which sent out the invitation. My only reservation was that we were able to get only a one way ticket, but Lewis said that it was probably because we did not know how many days our meeting would last.

The Chinese Airline's flight was very good. It only foreshadowed my problems which I already knew about. I am from Hungary and my eating habits stopped at age five. Obviously at that time there was no Chinese food in Hungary and therefore, it was not on my list of foods that I would like to eat. So my dilemma was that during an official invitation like this, if I am going to be served food which is not on my eating list, what to do about it.

I was curious what the PRC inspectors at the Beijing airport would say when they would check my luggage, in which

---

[278]Ramon Dominguez was with Solarex since 1975 and was working in various capacities and was one of the most competent person I worked with.

[279]John Johnson was the right-hand man of Gordon McKeague.

besides a few pieces of clothing, I had a big stick of Hungarian "Pick" salami, which interestingly was available in Hong Kong, several bottles of Evian water, and a bottle of Johnnie Walker Red Label scotch.

I never found out what the PRC inspector would ask when opening my luggage, because getting off the airplane some people greeted us and ushered us to a small bus and bypassing any inspectors we were driven to our hotel, which was next to the airport. The leader of that group told us that after we deposited our luggage in our rooms, we should go down and join them to discuss our program.

The hotel was modest and fairly rundown. I had a very big room with a bed, closet, and many chairs and a sofa. All of the furniture seemed to be very old. The bathroom was adequate. I had to fix the toilet to get it operating. My room had two windows. One could see in the distance the lights of a runway as it was already dark.

After we assembled downstairs in a meeting room next to the lobby, the leader—speaking good English—introduced himself. He was from the ministry and introduced the six other persons. They were scientists and engineers.

I introduced myself and the other five people.

The person from the ministry was very well dressed. He wore a dark well-tailored suit and a nice necktie. However, the others' clothing was poorly tailored; on some of them, the jacket sleeves were too long and on the others tooshort. The person from the ministry said that unfortunately they could not reserve rooms for us in the very good hotel downtown Beijing, because it was full; therefore, we were now not in the best hotel in Beijing. He hoped we would find it good. He outlined our program which would take about 2 weeks and included several sightseeing trips.

It was really very nice that they planned not only meetings but also sightseeing trips for us, but my problem was that I did not expect this and had a very tight schedule. October 6 was a Thursday when we arrived and I planned to be back in Hong Kong on October 10, Monday, as I was scheduled to go for 1 day to Taiwan and after that 2 days to Bangkok and spend a few days at our office in Hong Kong and wind up in Hawaii on 20th.

There I was planning to visit a sugarcane plantation, which was our customer as they bought a large number of our small-size PV modules. They used them to charge batteries that power a timer and valves to automatically back-flush large water filters for drip irrigation. At that time, the system was being used in 13,000 acres. They invited me to show me around. Anyway I planned to take a week vacation finally, before returning to Washington, DC. Now my problem was how I could tell them politely that I could not stay longer than a few days. Finally I tried to be very diplomatic.

I thanked him for the magnificent program they arranged for us and I believed that my colleagues agreed with me that this was an incredible opportunity for us not only to complete our business but also to have the opportunity to visit the world famous sights China had to offer. I did not want to take that much time away from them, who were very busy to do their work, and therefore, I suggested that we should come up with a shorter plan in which we could complete our discussions and we could see the most important attractions. I suggested that we start with their proposal of day 1, tomorrow, and before lunch Mr. Dominguez would give a talk about Solarex's technology. Naturally, he would be open for questions. After lunch I was going to describe Solarex's structure and marketing and sales issues, also I would answer any questions they asked. Furthermore, I thanked him that the vice-minister invited us for dinner, for which he should convey our thanks to him.

The next morning, at breakfast, Lewis gave us some money. He explained that two different kinds of money were being used in the PRC, one for the citizens and the other for foreigners. He told us the conversion ratio between the two types of money. He explained that the reason for these two types of money was that the one for the foreigners could be obtained only by converting foreign currency. There were several stores where the merchandise could be purchased only by paying with this money.

Shortly, the person from the ministry arrived and told us to come with him. We left the hotel, in front of which were a small black car with a driver and two taxis. He motioned us to the two taxis and he got into the black car on the rear seat. We got into the taxis and the cars took off at a very fast speed. We apparently went in the direction of downtown. As we left the hotel and got on the main road obviously leaving the airport toward the city, the streets and the big avenue got totally filled up with people on bicycles. They all cycled toward the city. I saw several buses and street cars, but I did not see a single automobile. Obviously our cars were slowed down by the bicycles. The taxi drivers must have had a lot of fun as they tried to scare the hell out the cyclists. They knew at least a hundred different ways to scare the cyclists, one of which was to speed up and drive the taxi a fraction of an inch next to the cyclist's leg.

We drove through the big Tiananmen Square, also filled with bicycles, and the walls and the gate to the Forbidden City could be seen from our taxi. We finally stopped at the entrance of a multistory gray building. The ministry person received us and paid the taxi drivers, and between military guards standing at the entrance, we entered the building. We went up some floors and he escorted us to a door. He opened it and we walked in.

Many people were in the room. I counted 20 Chinese people. Besides the well-dressed man from the ministry, there was another well-dressed man, I assumed that he should also be from the ministry. The rest of them had the same attire described above. The tables were arranged like a T. Six of us were sitting on the shorter table, the top of the T, while the others on the lower part of the T. On the side of our table sat one person, he was the person who translated our talks. The two ministry persons were sitting closest to us. Behind us was a screen and on the far end of the T was a projector.

The ministry person introduced me, but the large audience was not introduced to us. After that I greeted the audience and told them what a privilege it was for us to visit Beijing and have this discussion. I introduced everybody in our group. The last one was Ramon and I also said that he was going to make a presentation and if somebody had questions he would be glad to answer. My talk was translated and Ramon started his presentation.

The day went well and I realized that the audience was well-educated and quite knowledgeable about PV.

For lunch we all left the building and taxis waited for us. We all got into the taxis and went to a big hotel, probably that was the one which was full. On the way, we saw again the entertainment of the taxi drivers scaring the millions of cyclists. The food in the hotel's restaurant was standard "American"-type food. During lunch, I sat next to Lewis and asked him what we were going to do about our return flight to Hong Kong. He knew that I had to be back there Monday and this was Friday.

The afternoon also went very well. I gave a presentation and after that we started to discuss a program of cooperation.

At some point, the ministry person said that we had to leave for dinner. We all got into taxis and the taxi drove quite a distance, where we stopped in front of a big restaurant. We all went in and were escorted to a big room, where a huge round table was in the center and chairs around it. The well-dressed ministry person escorted Lewis and me to chairs and motioned to us to sit down. Lewis was on my left. He motioned to the translator fellow to sit down on my right but there was an empty chair between him and me. He escorted Ramon and David to one side of the table and after that John and Joseph to the other side. After this was accomplished, everybody sat down. One of the well-dressed fellows sat next to Lewis and the other next to the translator. After all that was arranged, the two well-dressed persons got up, left the room and came back a few minutes later with a tall also well-dressed person. Everybody jumped up. I assumed that he was the vice-minister.

Lewis and I got up too. He came to me, shook my hands and Lewis said something—the translator said that he was happy to see me and Lewis and the other people from Solarex, that he had heard a lot of complimentary stories about us and Solarex and it was a privilege for him to have dinner with us. I answered him and Lewis spoke a few words in Mandarin. After Lewis's talk, he said he was impressed that Lewis spoke Chinese.

As soon as everybody was seated, waiters came in with dishes, deposited them on the "lazy Susan," on the middle of the table. Other waiters came with bottles. Two of them came each with a bottle which was filled with a clear liquid. One filled up a small shot glass in front of me, while the other at the same time filled the shot glass in front of the vice-minister. After that each of them started to fill the shot glasses in front of people sitting to the right of the vice-minister and to the left of me. The vice-minister told me that the liquid in the shot glass was "Maotai," which was a Chinese specialty.[280] As these two waiters departed, two others came each carrying a bottle in their hands. The vice-minister pointed to the bottles and said (naturally it was translated) beer and wine. I selected beer; the vice-minister also selected beer. The waiters started to go around, like the Maotai waiters.

What happened then was much worse than I imagined on the airplane. I thought that I would pick from the food a few pieces, which I would eat, but what happened was that the vice-minister picked up my plate and started to load it with all kinds of things. After he finished, he did the same with his plate and told the interpreter to tell me what was on my plate. Facing the problem of the variety of unknown foods, the solution flashed through my mind in a second. I stood up, picked up the shot glass, lifted it, and thanked the vice-minister to invite us to come to Beijing, thanked also for the dinner and wished everybody well. Lifted the glass again and drank its entire content. The taste resembled somewhat the Italian Grappa, which I liked and it was at least 90 proof. Everybody lifted their glasses and drank the content. I sat down. Our glasses were refilled in a second. Now he got up and toasted to us. Bottoms up again. Glasses got refilled.

After that I picked up my chop-sticks. I used them before during my previous trips to Hong Kong. I turned to Lewis and asked him to eat fast, so we could exchange plates. I knew that Lewis, in spite of being a slim athletic person, was able to eat two or three full dinners. He nodded.

---

[280]During the *Qing Dynasty* (1644–1911), Maotai became the first Chinese liquor to be produced on a large scale. Maotai, the world-renowned liquor in China, brewed in Guizhou Province, is the honored National treasure in its kind. Maotai has been used on official occasions in feasts with foreign heads of state and distinguished guests visiting China (http://en.wikipedia.org/wiki/Maotai).

The procedure was simple. I ate a few pieces, which I did not know what they were. After each piece, I washed it down with some Maotai and some beer. When the vice-minister looked the other way and talked to somebody, I quickly exchanged my plate with Lewis. We performed this routine several times.

I also had a good discussion with the vice-minister. He saw very clearly that in this immense country with that many people, electricity was very important and should be provided for everybody, but to extend the grid to all of the remote areas was impossible; therefore, he believed, the PV electricity was very important for them and they would like to develop cooperation with a very reputable company such as Solarex. He thanked me that at short notice Solarex responded so fast. He wanted to continue this relationship.

As of the food, I only remember one item which I ate and was told it was jellyfish. It was a 3″ circular ring, looked like a white colored elastic band. It also tasted and felt like an elastic band. I compare it to an elastic band, which I never tasted before. As of Maotai, the vice-minister and also the two well-dressed men were astonished by the quantity I drank. I assured them that it was one of the best drinks I ever tasted. As a matter of fact, I later found out that the specific techniques producing the beverage make the strong liquor suitable for drinking and beneficial to the health. When they brought us back to our hotel, we agreed that we were going to be picked up at 8 a.m. the next morning.

When I got back to my hotel room, I was actually hungry as I had eaten very little. I cut myself a thick slice of Pick Salami and drank a full glass of Evian water with a little Scotch.

The next day was October 8, a Saturday. In those days, in many places a half day work was customary. We were told that it was a full working day in China. We were picked up at 8 a.m. and went to the same place we had our meeting the day before. Until noon we had a general discussion. Lewis invited them to the same hotel where we were the day before for lunch. So we proceeded to have lunch there.

After lunch, we continued our business meeting and it became evident that we were going to be able to write and sign an agreement Sunday morning. The well-dressed person at the end of the meeting summarized the 2 day's results and after that invited us for dinner to the same restaurant we were the day before. He also said that although all flights from Beijing to Hong Kong were fully booked, he was able to get for me and John Johnson Monday tickets as we needed to be back, and for the rest of the group, he was able to get tickets for Wednesday and told them that they would be glad to take them on sightseeing trips. Lewis thanked him for this.

The dinner at the restaurant was more casual than the day before. The vice-minister was not present and therefore, I was able to select from the "Lazy Susan" whatever looked to me I could eat. But first our interpreter, who now sat next to me, explained what I was picking out. This lowered the amount of my Maotai but increased my beer consumption. The interpreter was a very nice and friendly person, the others were also, but I could not talk to them. The well-dressed man, at least who was mostly dealing with us, was also a nice person, but very formal. The interpreter had a good sense of humor and sitting next to him gave me the opportunity to discuss with him a few questions I was interested in, but could not ask anybody else because of the language problem. During our short discussions, I had the impression their problem was that in their 5 year or whatever plan, they were supposed to produce a certain amount of PV modules, but in spite of the group being very knowledgeable and having a reasonable research facility, they were far from what the plan demanded and that was the reason they wanted to team up with Solarex. On the other hand, I told him that we had no chance to sell in China under the present conditions and we would be glad to work with them. Furthermore, the company belonged now to AMOCO— interestingly that it happened only a few weeks ago he knew about it—and AMOCO would also be glad to develop good relations with China. I knew he had to report our conversation and at least they could see why we were there.

At the end of the dinner, I picked up my shot glass with the Maotai, thanked them for everything, and told them that as a part of the agreement we were going to send an invitation for them to visit Solarex at a time convenient to them. They applauded this invitation.

Sunday morning the entire group was back in the meeting room. Several of them came only to say good-bye to me and John, because they did not participate in the discussion to finalize the agreement. We shook hands and they left.

After they left, only 12 of us remained: the six of us from Solarex, the two well-dressed persons, the interpreter, and three of the less well-dressed ones. By that time we all became well acquainted with each other. Their interpreter was a really nice person and according to our two Chinese members, David and Joseph, he made a very accurate translation from English to Chinese and from Chinese to English, including the technical matters. The interpreter was always wearing a cap, as I found out later, because he was bald.

We completed a one-and-a-half page agreement in about 2 hours when we ran into the following problem: They had no typewriter and the copying machine was almost unknown to them. As a result, a part of the group, like monks in the Middle Ages, started to hand write the agreement, we needed two copies. Ramon and I became the monks writing the English version and David, Joseph and the interpreter translated it into Chinese and wrote the two copies.

We finished it by noon, when I suggested going to the hotel for lunch, sign the agreement, and toast it with a glass of champagne. We signed only the English version and attached the Chinese version to it. Lewis invited them to dinner in the evening at the same place, which all of them accepted.

During the afternoon we went shopping at a store where only foreigners could shop. They had very nice cashmere and silk products which were much cheaper than in stores in Hong Kong. During the dinner they gave us little gifts. But I got a heavy package. I opened it and found a bottle of Maotai. That was very nice of them, I thanked them, but my problem was how I was going to take it to Hong Kong. The customs was not the problem. The problem was that I did not trust the cork in the bottle. It might leak Maotai into my luggage. I had to carry it in my hand with the cork always uppermost.

Monday, several of the colleagues showed up and Lewis, Ramon, David, and Joseph went sightseeing with them. Shortly after they left, our well-dressed friend came and told us that he was taking us to the airport. It was a very modern, small airport. The Washington National airport in comparison was also small, but not modern. In order to save electricity, the escalators worked only when somebody stepped on them. That was a quite modern feature.

He escorted us to the check-in counter and presented our tickets. He had a short discussion in Chinese with the woman attendant, who told us that the flight was full and she could not give us seats next to each other. John wound up with a window seat and I with an aisle seat, somewhere in the middle of the plane. After we checked in our luggage and got our boarding passes, two Germans came and gave their tickets to the attendant. The attendant checked them and gave them back saying that unfortunately the plane was full and she could not give them a boarding pass. The two Germans got very upset and told the woman that they had confirmed their reservation. The attendant responded that she was sorry, but there were no more seats available and she could not give them the boarding passes. The two were now extremely upset and asked the attendant to what airplane she was going to transfer them. She said they should go to an office—and told them where the office was located—and there they would be taken good care of.

Our well-dressed escort, while this went on, told us that he is going to take us to our gate and we should follow him. We did not go through where all tourists go. We went through some corridor to an office. He asked for our passports, we gave them to him. He gave them to a person in military or police uniform. He did not even look at them, just opened the page where our visa was and stamped it. We never saw any other official and ended up at our gate. He thanked us for coming to Beijing and working out a very good agreement. We thanked him also for everything. He wished us a good trip, shook hands, and left.

The airplane was really full. We were already flying at least an hour when John got up, came to me and said that I should look at one of the two engines on his side—it was a four engine Boeing 707—because—he said—its position somehow looked strange. I got up, looked out through his window and said that

I did not think the position was strange the window was just distorting the image.

We landed in Hong Kong. A bus waited for the passengers to transport them from the plane to the terminal. We got on the bus and as it went by our plane we saw that the engine John was pointing out to me was a different type from the other three. As a matter of fact it not only looked different but also was also mounted differently.

We checked in at the Mandarin Hotel and before dinner we sat at the bar. I had a Johnnie Walker red label without ice and a small Perrier in another glass without ice. John had bourbon with ice. I asked him:

"If you would have seen that one of the engines was strapped on probably to bring it here for repair, would you have offered our seats to the two Germans, who were thrown off of this plane probably because of us?"

"That is a mute question," he answered, "because we are having now a nice and relaxed drink in the Mandarin Hotel."

After I returned from our visit to Beijing, the president of Solarex kept me quite busy with various projects. I kept my title "vice-chairman" as many of the projects I was involved in gave me a free hand to direct them. One of these was the acquisition of Solar Power Corporation (SPC), Solarex's big competitor, which was an Exxon division and Exxon decided to close it. SPC had excellent technical staff and made good products but was more or less the casualty that they had to march toward the central power station idea and most of their business was government demonstration projects. When the Reagan administration, which followed the Carter administration, decimated DOE's PV budget, they were desperately trying to find commercial business, but commercial business cannot be built overnight. So Exxon closed SPC and on June 15, 1984, I signed the papers in New York buying for a lump sum SPC's assets, including inventory of wafers, cells, modules, etc., for Solarex.

The Chinese informed Solarex that their delegation would like to visit Solarex in July. The return visit by the Chinese delegation to Solarex started on July 9, 1984. The preparation and the schedule for this visit were organized by Ramon

Dominguez under the direction of the President of Solarex. I participated in the preparations, and contacts with the Chinese, but my role was mostly ceremonial.

On July 9, the first meeting at Solarex with the Chinese took place, in which I also participated. The delegation's six people were mostly the same who participated in our meeting in Beijing. The difference was that all of them were quite elegantly dressed. The interpreter was the same person still with his cap on. I welcomed them, introduced the Solarex people of whom they already knew Ramon. After they introduced themselves, I asked our president to take over the meeting.

After the working day was over, Solarex organized a dinner, where more Solarex colleagues joined, too. The interpreter was seated between me and our president so that he could translate what we were saying and what their people were saying. I talked with the interpreter about a few matters. One was my interest whether they were able to fulfill the production of the required amount of PV modules. He said no. I asked him if they would be interested to buy from Solarex the solar power modules, which we just acquired when we purchased SPC's assets. If they bought the entire amount of 92 kW, I assumed they would get it at a very low price. He said he would let me know.

Next day, I joined them only at lunch. The interpreter told me yes they would be interested to discuss the matter. After lunch I went to our president's office and told him the possibility that we could sell the SPC module inventory to China. He was surprised and very pleased. SPC produced good-quality modules, but for Solarex it would not be very good to start to sell them. We discussed the price and agreed that I would conduct the negotiations. Before I left I asked my secretary to find the translator and ask him to come out from the meeting for a minute. I waited at the door. He came with my secretary. I told him that the Solarex president agreed to sell them the entire quantity and told him the price, which was much below the normal market price at that time. I told him that they should talk it over and the next morning I would be back and we could discuss the details.

Next morning the president, Ramon, and I met them. They told us that in general they were interested but they would like to see the modules.

I told them they could see them immediately and we were going to transport them to the warehouse. I also told them that most of the modules were in individual boxes, but some of them were not. Solarex would properly package them. I also told them that we tested several of the modules and they all were OK.

With this Ramon arranged transportation and they all left for the warehouse. They were back before lunch and the interpreter said they would buy all of them, but wanted them at a lower price. I suggested that the entire transaction go through our Hong Kong office, they should pay the price that we asked, but Solarex would pay the shipping to Hong Kong. They should send the letter of credit to our Hong Kong bank and they should arrange shipping from Hong Kong, which they could do by trucks. I also told them, if any of the modules were broken, or not according to specification, they could return them to Hong Kong and the money for those modules would be refunded.

After a short discussion, the interpreter said that they agreed, especially as they did not have to do business with a company in the USA, but only with Hong Kong. This would make the transaction much easier. I thanked them and told them that as soon as the letter of credit arrived at our bank in Hong Kong, we would ship the modules in crates to our Hong Kong warehouse where they could pick them up. The SPC 92 kW PV modules wound up in China and we received no returns.

I found out only later[281] (Fig. 36.3) that in China the total 1979 yearly production of PV was only 5 kWp (kilowatt and not megawatt) which makes it understandable that a Chinese delegation visited Solarex to learn something and try to institute changes. But for some reason it seems that it did not work out because 4 years later in 1983 and 1984 they were interested

---

[281]Junfeng L (1998) *Commercialization of Solar PV in China*, China Environmental Science Press, Beijing.

in technology transfer from Solarex, as the total PV production in China was, respectively, only 30 and 50 kWp (kilowatt, not megawatt).

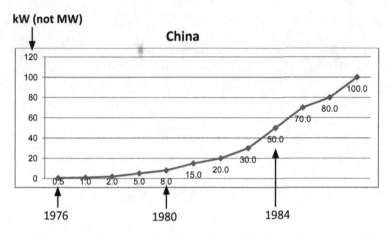

kW (not MW)

Figure 36.3 PV Module production in China (in kWp, not in MWp) per year from 1976 to 1987.[282]

In order to put the Chinese PV production in perspective, one should consider that the 92 kWp which they purchased in 1984 from Solarex (which as mentioned above was in the inventory of SPC when Exxon closed it and Solarex bought the assets) was almost twice the 50 kWp which was China's entire PV production of that year. The global production of PV modules in 1984 was 18 MWp out of which Solarex was producing probably 6 MWp.

---

[282]"Center for Renewable Energy Development, Energy Research Institute" published an extremely candid, impartial and factual Report entitled: *Commercialization of Solar PV Systems in China*, p. 117.

Chapter 37

The Chinese Miracle (中国奇迹)—Part 2

In 1999, the Center for Renewable Energy Development—Energy Research Institute published an extremely candid, impartial, and factual report entitled *Commercialization of Solar PV Systems in China.*[283] This publication (*Report of 1999*) was the result of the Energy Efficiency and Renewable Energy Protocol between the US Department of Energy and the PRC State Science and Technology Commission and signed in June 1996. The purpose of this agreement was to promote cooperation on renewable energy business development between the USA and China.

Allan R. Hoffman[284] in the foreword to the *Report of 1999* wrote, "With a degree of honesty not often seen, it critically examines China's past and current efforts to use photovoltaic and develop a PV industry and draws useful lessons for the future." The foreword written by Hoffman in 1999 concludes

[283]Chief Editor: Prof. Li Junfeng; writers: Ms. Hu Runqing, Ms. Zhu Li, Ms. Miao Hong, Professor Wang Changgui, Mr. Wang Zhongying, Mr. Song Yanqing, and Ms. Shi Jingli, China Environmental Science Press, Beijing, 1999.

[284]Office of Energy Efficiency and Renewable Energy, US Department of Energy.

Sun above the Horizon: Meteoric Rise of the Solar Industry
Peter F. Varadi
Copyright © 2014 Peter F. Varadi
ISBN 978-981-4463-80-5 (Hardback), 978-981-4613-29-3 (Paperback), 978-981-4463-81-2 (eBook)
www.panstanford.com

with a prophetic statement: "China's resources, population and need potentially could make it the world's largest developer of PV in the future, if appropriate policies are implemented and the investment environment is attractive."

In the preface to this *Report of 1999*, Zhai Qing[285] writes, "It can be seen from the report, even though China has a very large market potential, the PV industry is still at the initial stage, and the technology is behind the international level. Higher cost and lower quality of PV is a big constraint for the continued development of the PV industry."

A similar assessment of the PV situation in China is also candidly expressed at the beginning of the *Report of 1999* on page 103,[286] "In spite of the rather large market potential, the PV industry in China is a latecomer, and the technical level is rather backward. The PV products are poor in quality and high in price compared to other countries."

The information in the *Report of 1999* indicated (Fig. 37.1) that the PV production in China compared to the production in other countries was very small. In 1998, the total Chinese production (a-Si and c-Si) was 2.3 MWp and as described in the *Report of 1999* it was also of bad quality. The *Report of 1999* (page 118) describes the reasons for the bad quality. One of the reasons is that the mentioned 2.3 MWp included 0.5 MWp a-Si solar modules which it is described to "have poor stability and low efficiency." Compared to the Chinese production, Japan produced 61 MWp (Annex. 8) of excellent quality in the same year. Compared to the global production, the Chinese production was only 1.5%.

In 1989, when the total PV production in China was only 0.55 MWp, a part of the WB Group, the International Finance Corporation (IFC), had its first involvement in China, when it invested US $3 million in Shenzen YK Solar PV Energy Co. Ltd, a solar PV manufacturer with the result, modestly described in

---

[285]Resource Conservation and Comprehensive Utilization Department, PRC State Economic Trade Commission.

[286]The first part of the *Report of 1999* is written in Chinese and the second part in English; therefore, the beginning of the English part is on page 103.

an IFC publication[287]: "The investment did not meet the original expectations."

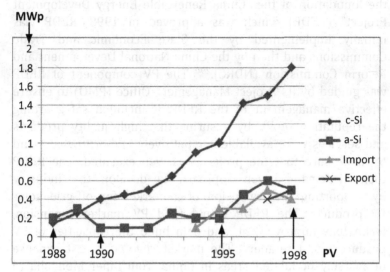

Figure 37.1 Module production in China MWp/year (1988–1998)[288] (c-Si, crystalline silicon solar cells; a-Si, amorphous silicon).

As described in the same publication, "by the mid-1990s the World Bank began supporting PV systems as a least-cost alternative to grid extensions for governments to deliver on promises of energy for development." In those days, China was still considered a developing country with a very large population without electricity. In March 1994, the WB sent Anil Cabraal and Ernesto Terrado to China to assess the support the WB could offer for renewable energy development. They identified off-grid solar PV as a priority area, among others. A more extensive study[289] co-lead by Cabraal and Terrado ensued. It detailed how such support could be structured. As a result, the Government of China sought WB assistance

[287]http://www.ifc.org/ifcext/enviro.nsf/AttachmentsByTitle/p_Catalyzing PrivateInvestment_SellingSolar/$FILE/SellingSolar.pdf.

[288]"Commercialization of Solar PV Systems in China" (*Report of 1999*).

[289]http://www-wds.worldbank.org/external/default/WDSContentServer/WDSP/IB/1999/08/15/000009265_3961214185459/Rendered/PDF/multi_page.pdf.

to implement decentralized PV power for households and community use (and grid-tied wind farms). This report became the foundation of the "China Renewable Energy Development Project" (REDP) which was approved in 1998. REDP was initially implemented by the State Economic and Trade Commissions and then by the China National Development and Reform Commission (NDRC).[290] The PV component of REDP was guided by its Project Management Office (PMO) to ensure effective management of the REDP. It included safeguarding the reputation of PV by ensuring that high-quality products and services were available competitively and at low cost and to make sure that the project was well executed. The REDP took a demand-driven approach to develop the industry by supporting the expansion of a market for off-grid solar PV products. The REDP would fund PV market expansion, technology improvement, and help build the capacities of PV producers and retailers. The market covered a vast expanse of sparsely populated areas in China from inner Mongolia to the north to Xinjian and Qinghai Autonomous Provinces in the west to Tibet in the south. The REDP was provided a grant of US $25.5 million from the Global Environment Facility (GEF) to support the PV project. The project was enormously successful. By the end of the project in 2007, the participating PV companies had sold 625,000 PV systems, the installed cost of SHS had dropped from $16/Wp to $9/Wp, the number of participating PV companies had grown from 17 to 25 as well as large number of companies producing PV components and systems to quality standards, the REDP PV project standard became the national standard and companies had invested US $188 million in technology improvement projects.[291]

---

[290]When REDP was set up, it was done under the leadership of SETC, State Economic, and Trade Commission. Mid-way into the project implementation, the major state commissions were reorganized and NDRC were set up and the REDP moved under its leadership.

[291]See: The WB, Implementation Completion Report of the China Renewable Energy Development Project, May 2, 2009, Report No: ICR0000880. http://documents.worldbank.org/curated/en/2009/05/10621768/china-renewable-energy-development-project.

By the mid-1990s the WB realized, based on experiences not only in China but also in other developing countries, that one of the more important prerequisites for PV systems was to deliver quality and reliable service. That was one of the reasons that the WB prepared and issued in June 1996 a study, *Best Practices for Photovoltaic Household Electrification Programs: Lessons from Experiences in Selected Countries* (WB Technical Paper), by Anil Cabraal, Mac Cosgrove-Davies, and Loretta Schaeffer.

The problem of the quality of Chinese PV manufacturing was a center point of the *Report of 1999*. It was a serious problem that the WB also wanted to solve—hence the singular focus on quality in REDP—not just the quality of components, but also systems and services. The guidance to REDP and to the PMO from the WB was given primarily by Anil Cabraal, who was helped by Enno Heijndermans. They also realized that a major problem which had to be solved was as Cabraal wrote[292]: "Quality *(of PV systems)* must not be compromised."

The first Chinese National Module Standard was established in 1988 but only in 1998 did the Chinese Module Standard become equivalent to the IEC Module Standard (IEC 61215), but there was no PV module testing laboratory in China. The WB supplied money to establish a Chinese PV testing laboratory.[293] In the absence of a quality standard for SHS, the WB—building on its experiences from Indonesia and Sri Lanka SHS projects, worked with the REDP PV Standards Committee to establish quality standards for SHS products to be used for REDP sponsored projects. The projects also provided cost-shared assistance to manufacturers to have their products tested and certified according to the REDP standard. Eventually, the REDP standard with some modifications became the national standard for SHS and hybrid systems in China.[294]

---

[292]Cabraal A (2006) *Selling Solar: Lessons from a Decade of IFC and World Bank Experience*, 21st European Solar PV Conference, Dresden (updated January 2007).

[293]By 2007, three PV component and systems testing institutions received ISO 17025 quality certification.

[294]http://siteresources.worldbank.org/INTRENENERGYTK/Resources/5138237-1239644200204/Solar0Photovol1nd0Photovoltaic0Wind.pdf.

As mentioned in Chapter 27, the WB, Anil Cabraal, and the head of the bank's ASTAE program, Loretta Schaeffer, supported the PV GAP program very much, and Cabraal in 1997 already believed that the PV GAP quality system could help in solving the quality issues in China. He asked the writer of this book, who was at that time the chairman of PV GAP to assemble the questions to be solved for the solution of the quality problem in the Chinese PV industry.

One of the suggestions was that by that time there were many manufacturers of PV products, cells, modules, systems in many countries, and that these companies should be educated about quality management. It was also needed to increase the number of testing laboratories able to test PV modules and systems; therefore, a manual should be created for testing laboratories what they need to enter the PV testing business.

In 1999, the WB with funding from the Government of the Netherlands, and also the Energy and Atmosphere Program of the United Nations Development Program (UNDP), sponsored PV GAP to write a *Photovoltaic Test Laboratory Quality Manual*[295] to help testing laboratories in developing countries so that they would be able to become accredited PV testing laboratories. For PV manufacturers, the WB sponsored PV GAP to establish a *Quality Management in Photovoltaic—Manufacturers Quality Training Manual*,[296] to help manufacturers in developing countries to institute a quality management program in their factories. The first issue of these manuals was presented at a meeting in Geneva on May 3 and 4, 1999, to which several people from developing countries were invited. From China, the WB proposed and sponsored Mr. Wang Sicheng,[297] who was the manager of the state owned Jike Energy New Technology

---

[295]Fitzgerald M (July 1999) *Photovoltaic Test Laboratory Quality Manual*, PV GAP.

[296]Varadi PF, Dominguez R, Schaeffer L (February 2002) *Quality Management in Photovoltaic, Manufacturers Quality Control Training Manual*, The World Bank, Washington DC.

[297]Wang Sicheng winner of the Robert Hill Award for the Promotion of PVs for Development, presented at the 23rd European Photovoltaic Solar Energy Conference (EPVSEC) in Milano, September 2007.

Development Company in Beijing and who did a lot of work for REDP.

Based on the recommendation of the participants of the meeting in Geneva, the *Quality Management in Photovoltaic— Manufacturers Quality Training Manual* was modified and reissued because the WB found it important to sponsor a training course in India and at a later date in South Africa.

The training course in India was held on October 11–15, 1999 in Jaipur and was conducted by Ramon Dominguez, one of the writers of the training manual. PV GAP was represented by Markus Real, a board member, and Enno Heijndermans of the WB, who was also on the WB team overseeing the REDP, also participated.

The training course held in India was attended by people involved in PV in India as well as in Sri Lanka. Two people delegated by REDP[298] also attended the training course. One of them was Wang Sicheng. He was at the beginning very skeptical about the usefulness of the quality training program for PV manufacturers. Interestingly, as a result of the training program, he became very enthusiastic about it realizing the great advantage of utilizing a quality program to improve not only the quality of the PV products but also the manufacturer's cost savings. He suggested that the same QM course should also be conducted in China. After returning to China in February 2000, Mr. Wang Sicheng had the PV GAP training manual for PV manufacturers translated and printed in Chinese (Annex. 10 shows the copy of the cover page) and utilizing the PV GAP manual, conducted a course (Fig. 37.2) at which 54 representatives of the PV industry and management participated. On December 4, Mr. Wang Sicheng became a member of the PV GAP Advisory Board. By 2002, the PV industry in China became larger and Mr. Qin Haiyan[299] set up the China General Certification Center (CGC) with the purpose to create a third-party certification system for renewable energy products in China.

---

[298]Wang Sicheng of "Jike Energy New Technology Development Company," Beijing, and Su Jianhu, Hefei University of Technology, Energy Research Institute, Hefei Anhui.

[299]Qin Haiyan (2011) Leaders of the Early Days of the Chinese Solar Industry, in Palz W, ed.: *Power for the World*, Pan Stanford Publishing, Singapore.

Figure 37.2 Mr. Wang Sicheng conducting the PV manufacturers' training course in 2000.

In a 2004 presentation to the PV GAP Board, Enno Heijndermans[300] reported that the REDP supported companies are considered to produce good-quality SHS systems, but several quality problems were still experienced. For example, some of the companies were misleading customers because although they were not participating in the REDP quality requirements, they put signs on their systems that they were approved by REDP. Some modules which were made by companies approved by REDP delivered less than 75% of the rated power. They had problems with batteries which had far below rated capacity (as low as only 25%). Accordingly, even in 2004 the quality was an issue.[301] On the recommendation of Dr. Wolfgang Palz, PV GAP elected on September 17, 2004, Mr. Qin Haiyan to the PV GAP Board. He was by that time the general director of the CGC.

---

[300]Heijdermans E, Cabraal A (17 September 2004) *Presentation at the PV GAP board meeting*, Brussels.

[301]Cabraal A, Heijndermans E (2004) *The Chinese Quality Situation, Lessons from REDP*, presented at the PV GAP board meeting, Brussels.

   Anil Cabraal of the WB approved the suggestion that another training course in QM should be given in China and Mr. Dominguez was asked to give the course (Fig. 37.3), which he had imparted by then in many countries. The training manual was revised based on the new ISO 9001:2000 standard (Annex. 10—Cover pages of the English and the translated Chinese training manuals). The updated training manual and the viewgraphs were translated to Chinese and the course was held in Chengdu, Sichuan Province August 1–3, 2005. The course was conducted in English and Mr. Wang Sicheng translated it. There were 31 participants and also several members of REDP and from CGC were present. All of the participants received a copy of the Chinese language manual and a copy of the viewgraphs (a total of 220 viewgraphs). The translation by Mr. Wang Sicheng helped very much by interfacing the lecture and the questions and answers between the various participants during the class exercises. Mr. Dominguez visited several PV companies and as he states in his report, "There are significant indications that all the companies visited are progressing in the pursuit of a quality product."

Figure 37.3 In 2005, PV Quality Management training course utilizing the new ISO 9001:2000 standard conducted by Mr. R. Dominguez (center) translated by Mr. Wang Sicheng (far right side).

An example that Chinese companies were anxious to introduce quality to their PV production can be seen from the following story. Mr. Dominguez, as mentioned, visited several PV factories. In one of them, he was shown the facilities. He was shown the up-to-date machine for testing the power output of the modules. Such a machine consists of a very big flash bulb, similar to which is used in photographic cameras, but much, much bigger, which flashes the light on the PV module to be tested. The electricity generated as the result of the flash is then measured by a computer which calculates the quality of the module.

The machinery according to Mr. Dominguez was a good one, but he pointed out that such measurements should not be conducted in a room the walls of which are painted white, because the reflected light would affect the results. The wall should be painted black, not to reflect light. When Mr. Dominguez ended his visit in that company, they asked him to visit the testing area again where they pointed out that while he was visiting, they repainted the walls with black color.

The PV GAP certification program management was transferred to, and was carried out at that time by IEC's *Worldwide System for Conformity Testing and Certification of Electrotechnical Equipment and Components (IECEE)* which because of political reasons was not able to establish the PV GAP program in China. Mr. Qin Haiyan, who was a PV GAP board member and was very familiar with the requirements of the PV GAP quality management system, based on his own opinion and also as a result of the success of the training courses, believed that a PV GAP type quality system would be needed in China. Therefore, he wanted to establish a similar certification program in China and shortly after the training course in January 2006 proposed to the project management office of REDP to establish a Chinese National Certification Program for PV products for REDP and establish the "Golden Sun Quality Mark." He invited the writer of this book to become a member of the group of mostly Chinese experts developing the certification program.

I had the opportunity as a member of the team developing the "Golden Sun" system to review the draft and comment.

The program was completed in 2006 and the requirements to obtain the right to display the Golden Sun quality mark on products were very similar to the requirements to obtain the PV GAP quality mark. The requirements included testing to IEC PV standards and if an IEC standard was not available a Chinese standard was developed for that product.

By 2006, major improvements were achieved in the quality of Chinese PV products. As Cabraal writes,[302] "Bank-funded projects now rigorously enforce quality requirements." He also observed that to achieve good-quality products "an additional support in China was the introduction of the 'Golden Sun' quality mark (Fig. 37.4) to help consumers identify quality certified products."

Figure 37.4 Introduction of the "Golden Sun" quality mark to PV manufacturers, 2006.[303]

[302]Cabraal A (2006) *Selling Solar: Lessons from a Decade of IFC and WB Experience*, 21st European Solar PV Conference, Dresden (updated January 2007).

[303]Press Conference: (From left to right) Qin Haiyan (China General Certification Center); Zhao Yuwen (Chinese Solar Energy Society); Zhu Junsheng (Chinese Renewable Energy Association); Wu Dacheng (Renewable Energy Development PMO); Shen Jun (Certification and Accreditation Administration of P.R. China); Li Junfeng (Energy Research Institute National Development and Reform Commission).

The requirements to be able to obtain the usage of the Golden Sun quality mark (Fig. 37.5) are listed in Annex. 11.[304]

Figure 37.5 The Chinese "Golden Sun" quality mark.

By the initial years of the Millennium REDP with the WB/GEF money and guidance was able to create the beginning of a quality-centric Chinese PV industry.[305]

The REDP team (Fig. 37.6) from its inception recognized enforcing quality without assisting manufacturers to improve quality was of limited value. The REDP also developed and implemented a quality enhancement program that offered grants to companies targeted to improve their own products and to increase their quality. It was modeled on a similar program in the Netherlands. The program was very successful, not only in bringing improved products to market but also helping Chinese PV suppliers to attract investment capital to their businesses. The WB's REDP Implementation Completion report issued in 2009[306] noted, "Measuring the direct impact of these activities is difficult but it is notable that several of the companies supported by the project have enjoyed substantial growth and increasing market share as the worldwide demand

---

[304]http://www.cgc.org.cn/eng/news_show.asp?id=3.

[305]Cabraal A (May/June 2004) *Strengthening PV Businesses in China*, Renewable Energy World.

[306]World Bank (February 2009) *Implementation Completion and Results Report* (IBRD-44880 TF-22462) http://www-wds.worldbank.org/external/default/WDSContentServer/WDSP/IB/2009/06/02/000333037_20090602004332/Rendered/INDEX/ICR8800P0468291C0Disclosed051291091.txt.

for PV systems has expanded. China now exports PV products proof that many companies are now meeting international standards; several of these companies were beneficiaries of the program." The success of the WB/GEF project can be measured not only by the large number of installed solar home systems but also by the success of the "Golden Sun Quality Mark," which was established for 11 PV product categories and as of October 9, 2012, a total of 234 PV manufacturers qualified to be able to display it on their products. In the category "PV modules" 125 companies qualified (Table 37.1).

Figure 37.6 In May 22, 2008 at the conclusion of the China Renewable Energy Development Project (REDP).[307]

The importance of the Golden Sun Quality Mark is also evident from the fact that TÜV Rheinland, which enjoys high reputation and great experience and is testing 80% of all worldwide solar module manufacturers' products, signed in 2011 an agreement with the CGC that the PV modules and

[307]From left to right: Wang Wei, PV Market Development Program Manager, REDP; Zhang Minji, Technology Improvement Program, National Proposal Solicitation Officer, REDP; Emil Ter Horst, Consultant, Technology Improvement Program; Song Yanqin, Technology Improvement Program Manager, REDP; Enno Heijndermans, Technology Improvement, WB; Noureddine Berrah, Previous WB REDP Team Leader; Anil Cabraal, PV Development, WB; Jim Finucane, PV Market Development Consultant; Miao Hong, Previous PV Market Development Program Manager, REDP; Richard Spencer, WB REDP Team Leader; Dai Cunfeng, Financial Manager, REDP; Wu Dacheng, Executive Director, REDP.

components tested by TÜV Rheinland can be recognized by CGC via the Golden Sun Mark Certificate. Even several non-Chinese (US and European) companies desired to obtain the Golden Sun Quality Mark. According to TÜV Rheinland's press release, "The Golden Sun Mark is becoming the most influential certification mark in the PV industry."[308]

Table 37.1 The success of the "Golden Sun Quality Mark"

Year	Approved PV module manufacturers
2007	2
2008	4
2009	19
2010	36
2011	38
2012	26
Total	**125**

The German FiT law of 2000 and its modification of 2004 created by 2006 such a high demand of PV modules that every tested or non-tested PV modules were purchased, and the demand grew day by day. As an example, the magazine *Photon International*[309] in 2006 published a list of the 555 module types over 140 Wp, manufactured in 24 countries by 111 manufacturers, 30% of them were not even tested (IEC 61215 or 61646). Nevertheless, the European developers of PV installations purchased everything.

PV manufacturers in China became aware of the business opportunities in Europe, especially in Germany, but at that time compared to the production in other countries still relatively little was produced in China. In spite that in 2000 when the German "PV rush" started, a relatively large number of PV manufacturers existed in China, but their total production capacity was extremely low. For example, when in 2002, Suntech Corporation was started in China, as described by Qin

---

[308]TÜV Rheinland Partners with CGC for Golden Sun Mark.
[309]Photon International (February 2006).

Haiyan,[310] its first 10 MW production line "capacity was equal to the total of the nation's (China) solar cell production over the past four years." The Chinese miracle was that while in 2002 the Chinese production of PV modules were only close to 0% of the total world's production, suddenly in 2011 the PV modules manufactured in China were close to 50% of the total production of the entire world. And this happened in only 9 years.

It is also interesting and amazing that in 2002 as mentioned above the Chinese PV module production was only close to 0% of the entire world's production, by 2011 five out of ten of the world's largest PV module manufacturers were Chinese.

How Did This Miracle Happen?

I am pretty sure that there will be a lot of case studies made at the various business schools, to study how was it possible to bring the Chinese PV cell and module production which was only close to 0% of the total world's production in 9 years, by 2011 close to 50% of the total production of the entire world. Those case studies will be very helpful to fully understand the process, but there are some facts which could give us some explanation.

After making the research for the chapter about the Chinese Miracle in PV, I was able to understand how probably that miracle happened, because I lived several years in Hungary under the doctrinaire communist system and now I am living for many years in the capitalist world and therefore, I have some understanding of how these systems work.

As we have seen until about the end of the 1990s only very small amount of solar cells and modules were manufactured in China in spite that the country had many experts in this field. I know they had many experts, because I met several of them who visited Solarex in 1979 and also when I had the privilege to meet them in Beijing in 1983 or when they visited Solarex in 1984 and I also met them at scientific meetings.

---

[310]Qin Haiyan (2011) Leaders of the Early Stage of the Chinese Solar Industry, in *Power for the World* (Palz W, ed.), Pan Stanford Publishing, p. 501.

The reason that only very small amount of PV was produced, was the result most likely that during that time under the doctrinaire communist system the manufacturing of almost everything had to be in state-owned facilities and as PV was a relatively small business, it was not suitable to establish a state-owned large enterprise to fabricate it. It may have been possible that an existing factory could have started to produce solar cells and modules. But as I learned in Hungary, the problem with the system was that if the development to manufacture PV devices were introduced inside of the state-owned company, and it proved to be successful, the person who introduced it received either no benefit from it, or perhaps received a reward. However, if the proposed new product did not work out, it was clear that the person was a saboteur, pushed the imperialist's cart, perpetrated a crime against the people, which everybody knew had the most terrible consequences. Therefore, it did not make any sense to change methods which already existed or to introduce new products. So whatever PV was fabricated was done probably in state-owned research and development facilities and not in dedicated manufacturing operations.

Also, as a command and control economy, in China, nearly every province wanted a little piece of the action. So small PV manufacturing plants were scattered throughout many provinces—too small to be viable and with small investments insufficient to improve production or scale up. So costs were high, quality was low, and so was output.

The Chinese Government in the early 1990s started to deviate from the doctrinaire system and has introduced wide ranging reforms into the state-owned sector that dominated the economy. Since 1998, a policy of "letting small enterprises grow" was introduced and the number of state-controlled industrial enterprises decreased to over one half in the following 5 years. The changes have permitted the emergence of a powerful private sector in the economy.

The WB, by incentivizing and helping to create a market, made possible for the local PV industry to develop. In the early days of REDP, Dr. Cabraal[311] observed was that the seeds of PV

---

[311]Private communication from Dr. Cabraal.

entrepreneurship emerged out of these state-owned enterprises and research institutes. They had the expertise and when given the opportunity, those who could became entrepreneurs and those who were more capable succeeded.

As the SHS and especially with the export market the need for PV expanded, more and more manufacturers entered the PV business. In 2010, as written by Jager-Waldau:[312] "More than 50 solar cell and more than 300 solar module companies exist in China."

The PV market grew in China because of the domestic need and added to this in 2000/2004 was the beginning of the great demand for PV in Germany fueled by the introduction of the "FiT." The production of solar cells and modules is relatively simple and at the beginning the Chinese solar cell and module manufacturing was fairly primitive and relying on the inexpensive labor. But by 2003/2004 a lot of automatic production lines were available from manufacturers in Europe, Japan, and the USA.

It seemed that everything was ready in China which was needed to supply good-quality PV products. Labor force knowledgeable to manufacture solar cells and modules complying with the quality requirements enforced by REDP/WB existed. By that time the Government in China approved the establishment of privately owned large factories to be able to mass produce PV. The labor force was still cheap, but automatic production lines could be obtained. Only three items were still needed to increase the production of PV and those were: Money, money, and the last thing was also money.

Money was needed not only to establish the solar cell and module manufacturing operations but also for the production of the basic raw material Si, Si wafers etc., and also to finance the incredibly growing quantities of solar modules and receivables.

Those people who will be involved in writing case studies about the financing of the Chinese miracle from nearly 0% to 50% of the entire world's PV cell/module production will have

---

[312]Jager-Waldau A. *Photovoltaic Industry in China*, http://www.evwind. es/2009/09/29/photovoltaic-industry-in-china-by-arnulf-jager-waldau-jrc/1549/, 2010

a lot of fun to trace where the money came from, because the money came from various places. By that time some Chinese individuals had money which they could invest and also in some instances local governments were investing. Obviously it was also the Chinese banking system where the rules were different from the rules of other countries.

At the beginning the money for all of the small Chinese factories came as equity investment from private sources—mainly family and friends—and also to create more jobs from local governments. It is not clear how much came from which of these sources. The interesting element of raising money was that these Chinese startups, some of them perhaps only 1 year in business were able to raise awesome amount of money on the various stock exchanges outside of China. Suntech Holdings Company registered in the Cayman Islands was the very first Chinese PV company to be listed on the New York Stock Exchange (NYS) in 2005 with an Initial Public Offering (IPO) $365,700,000. This represented only 18.2% of all of the company's outstanding stocks. The founder of Suntech was Dr. Shi Zhengrong, who studied solar energy under world famous Prof. Martin Green at the University of New South Wales in Australia. After returning to China, with the support of the Wuxi municipal government in January 2001 established Suntech Power. Following Suntech, several other Chinese PV manufacturers were able to raise considerable amount of money on the NYS. The same companies and also others may have raised additional money in second offerings, or on other stock exchanges, for example on AIM London, Hong Kong, or Chinese exchanges.

Table 37.2 lists some of the Chinese PV companies IPOs at US markets.

Several of these companies are registered at the Cayman Island. The lead underwriters were very prestigious companies: Credit Swiss First Boston; Morgan Stanley; Goldman Sachs; CIBC World Market; Merrill Lynch; Piper Jaffrey; Lazard.[313]

---

[313]Some of the information in Table 37.2 may not be accurate because of the sources I had to use.

Table 37.2 Chinese PV companies' IPOs on US stock exchanges

Company	Symbol	Product	Date	IPO Date	Market	IPO total ($)	% US ownership
Suntech Holdings Co	STP	Cell-module	2001	December 14, 2005	NYS	395,700,000	18.2%
Canadian Solar Inc.	CSIQ	Cell-module	2001	November 9, 2006	NASDAQ	115,500,000	27.8%
Trina Solar Inc.	TSL	Wafer-cell-module	1997	December 19, 2006	NYS	98,050,000	90.2%
Solar Fun	SOLF	Cell-module	2004	December 20, 2006	NASDAQ	150,000,000	5.0%
JA Solar Holding	JASO	Cell-module	2005	February 7, 2007	NASDAQ	225,000,000	34.0%
China Sunergy Ltd	CSUN	Cell-module	2006	May 17, 2007	NASDAQ	93,800,000	57.2%
LDK Solar Co	LDK	Wafer	2005	June 1, 2007	NYS	469,368,000	14.5%
Yingli Green Energy Hldg	YGE	Wafer-cell-module	2003	June 8, 2007	NASDAQ	319,000,000	23.0%
ReneSolar	SOL	Wafer	2005	January 29, 2008	NYS	130,000,000	8.0%
Jinko Solar Holding	JKS	Wafer-cell-module	2006	May 14, 2010	NYS	64,185,000	25.5%
DAQO New Energy	DQ	Poly Si-cell-module	2010	October 7, 2010	NYS	76,000,000	5.0%
					Total	2,136,303,000	

Note: "Date" means the year in which the company was formed; "% US ownership" is the approximate percentage the number of shares purchased of the outstanding stocks.

After raising that much money, all of the Chinese "newcomers" to the PV business decided to manufacture the conventional c-Si PV cells and modules. The new Chinese companies were able to buy the newest fully automated and efficient product lines with turnkey operation to produce quality PV products, wafers, cells, and modules. They could buy these newest systems at competitive prices from manufacturers in the USA, Europe, Japan, and even in China. In the PV industry, as is normal in every other industry, because of the extremely increasing demand for modules, manufacturers' world-wide added large production capacities. The Chinese had a big advantage over most of their "older" competitors in other countries because production lines of the competitors were assembled during their many years of operation; they were not the modern automated systems that the Chinese companies were buying.

This book is about the history of the terrestrial PV industry. "Follow the money" was the advice given to the reporters investigating the "Watergate break in" in the movie *All the President's Men*. In Acts 1 and 2 of this book, to write the history of the terrestrial PV industry was helped very much by "following the money." But to write the history of the PV industry, in addition, one also had to follow the politics/government actions. To narrate the story of Acts 1 and 2 was easier because we can call it ancient history, it happened in the last century. To follow the money and politics in the first 12 years of this century is not as easy; the fog had not lifted yet, and that makes the narrative of the turbulent PV industry more difficult.

It was described above that the seed money for the Chinese PV industry most likely came from private Chinese people and from municipal governments. For example, Mr. Xiaofeng Peng, who realized that probably nobody was manufacturing Si wafers in China, invested $30 million of his own money and started LDK Solar Company in July 2005.[314] He also received some 80 million in venture money. The first sale of multi-crystalline wafers was in April 2006. LDK was only about 2 years old when in June 2007 it was able to raise $469,368,000 with an IPO on the NYS (lead underwriters were Morgan Stanley

[314]http://en.wikipedia.org/wiki/LDK_Solar_Co.

and UBS Investment Bank). As a matter of fact, as was shown above, the Chinese PV industry was financed with over 2 billion dollars raised on the US stock market from probably American investors. *Without that money obtained in the USA, there would have not been a Chinese PV miracle.*

In comparison, during the same period, only two US companies producing PV cells and modules raised money on the NYS: SunPower Corporation[315] raised $145 million in November 2005 and First Solar $400 million in November 2006. In Germany, SolarWorld raised money in an IPO in 1999, and Q Cells, which was founded in 1999, raised on the German stock exchange $350 million (Euro 313 million) in October 2005. It seems money for solar investment was available. Then why was there difference in the number of companies which benefited in China, Europe, and the USA?

From my experiences at Solarex, I can fully understand what happened in the following years. As I wrote in a previous chapter: "Because Solarex was a 'rapidly expanding' company, it needed a lot of cash for its operation." But this is an understatement compared to the situation that happened recently. During the years from 2005 to the present, the demand for PV modules was not "rapidly expanding" but was "exploding." All of these companies were profitable at the beginning, but as with Solarex the profit was insufficient to provide all the money needed for such an expansion. To finance their rapidly expanding production and sales, in addition to the equity investments, Chinese PV manufacturers have also received very large amounts of loans from many sources: private, banks, local municipalities, and even the WB's IFC. These PV companies had to take loans to finance the explosive expansion and their profit seemed to be sufficient to service those loans. LDK, for example, winded up with much more than a billion dollar loans. It is estimated that in some cases the debt became several times the company's total equity.

Because of global recession, changes in the FiT system, the demand for PV modules from 2011 did not increase as expected.

---

[315]An excellent review of the history of SunPower was written by its founder Dr. Robert M. Swanson (2011) *Palz: Power for the World*, Pan Stanford Publishing, pp. 531–553.

The over-capacity resulted in price erosion and accumulation of inventory and substantial financial losses. Many of the small companies were closed.

Among the very large ones, some of them were sold, some of them would go bankrupt, and others would find money to survive. A solution for some of them was that if they received more orders than they were able to handle, the essence of Finlay Colville's description in his paper[316] was that the possibility existed that other c-Si manufacturers having excess capacity or even being temporarily shut down because of lack of work could be used by leading c-Si manufacturers as "virtual capacity" to save time and outsource the work to them instead of adding another production line. The product would be obviously labeled with the leading PV manufacturer's name and sold under its name.

The problem occurs when the leading manufacturer advertises or sells the product under his name stating that the leading manufacturer has ISO 2001 certification and the product was tested according to the relevant IEC standard, but the "virtual capacity" does not use the same Quality Management Program as the leading manufacturer and samples from the "virtual capacity" have not been tested according to the IEC standard. This can be avoided as follows:

- Either the manufacturer of the "virtual capacity" uses the same QM program and materials as the leading manufacturer and gets an ISO 2001 certification from a reputable registrar, and furthermore, samples from the products from the "virtual capacity" would be also tested;
- If that is not the case, display the PV GAP Quality Mark (or the Golden Sun Mark) on the leading manufacturer's product and nothing on the other, providing thereby a difference which customers can realize.

Most of these companies will obviously survive because of their excellent products, having competent management and also as a result of the unfolding extremely large global as well

---

[316]NPD Solarbuzz, http://www.pv-tech.org/guest_blog/pv_for_sale_or_rent_chinas_12gw_virtual_fab.

as domestic PV market in China. Starting from a very low level, by the end of 2012 China has installed 8.3 GW of solar PV capacity. Most of these companies will be able to solve their financial problems and will stay as major producers of PV products.

It seems that the downhill ride may be over as predicted by the market research group IHS. They reported that the first quarter of 2013 marked the end of more than 2 years of sharp declines as pricing for solar PV modules achieved stability, paving the way for the return to profitability for some leading suppliers.[317]

One interesting note about the quality of the Chinese PV modules: Based on the Chinese PV story described here, it is worth noting that as a result of the WB insistence on the importance of quality and the recognition of the value for the sale of their products and also the financial benefit of a QM program based on the PV GAP system, the "Chinese Golden Sun Quality Mark" was introduced for qualified PV products. This was also the reason that the Chinese PV modules except some of the minor manufacturers are very good.

Another result of the Chinese miracle was that the mass production achieved by these Chinese PV manufacturers proved Palz's First Axiom (Chapter 30), which was predicted by T. Bruton's work that PV production using the existing technology and utilizing c-Si wafers, by reaching the production level of 500 MWp of PV modules, could achieve prices that could go under $1.00/Wp. The PV module price of under $1.00/Wp is now a reality.

---

[317]http://www.pv-magazine.com/news/details/beitrag/sunny-days-ahead-for-module-suppliers_100012015/#axzz2apKrPfCc.

Chapter 38

Concentrated, Tracking, and the Solyndra PV Systems

Concentrated PV Systems

Concentrated sunlight was used in ancient times. Greek inventor Archimedes is said to have used mirrors to burn ships of an attacking Roman fleet in Syracuse, Sicily during a siege when Syracuse was a Greek colony (BC 224–222). The legend says that he used mirrors to concentrate sunlight on Roman ships and managed to burn them down.

To use sunlight concentrated by reflecting surfaces (mirrors) or lenses to produce electricity from crystalline silicon solar cell was considered as soon as solar cells were used for terrestrial purposes. In a concentrated solar cell system, the solar cells always have to be in the focal point of the mirror or lens and therefore the system has continuously to track the sun so the sun rays arrive perpendicular to the mirror or the lens. A stationary solar cell system receives the sun rays perpendicular to the cell surface only at noon. Because the mirror, the lens, or the solar cells have to be always perpendicular to the sun rays, the system has to have a tracking mechanism. The tracker is an electro-mechanical device which is programmed so that the

Sun above the Horizon: Meteoric Rise of the Solar Industry
Peter F. Varadi
Copyright © 2014 Peter F. Varadi
ISBN 978-981-4463-80-5 (Hardback), 978-981-4613-29-3 (Paperback), 978-981-4463-81-2 (eBook)
www.panstanford.com

sun ray should always be perpendicular. This means that the system should be turned to follow the sun.

There were several reasons why concentrated PV (CPV) systems were considered to be used. One of the reasons was that mirrors or lenses were cheaper at the beginning of the solar cell age to direct light onto solar cells than having the same area of solar cells to produce the equivalent amount of electricity. Another reason was that it was also found that the solar cells' efficiency is increased, thereby producing more electricity, by irradiating the solar cells with light the amount of which is the multiplicity of the sun's light.

The limitation of the CPV systems compared to a standard PV system is that, while the non-CPV system is producing electricity at any light levels, independent of whether it is a scattered light or direct sunshine, the CPV systems require direct sunlight to function.

When Solarex was started, the price of solar cells was very high. It was considered that using mirrors to concentrate or reflect sunlight on the solar cells could make the price of the produced electricity cheaper. The first experimental PV systems used mirrors as reflectors and not as concentrators which made them useful during the times when there was no direct sunshine.

Several experimental PV modules were made utilizing reflecting mirrors (Fig. 38.1). These experimental modules were tested and found acceptable but in spite of their lower price, in those days customers rather purchased the somewhat higher priced but not concentrated type.

High concentration (100–over 1000 suns) PV systems (HCPV) could have one more advantage. Because of the concentration of sun rays on the solar cell, the solar cell will become hot therefore, either air or water (or oil) cooling had to be used, to keep the cells at the optimum operating temperature, where they produce the most electricity. The hot water produced as a result of cooling the solar cells could be used, for example in bathrooms or other purposes. Therefore, a PV concentrator system producing electricity as well as hot water could be very useful for private houses or for housing developments and was thought that could be an important PV market segment. In the mid-1970s, a company in Switzerland was developing a dual axis (means that it follows the sun from the east to west horizontally as well as vertically because of its seasonal move), hybrid (producing electricity as well as hot water) concentrator system with parabolic mirrors, and in the focal plane a PV receiver was mounted to convert the concentrated light to electricity (Fig. 38.2). They contacted Solarex's European subsidiary in Switzerland to make for them such a strip PV receiver. It was decided that the system would need a water-cooled PV strip so the mounted solar cells should convert the high concentrated sunlight efficiently.

Figure 38.1 Low concentrated PV modules utilizing reflecting mirrors (1975).

PV Receivers in the focal point of the mirror

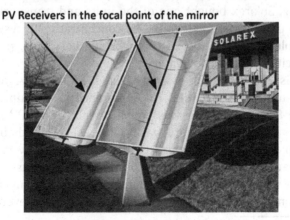

Figure 38.2 Dual axis concentrated PV system (1977).

Solar cells mounted on a water-cooled backing were designed. The solar cells were made in Solarex's Rockville, MD, facility, and the water-cooled PV strip (Fig. 38.3) attaching the solar cells were made for them by Solarex in Switzerland.

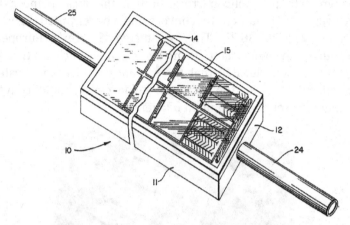

Figure 38.3 Water-cooled strip PV receiver[318] (10: copper enclosure; 11 and 12: walls of the enclosure; 14 and 15: solar cells connected in series; 24 and 25: water inlet and outlet).

It was believed that this system would be very good for the USA market, too, especially in the southern states, where there is abundant direct sunshine and where there are many private houses, housing developments, and schools. When the Swiss company completed the development of the HCPV system in 1977, Solarex bought one of the units and transported it to our US plant in Rockville, MD. The concentrator PV system was assembled and tested, and was working very well, producing electricity as well as hot water.

This HCPV system was marketed in the USA and was also exhibited at the IEEE PV Conference in Washington, DC, June 5–8, 1978. Solarex was trying to sell these CPV systems, but it was found that there was no interest to buy them. There was no market for the small or the hybrid one. By that time, I learned from my "business experience" that if customers do not buy a product, then one should not continue to make it. We abandoned the CPV system market. In those days several other

---

[318]US Patent No: 4,056,405 (Filed May 10, 1976).

companies tried to make and market PV concentrator systems. We made the special solar cells for them, but they were not very successful either.

These experiences were the reason that the author of this book, in a paper presented in 1993[319] at the ISES conference in Budapest, said, "The story of concentrators is the most puzzling non-development of the past 20 years. PV cells for 1 to 1000 times concentration of solar energy were developed practically 20 years ago. Tracking or stationary concentrator systems were used also practically 20 years ago. The efficiency of hybrid (PV and hot water) concentrator system could be 30–50% efficient, making them useful in many applications. In spite of this, CPV systems never became factors in this business. Will they become factors in the second 20 years? In spite of the history of the first 20 years, I would not count them out." But now the second 20 years passed, and I forecasted that I would not count out that PV concentrator systems are going to be used. Well, I was right.

As of 2012 they were about 100 MWp CPV systems installed globally, which is about 0.1% of the total PV installed in the world. It is probable that in 2012 CPV and HCPV systems installation could reach 90 MWp and it will have a rapid growth and in 2016 the total installed capacity will reach 1.2 GWp.

As mentioned before, CPV systems could be "low concentration" (LCPV) means 2–100 suns, or high concentration (HCPV), which means over 100–1000 or more suns. HCPV systems, besides requiring cooling so that the solar cell should operate efficiently as mentioned before, also require a precision two-axial tracking system so that the sunlight hits the cells exactly and perpendicularly. The precision for LCPV systems should be 1–3°, while for high-concentration systems the requirement is at least 0.5°. For highly efficient systems either mirrors or lenses may be used. Si solar cells are usually used for LCPV systems. For HCPV systems, Si as well as high-efficiency *multi-junction PV cell*, which are made for space, are also used.

---

[319]Varadi PF (August 1993, 23–27) *Photovoltaic Industry: Past, Present, and Future*, International Solar Energy Society meeting, Budapest.

Several companies are manufacturing HCPV and also LCPV systems (Figs. 38.4 and 38.5). One of the limitations about where CPV can be used—as mentioned before—is that they require direct sunshine. Other limitations are that in areas where there is plenty of direct sun, but sandstorms may occur (for example, parts of the Sahara), the mirrors or lenses could be damaged. Also, the tracking mechanism requires maintenance.

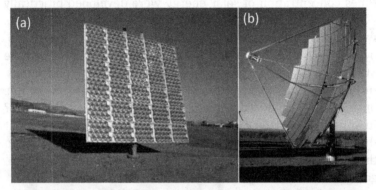

Figure 38.4 (a) SolFocus (Spain).[320] (b) Solar Systems Australia.[321]

Figure 38.5 Amonix (USA).[322]

---

[320]http://www.solfocus.com/en/.
[321]http://solarsystems.com.au/.
[322]http://amonix.com.

Because of very low PV module prices, no CPV projects were completed in the USA in 2011 and several announced CPV projects were converted to fixed PV systems.

Tracked (but Not Concentrated) PV Systems

The electric output of a fixed PV array (system) can be increased if the array is mounted on a stand which is able to track the sun. The tracking system can be one-axis tracker which follows the sun from the east to the west. In this case the PV system is mounted at an angle where it is most efficient during winter. Or a two-axis tracker can be used, which follows the sun during the day horizontally and also adjusts for the yearly motion of the sun, vertically. If one compares two identical arrays, one is in a fixed position, the other is mounted, for example on a two-axis tracker the tracked array will outperform the fixed one (Fig. 38.6). In the USA, the annual difference between the two arrays in favor of the tracked array, dependent on location, will be between 25% and 45%.

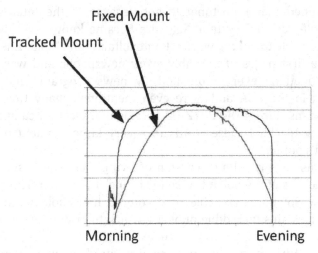

Figure 38.6 Comparison of fixed vs. tracked mounts (27%).[323]

It is obvious that the PV system mounted on a track mount will deliver more electric power than the identical one on a fixed

---

[323]John C. Sagebiel.

mount. The decision to design and build a PV system that it should be fixed mount or tracked mount will depend on which of them is less expensive.

To calculate the price difference is very simple. For example, a 1000 Wp fixed mount system requires, for example, an additional 400 Wp to produce the equivalent amount of electricity than the track mount system. This cost will be counter balanced by the cost of the track mount system and the electricity it uses for operation. The fixed mount system requires no maintenance, while the track mount system requires maintenance and probably repair, and its life will depend on the environment where it is deployed.

The Solyndra "Self-Tracking" PV System

The unique PV tracking system, developed and manufactured by the California-based Solyndra Corporation, which called it "self-tracking,"[324] has to be mentioned in this chapter. The power output curve of the Solyndra module has some resemblance to the curve of the "track mount" PV module. It has also to be mentioned that unfortunately not because of the interesting PV "self-tracking" system, Solyndra became known worldwide and was able to put the words "Photovoltaics" and "Solar Energy" on the first pages of probably every newspaper, and was also mentioned on every radio and TV news programs and talk shows in the USA and were even mentioned many times by politicians during the 2012 US presidential election campaign, thereby teaching these words to a very large segment of the world's population.

This requires the discussion of two issues. The first is the uniqueness of the Solyndra "self-tracking" PV system. The other is how Solyndra was able to become a household word and got for PV this incredible publicity which the PV industry could not achieve during its 40 years of existence.

Solyndra designed and manufactured a new type of unconventional PV module which consisted of a hermetically sealed cylindrical glass tube having a copper indium gallium diselenide (CIGS) thin-film solar cell on a second glass tube

---

[324]http://www.solyndra.com/technology-products/cylindrical-module/.

inside the hermetically sealed glass tube. The inner glass tube with the CIGS thin-film solar cell on its surface was mounted in the outer glass tube and the two were coupled with an optical coupling agent. The CIGS thin-film captured the sunlight across a 360° surface capable of converting direct, diffuse, and reflected sunlight into electricity. The CIGS thin-film solar cell has been used successfully as solar cells before (as described in Chapter 34), but only in a conventional flat module. The unusual nature of the Solyndra PV module was that it was a cylindrical glass tube, while all other PV modules (c-Si and thin-film) were flat.

Solyndra was founded in 2005 and was able to raise approximately $700 million from private investors and an additional $535 million US Government–secured loan. This adds up to a staggering $1235 million. Despite that this huge amount of money was invested, in a relatively short time on August 31, 2011, Solyndra went into a spectacular bankruptcy.

At the same time, as stated in Chapter 37, 11 Chinese PV startups raised a total of at least $2140 million, which per company was much less, in average about 10% than the capitalization of Solyndra. They raised the money not from the Chinese Government, but from US investors who bought their stocks listed on the US stock exchanges.

There was, however, a very big difference between Solyndra and the Chinese PV manufacturers in the utilization of the money. The Chinese companies were not gamblers. They used their money to purchase off-the-shelf modern automatic production lines available from several companies at competitive prices, making guaranteed quantity, and quality of the conventional c-Si solar cells and modules.

Solyndra, on the other hand, started to develop a brand new type of solar module which required the development of a new technology from the scratch and a new special manufacturing line in which practically nothing could be used from the conventional PV technology and manufacturing lines. The newly developed production line output to produce these new and unusual modules was planned to be so large that it would have put the company's capacity close to the top 10 PV manufacturers in the world.

The elementary mistake of the business plan was that it was believed that to achieve "competitive" prices for these totally different looking solar modules, a very large factory should be built. This fabrication plant, "Fab 2," was a $733 million state-of-the-art robotic facility in Fremont, CA, which opened in September 2010. The capacity of this plant was 300 MWp, to become 610 MWp by 2013.[325] It was probably another assumption that these tubular solar PV modules, in spite of being totally different looking and with a minimal field experience, at "competitive" prices would take a great market share from c-Si modules having 40 years of field experience. In reality, the size of a factory has to be built according to the demands for the product.

For a totally new type of product, such as Solyndra's cylindrical solar module, what is required is proper marketing and long time. This is true for every business, but one can take an example from the PV business, when the multicrystalline solar cells were developed. Solarex had to develop only from scratch the fairly simple automatic manufacturing machinery to produce the cast Si multicrystalline wafers, but the Solar cells from the wafers were made on the same production machinery used for the single crystal wafers. The only difference in the end product was that the surface of the solar cells made from the multicrystalline wafers looked different from the ones made from single crystals. When Solarex started to produce multicrystalline wafers in quantities, several people urged to switch the entire production from single crystals to the new multicrystalline wafers. Luckily we realized that people are very conservative in everything, except in women's clothing design and dancing. People, especially engineers, buy what they know works and therefore, the new products selling even at the same price or maybe even a little cheaper, have to be properly introduced. This was the case with the multicrystalline modules, which were only very little different from the single crystal ones, compared to the tubular Solyndra modules which were totally different from the flat

---

[325]http://en.wikipedia.org/wiki/Solyndra.

conventional PV modules. After lots of in-house testing we started with a 200 kWp (which in 1982 was a large system) which was installed on Solarex's factory roof. Based on that, DOE financed in 1983 a 300 kWp (which was about 3% of the world's total production in that year) PV roof at Georgetown University in Washington, DC, which was followed by installations by customers who knew Solarex very well and ventured to acquire modules with the new type of PV cell and slowly, in about 5–6 years after the production of the multicrystalline wafer was introduced, Solarex was able to switch over almost the entire production to multicrystalline. It is only now, about 30 years later, that somewhat more than 50% of all c-Si systems are using multicrystalline wafers. This story is only to indicate that the acceptance of a new product takes many years.

In comparison, Solyndra, after developing the cylindrical PV modules—drastically different from the flat ones including the mounting hardware, which people had used for 30 years, built a factory immediately and hired 1000 people to produce 300 MWp/year, and in 2010, the company's largest installation was only a 3 MW[326] Solyndra "self-tracking" system in Belgium, which was 0.02% of that year's PV production.

The elementary mistake of Solyndra's business plan was that it assumed the new tubular PV modules would overnight replace the flat PV ones, which neither the CPV nor the simple tracking systems could do in 35 years. Nevertheless, the "self-tracking" concept is interesting and after people forget about this fiasco, the concept may start to be used for certain applications.

---

[326]http://www.solyndra.com/about-us/timeline/.

Chapter 39

The PV Rollercoaster

It can be considered a real controversy to state that the demand for PV systems is growing, that the PV business is "unstoppable," and at the same time that the PV industry is going through a crisis in which not only the small manufacturers go out of business, but some of the largest ones may go into or teetering around bankruptcy, or are sold or merged at bargain prices. Some people consider that it will be detrimental for PV. I, however, think this is a normal cyclical phenomenon as happens in every other industry. It happened in the PV industry before, we just forgot about it.

The PV industry experienced it in the 1980s when the Carter administration in the USA strongly promoted PV, which resulted in an overcapacity and then came President Reagan, who decimated the RE budget. The government-fueled demand went down and some companies heavily involved in government business, for example the Exxon-owned Solar Power Corporation, which was at time one of the largest PV cell and module producer, and some others were shut down. Solarex and Arco Solar survived, because they were more diversified, and their international and commercial business was far more than their government one.

Sun above the Horizon: Meteoric Rise of the Solar Industry
Peter F. Varadi
Copyright © 2014 Peter F. Varadi
ISBN 978-981-4463-80-5 (Hardback), 978-981-4613-29-3 (Paperback), 978-981-4463-81-2 (eBook)
www.panstanford.com

PV started in 2001 with increasing big demand. Insufficient amount of products were available due to the Poly-Si shortage until 2007, which started to push prices and profit up (Annex. 8). What happened was that because of the expectation of continued growth, all of the PV cell and module manufacturers embarked on an expansion of manufacturing operation and several new manufacturers entered because of the great demand. When Poly-Si became available again, the resulting expansion of the production capacity for c-Si and also for the thin-film PV modules was tremendous. The automatic c-Si cell and module as well as for thin-film PV module turnkey mass production lines were available from several equipment manufacturers. As companies were able to raise large amounts of money issuing stocks, bonds, or taking loans, they could easily increase their PV cell and module production capacity and as a result practically overnight every manufacturer winded up with "overcapacity." This resulted in a price war pushing prices lower and as a global recession and some changes in the FiT laws in various countries took place a slackening in the demand which happened to cause a large increase in the inventory. It was described before the PV module prices started a nosedive in 2007. That year, the price was about \$3.50/Wp and wound up around \$1.09 in 2011. In 2012, the global PV production was about 30,000 MW. During the second quarter of 2012, the price of a Wp went down to \$0.98 and in the fourth quarter it was only \$0.80. The results of these price reductions were manifold.

It obviously increased the demand for PV systems but caused financial losses to PV manufacturers, and as a result, some case PV manufacturers wound up in a bad financial situation. The world's financial crisis also affected banks and their capability to provide loans to these companies. Some of the PV companies closed, went bankrupt, or were sold.

Cyclical "overcapacity" and large inventory is a normal event in most industries and it is not a new phenomenon in the PV business either.

The building industry is one example. The cycle starts with the need for building space for offices and/or housing. People start to build, and the buildings are rented or sold fast. Builders see the opportunity to build and banks see the opportunity to lend money. This goes on for years and more and more companies or people and banks see the opportunity and get in. Price for space goes up, up, up, people or companies speculating that rental or space prices will continue to rise and build more and more, but at some point something happens and the need for space and housing decreases. Builders have overcapacity and inventory and as the result of lower demand, prices start to fall. Many builders and people who speculated in rising prices wind up with unsalable inventory and go bankrupt. This happened many times. For example in 1976, when the landlord, who owned the building in which Solarex rented only half of it, was not able to rent the second half and was not able to sell the building either, gave an offer to Solarex to take over the entire building and he would sell it to Solarex at a very low price. This was the bottom of the cycle. Solarex needed space and acquired the building in 1979. Four years later in 1983 Solarex outgrew the building and built a new factory, the building cycle started to go up and Solarex sold the building which was acquired in 1979 with profit without any problem.

The last time the US building cycle was on top was in 2007/2008, when it suddenly collapsed which everybody remembers. This rollercoaster cycle in the building industry is about 5–7 years up and 5–7 years down. The US building cycle was in early 2012 at the lowest point and now it is starting to go up and house prices are slowly rising and some builders venture to build. The automobile industry is also cyclical: Even the mighty General Motors had to be rescued from total bankruptcy in 2009.

What is happening now in PV is normal. The German FiT caused everybody to jump in to produce Si, wafers, solar cells, modules, and manufacturing equipment. This cycle peaked in 2011. Changes in FiT and recession, problems with banks and so on changed the landscape. Some of the companies will disappear, but many of the present ones will survive and the up-cycle started.

PV for several reasons is not stoppable anymore. The mass production achieved the low prices that were needed to be competitive. PV provides electricity during the day when most is needed. Because it is of a decentralized nature, it can be used anywhere and will fit into the new wireless environment. But there is one more reason it is unstoppable. It created many jobs and more jobs will be created as time goes on. The German Government realized that and nothing can be done to abolish PV.

Many papers published in magazines are forecasting doom for PV. But they are all wrong. This rollercoaster for any industry is a normal event. Those forecasting doom for PV are all newcomers. We old timers have seen this rollercoaster effect before, but PV was never as unstoppable as it is now. We are now at the beginning of an upswing. In a few years, the upswing will peak again, but during the upswing PV manufacturers who want to survive the next downhill ride have to learn two words: "smart diversification."

Chapter 40

The Last Barrier: "Bankability" = Proof of Quality/Reliability

In the past years, I have been reading more and more about the "bankability of PV systems." I found out from the *Merriam Webster* dictionary that the first known use of the word "bankable" was in 1818. I am writing a book about the PV industry and wanted to find out what the experts in word usage say what the word "bankability" mean and how does it relates to solar energy?

Is "bankability" being used in other businesses? Why did this question never come up buying a condominium or an automobile? Being a born paranoid, I immediately assumed that for PV, "bankability" is a new barrier to enter the market established by the anti-PV forces. But I found the use of "bankability" also in another industry. It is mentioned in the Thesaurus of the Word for Windows®: It means "likely to bring in money." This brings up an example: "a bankable movie star."

I came to the conclusion that it is obvious a project is considered bankable if lenders are willing to finance it. But in order to be willing to finance it, they need some assurance

Sun above the Horizon: Meteoric Rise of the Solar Industry
Peter F. Varadi
Copyright © 2014 Peter F. Varadi
ISBN 978-981-4463-80-5 (Hardback), 978-981-4613-29-3 (Paperback), 978-981-4463-81-2 (eBook)
www.panstanford.com

to believe that they can get their money back possibly with a suitable profit. I realized what "bankability" means from a bank's point of view when I compared the movie and the PV business. This can be understood from these two stories.

On one side of the DVD disk: Two fellows go to the bank for a meeting with the top director who is responsible for large loans to tell him that they need a $50 million loan for a planned movie and would like to make a presentation. They present him with a book containing the movie script and make a fantastic PowerPoint presentation about the proposed movie. Being in Hollywood, the Bank director probably sees every day at least two of such presentations. He asked them why they think the movie will be a success. They did not answer him, just turned on the last slide which read, "The movie will be directed by *Steven Allan Spielberg*." The "bankability" of the movie project was instantaneously established. The director picked up the telephone and told his secretary, "Please bring the forms needed for the loan to a movie."

On the other side of the DVD disk: Two fellows go to the Bank to have a meeting with the top director who is responsible for large loans to tell him they are looking for a $25 million loan for a planned PV system. They show him one of the PV modules to be used and make a fantastic PowerPoint presentation about the proposed 10 MW PV system to be erected on top of a landfill area, which is not used anymore for landfill so the land is not costing money. The presentation showed several similar systems, some of them even much larger. The presentation included a rendering of the proposed system and also that the local very reputable large power company is going to buy the electricity generated by the PV system for the next 20 years at a set price, which will be sufficient to service the loan and pay it back.

The director thanked them and said that it is a very interesting project, but he would like to have one of the bank's experts also see this presentation and they should come back a week later. A week later, a younger guy was also there in a dark blue suit and blue necktie. After the presentation he said:

"The banks are usually concerned about the return of their investment and how can it be proved to the bank that the PV system is dependable and reliable. The PV system has a mechanical structure and a number of components. The sturdiness of the mechanical structure can be easily verified by an engineer checking it. The components can be easily repaired or replaced. But how can the bank be assured that the PV modules are going to provide the required electricity for 20 years, when, for example, the manufacturer of the modules is only 7 years in business. This is the major issue. Come back with some proof of this to establish the bankability of the project."

This example explains bankability. A movie directed by Spielberg based on his successes provides a very good bet that the money will be returned to the bank in a short period of time after the movie is released. But how could anybody believe that a system with those PV modules exposed to the sun will supply electricity for 20 years and the only thing may be needed is to cut the grass occasionally between the rows of the modules? On what basis can these two people or anybody else prove and assure the "bankability" of the PV modules, which means that the PV system will provide the predicted amount of electricity for 20 years?

The conclusion is that "bankability" in every industry has to be defined differently. In the movie business, bankability = previous success. In the PV business, bankability = quality and reliability primarily of the PV modules have to be proven.

If we define that the bankability of PV systems mostly hinges on a proof of quality and reliability of the PV modules in the system, I am not surprised that the bank needs reassurance, some well-defined yardstick to measure these attributes especially that in the manufacturer's advertisements and brochures confusing statements may exist. Also that in a large number of solar reports published in the media in 2011–2013, one can hear a tocsin concerning PV modules.

Let's start with the essence of Paula Mints paper discussing that previously the PV industry participants during the first decades knew that delivering high quality, clean, reliable electricity, and profitability was advantageous for the industry and highly possible.[327]

I assume "previously" means in this case "8–12 years ago." What happened during those 10 years let's quote Mints, "Poorly installed systems and even counterfeit module products that carry a brand name and are really an off brand—good grief, what's an industry to do? Any day now we may encounter a guy on the corner with a pile of counterfeit modules, an armful of fake Rolexes and a pile of faux-Gucci purses all to be had at "$9.99."

I disagree, however, with the person quoted in Susan Kraemer's paper, "concerns about long term quality blowback from the price wars." The quality "blowback" started long before the 2011–2012 price wars, it started during the years 2004–2009 when PV was a seller's market and the customers bought everything which had a "solar look" and the word "quality" was erased from the dictionary. I, however, fully agree with Mints: "The PV industry needs to get back to its quality roots."

This would not be the first time that PV module *manufacturers needed directions to improve or maintain* quality. In Chapter 9 of this book, we compared the terrestrial PV industry's very first PV modules to those "magnificent flying machines" made at the beginning of aviation. Then as described, in 1975 the Jet Propulsion Laboratory (JPL) quality program was initiated, which established the basis of the quality manufacturing to produce

---

[327]Mints P (September 6, 2012) *A Solar Panel Manifesto*, Renewable energy World. http://www.renewableenergyworld.com/rea/news/article/2012-/09/a-solar-panel-quality-manifesto.

Wong U (15 November 2012) *Solar Struggle: A Rise of Poorly Made PV Modules?* Renewable Energy World, http://www.renewableenergyworld.com/rea/news/print/article/2012/11/solar-struggle-a-ri.

Kraemer S (20 November 2012) *Is the Solar Industry Prepared for a Quality Blowout?* PV Insider.

Kaften C (1 June 2012) *PV Modules: Branding and Bankability do not Always Correlate to Quality*, PV Magazine.

Hall M (24 May 2013) *Call for Tougher Solar Module Standards*, PV Magazine.

Woody T (28 May 2013) *Solar Industry Anxious over Defective Panels*, The New York Times.

long-lasting PV modules. It was estimated that manufacturers implementing and fully adhering to the JPL quality method were able to produce PV modules lasting at least 20 years, and in this book Chapter 9 describes this and if somebody needs more information and verification, please read Annex. 9.

The JPL program was completed in 1985 without any continuing program. Manufacturers who participated in the JPL program realized the advantages of quality products and continued the practice. However, several of the new manufacturers who entered into the PV cell and module manufacturing business probably knew nothing about the JPL Quality Management program and its requirements. As a result some of them did not even test the modules. Others, to save money, did not test them either. The quality of the modules produced by these manufacturers was—we can generously call them—"questionable."

That was the second time PV module manufacturers had to make corrective actions to "get back to the quality roots." The PV module manufacturers adhering to quality standards were still at that time in the majority, and to differentiate their good-quality products from the non-good-quality ones—as described in Chapter 27—with the support of the World Bank, UNDP, etc., in 1996 they established the PV GAP, which was basically the then up-to-date version of the JPL program requiring a similar quality approval system.

The PV GAP program was supported by the European (EPIA), Japanese, and USA PV industry associations. The PV GAP qualification requirements were simple, not expensive, and not difficult, and several manufacturers qualified their products without any problem. If a product, for example, a PV module, qualified, the company received certification and was also licensed to display a distinguishing PV quality mark on the product (see Chapter 27) which was provided free of charge by the PV industry associations so customers would be able to visually recognize and be able to distinguish them from other products of unknown quality. The World Bank accepted the PV GAP quality program and recommended to their clients the use of those PV products which were approved according to the PV GAP system.

In February 2003 issue of *Photon International*, Michael Schmela, editor-in-chief wrote, "The non-profit organization's (PV GAP) idea is worth supporting."

The support of PV GAP recommended by Schmela started to fade in 2004 and disappeared in 2006. It is rather easy to understand today why the PV industry associations which were founders of PV GAP to ensure recognition of quality PV components and systems suddenly abandoned it.

It was not that PV GAP became expensive. On the contrary, PV GAP was a non-profit organization approved by the Canton of Geneva, Switzerland. Its administration was located in the IEC building in Geneva in a partial office provided free of charge by IEC, a part time secretary was paid by PV GAP. PV GAP also needed some money for stationary and brochures. This was an absolute minimum financial support PV GAP received for its operation from the beginning (1996) from the industry associations and that amount was never changed. Persons (such as myself) involved in PV GAP did not receive any compensation, including travel or out of pocket expenses.

The reason of the abandonment of PV GAP was simply that the PV GAP system insisting on quality and reliability of PV products suddenly became an unwanted handicap as many modules—as it was briefly mentioned before and as we can see below—were not even tested and would have been bad for business to call attention to this fact by having some displaying the PV GAP Quality Mark and others not.

As described in Chapter 35, realizing the profit opportunity created by the FiT system, many promoters got into action at the beginning mostly in Germany and raised the huge amounts of money from the public to establish PV systems. Because of this demand and seeing all that cash, the "PV system integrator" branch of the PV industry attracted many newcomers. As far as the size of the PV systems and the money required were concerned, only the sky was the limit.

Because of the extremely large number of PV systems to be built the problem for the "PV system integrators" was that larger and larger quantities of PV modules were needed to fulfill the ever increasing demand. Their problem was where can they get that large quantity of solar modules?

The PV industry, which was fighting for survival for a decade until the FiT produced the "PV rush," was obviously not prepared for this sudden incredible large demand. It is relatively simple to set up a not very large-scale PV cell and module production, which required only a little capital investment especially at places where the labor was cheap. To fulfill this "PV rush" to produce modules, hundreds or maybe even a thousand solar cell and module manufacturer appeared overnight all over the world, such as mushrooms after rain.

This was the beginning of PV's "wild west" period. Every module independent of their quality could be sold. Every year in its February issue, *Photon International* magazine provided a table listing all of the PV modules on the market and also posted their data (Table 40.1). The data published in *Photon International*[328] magazine's February 2004, 2006, 2009, 2010, and 2012[329] issues showed the total number of PV modules and indicated whether the module was tested according to IEC 61215/61646.

Table 40.1 Large percentage of all of the PV modules on the market are not even tested (!)

Year	Number of modules listed	Not tested to IEC 61215/61646	Percentage of not tested
2004	342	85	25
2006	555	166	30
2009	2765	704	26
2010	884 "Leading suppliers only" (USA edition)	115	13

The so-called "untested" PV modules offered on the market were also purchased. Manufacturers of those untested modules at least told honestly that their product was not tested as they knew at that time the customers did not care whether the product was or was not tested. (I mention "honestly," because tested modules, as Mints says, could be "counterfeit modules"

---

[328]*Photon International* (February 2006).
[329]*Photon Magazine*, USA edition, 2, 2012.

as testing laboratories do not check routinely the origin of the modules they receive for testing.)

Photon explains the decrease[330] in the listed modules between 2009 and 2010, "Last year (2009), a large number of small vendors were added using data sheets collected at trade fairs, this year (2010), many of these companies didn't respond to our request for information. Moreover, many vendors have disappeared from the market." This is an indication of the stability of the module suppliers.

An unfortunate result of the explosion for the demand of PV modules caused by FiT was that everything which had only a "solar look" was bought up as nobody checked for quality. But regardless of whether the PV module was tested or not, the power rating as well as the 20-year guarantee was surely stamped on the modules. The only requirement was quantity, and everybody accepted whatever power ratings and 20-year guarantee was stamped on the module.

This explains that the PV GAP requirements became a very serious handicap in obtaining large amounts of modules. The best approach was to forget that it even existed.

Obviously utilities owned only very few PV systems and they did not care about the quality of the Independent Power Producer's (IPP) PV systems whether it works or not because they were only responsible to pay for the electricity which actually was fed into their grid by net metering or under FiT rules and not what was stamped on the module.

Under the FiT or net metering program, for an investor investing in a PV system, the return of investment was based on that the PV modules will perform for at least 20 years with a minimum of 80% power guaranteed. Obviously if the PV module will not last 20 years, investors will not be happy to lose money. The question is: On what basis is this 20-year guarantee given by the manufacturer, especially if the company's PV module manufacturing started only 7 (or not even 7) years ago? And will the manufacturer providing the guarantee be in business 20 years from now?

---

[330]*Photon International* magazine 2, 2010, page 136.

At the beginning, the FiT systems were relatively "small" and the majority of the systems were installed in small homes the investors were buying it on the advice of a promoter or like gambling in a casino they buy the PV system from a randomly selected vendor. Obviously private investors can take risks, but when the proposed PV systems became bigger and the investors were sophisticated money managers, the PV systems' reliability and durability started to become an issue. In other cases, large investors started to work with bank loans, but banks—if they are going to loan money—have to express a serious interest of satisfaction about the quality and durability that the PV system will last and perform for 20 years. This is the point where we are back to the requirement of "bankability."

After 1975 and 1996, the PV module industry in 2013 is back for the third time to clean up their ranks to provide for banks, "system integrators," individual customers, and insurance companies a globally accepted certification to identify the quality PV modules, which are "bankable" in order to distinguish them from those of unknown quality.

Considering the 40 years of experience of the PV module industry, this would have been easy. But what is the present situation?

I was amazed when I Googled "Solar Bankability Consultants?" and got a large number of results.

One consultant started its advertisement with big red letters: "Solar Module Bankability: Is it Time for Manufacturing Standards?" (Suggestion: They should read Annex. 3 of this book; it may have all they were interested in.)

Another one advertised that the consultant company developed a "Breakthrough PV module Rating System," a calculation to show the modules' 25-year lifetime energy production (LEP?), but the calculation is proprietary and to protect the company, the ratings have not been validated by a neutral third party.

Another consulting organization developed the PV module manufacturers "Tier System." The Tier was primarily defined by the manufacturer's size and balance sheet and by brand awareness (?). It did not look at the quality or reliability of the finished product. There was only a "very loose correlation" found between the performance and the Tiers. (One could ask the question: Was the company's balance sheet good, because they saved money by not testing the modules?)

Another one advertises its services: "Our bankability services offer one-stop quality assurance that covers your entire quality control needs—from raw materials to modules and PV systems."

There are banks or insurance companies which declare that they do not believe in consultants; they hire people and send them to the manufacturers to establish their own opinion. But on what basis?

PV "bankability consulting" is obviously a good business. There are thousands of prospective customers: PV manufacturers, "system integrators," individual customers, banks, and insurance companies. By reading the advertisements of the "bankability consultants," it seems that the great majority of them are trying to invent a system to rate the PV modules not necessarily according to their quality and reliability.

Obviously it is not in the PV modules "bankability consultants" interest to realize that terrestrial PV modules have by now been manufactured for 40 years and during that time the PV industry and their customers must have faced the problem of quality and reliability. The reader of this book may have realized that the PV module quality and reliability was one of the major issues from the outset of the terrestrial PV industry.

The "consulting system" would force PV manufacturers to obtain for each job a different "bankability" certification from a "consulting company" whichever the bank selects. This would slow down the implementation of the systems and would add to the cost.

Instead of this new untried and at times even amusing "consulting system" approach, it would be very simple to use the already globally accepted, tried, and utilized PV GAP proven quality certification system which was created by the PV industry and customers (such as the World Bank, etc.) to establish the quality and reliability certification not only for PV modules but also for other components and systems.

The following system, which is based on the best scientific and technical knowledge of today, could be introduced immediately and it would provide a useful tool to establish "bankability."

Let's repeat it once more and express it in simple terms:

(1) Products (e.g., modules) to be tested by an accredited PV testing laboratory to the latest Qualification Test Standards (IEC 61215 and 61646). The IEC 61215/61646 standards were not designed or covered reliability of PV modules. But today it is the best known method to forecast that the module will provide at least the 80% of its rated electric output up to a 20-year life (see Annex. 9.).

(2) The manufacturing facility has to have ISO 9001:2001 certification for the production facility certified by an accredited ISO registrar.

This step verifies that a real manufacturing facility exists and has a Quality Management system which assures that all of the manufactured products will have the same quality and the inspectors may take samples for testing, elimination, and cheating of the test data (see. Annex. 9.).

(3) Random unscheduled inspection may also be able to take samples for testing.

(4) Displaying a PV Quality Mark (PV GAP or Golden Sun). This would also assure that the manufacturer indicates that the product, for example, the PV module, was produced fulfilling all of the above requirements.

The PV GAP system is now administered by the IEC System for Conformity Testing and Certification of Electrotechnical Equipment and Components (IECEE). The IECEE PV program is described in detail in the IECEE Operational and Ruling Document.[331] At present, IECEE has in 14 countries 39 PV

---

[331]IECEE PV Program, Procedure for Certification of PV Products and the use of the IECEE PV quality mark and the OPV quality seal, OD-2051-ED. 1.0.

member organizations[332] to provide PV product certification and to authorize the use of the PV GAP Quality Mark and Seal for qualifying products.

Because the World Bank had a major influence in China by insisting on quality for PV systems, China adopted the PV GAP system and created the equivalent in the Golden Sun Quality Mark program (detailed in Chapter 37) for PV.

These could be an immediately available solutions, but a final solution detailed in Annex. 9 should be completed as soon as possible. Also, an "International Conformity Assessment" body needs to be established. The best option is that an IEC Renewable Energy (IECRE) should be formed to do the conformity assessment for PV, wind, solar thermal, and ocean wave energy.[333]

---

[332]http://www.iecee.org/pv/html/pvcntris.htm.
[333]Wohlgemuth J (NREL) private communication.

Chapter 41

The Effect of PV on the Transformation of the Electric Utilities

The Solar Electric Power Association's (SEPA) President Julia Hamm in her speech[334] in 2012 pointed out that "utility and the solar industry continue at odds rather than focused on finding win–win solutions." She has a very good point but I believe the utility and the solar industry did not focus on finding win–win solutions and this was the result of the utilities' corporate culture as the PV community from the beginning believed in a strong cooperation with utilities. One should recall from Chapter 4 of this book that 40 years before her presentation, in 1972 an expert panel believed that "central ground stations" will be one of the three major applications of PV. The US Department of Energy (DOE) and the oil company investors in PV companies continued to steer the PV industry toward this idea, the "big picture," the central PV power station. But this idea was from the outset not even considered or maybe to use a better word, ignored or not taken seriously by the utilities.

---

[334]Hamm J (2012) *Speech at the Solar Power Industry (SPI) Conference*, Orlando, FL.

Sun above the Horizon: Meteoric Rise of the Solar Industry
Peter F. Varadi
Copyright © 2014 Peter F. Varadi
ISBN 978-981-4463-80-5 (Hardback), 978-981-4613-29-3 (Paperback), 978-981-4463-81-2 (eBook)
www.panstanford.com

This is best illustrated from the Electric Power Research Institute's[335] (EPRI) recent (2011 and 2013) Websites. The utilities are maintaining EPRI and its research activities are funded by the utilities membership participation. (Members represent more than 90% of the electricity generated and delivered in the US, and International participation which extends up to 40 countries.) Perhaps it would have been EPRI's role to work with the fledgling PV industry from its beginning and focus on "finding win–win solutions."

Instead, EPRI—since the terrestrial PV systems were introduced in 1973—expressed only a marginal interest in PV and even in June 2011 (after the Japanese nuclear disaster) on its Web site the "Featured Research" included *Advanced Nuclear Fuel Cycles—Main Challenges and Strategic Choices,* but no research related to PV was listed.

It seems that EPRI did not grasp the extent to which PV was used in the world when in 2011 it published[336] its 2013 Research Portfolio forecast in which is stated, "EPRI will continue to track the development of all major solar technology options and provide insights on technology maturity, market trends, major manufacturers, and the likely scale and timeframe of market growth. In addition, the 2013 Solar Program will look to enhance performance and reliability through field testing, demonstrations, and targeted studies." For this, EPRI's estimated 2013 program funding (according to the 2011 Webpage) was scheduled $920,000 (less than 1 million dollar). It seems that EPRI in 2011 believed the calendar year was still 1980, when the total PV module sale in the world was only 4 MW[337] and not 17,400 MW as it was in 2010 (equivalent to a few nuclear power plants) and the investment in PV was not in the millions, but in the $100 billion range.

Reviewing EPRI's 2013 Web site,[338] it seems that in the beginning of 2013, when in Germany 32,300 MW and in the USA

---

[335]EPRI website: http://my.epri.com/portal/server.pt?.

[336]EPRI's 2013 Research Portfolio (http://portfolio.epri.com/ProgramTab. aspx?sId=ENV&rId=221&pId=6912).

[337]Starr MR, Palz W (1983) *Photovoltaic Power for Europe, An Assessment Study,* Reidel Publ. Co., Dordrecht, The Netherlands, ISBN 90-277-1556-4.

[338]http://www.epri.com/Pages/Industry-Challenges.aspx.

about 7600 MW PV systems were already installed, EPRI's report on "Renewable Energy and Integration" seems to be still in the 1980s, because it has the following statement, "Utilities need a clearer understanding of the opportunities and risks of relying on renewable generation and how this likely will affect the bottom line."

On the other hand, it seems that EPRI had no study planned on how to deal with decentralized electric power sources, in spite that in 2003 the IEEE 1547 "Standard for Distributed Resources Interconnected with Electric Power Systems" was ratified. As Dick DeBlasio, who is the chair of the IEEE 2030 working group developing the standard wrote,[339] "Utility electric-power systems (EPS) historically had not been conceived to connect with active distribution-level generation and storage technologies."

As described in Chapter 10, the Sacramento Municipal Utility District (SMUD), in 1983 after 11 years from the beginning of the terrestrial PV age, was the first and until not long ago the only utility in the USA which installed a utility-scale (3.2 MW) solar system. On the other hand, utilities did everything to ensure that the electricity generated by RE, including the fledgling PV industry, should not be connected to the electric grid.

Many of the utilities—like an ostrich burying its head in the sand—even until recently tried to ignore the existence of PV, and now when globally already over 100,000 MW of PV systems are installed, only some of the 3269, US utilities realized that PV systems do exist.

This can be seen from SEPA's recent report "2012 SEPA Utility Solar Rankings" (June 2013). In January 2013, the utility solar rankings survey was sent out by SEPA to about 400 utilities in the USA. From the total, only 265 utilities responded. This indicates that in general a part of the US utilities are still not cognizant how far PV has advanced. In many cases, they were forced by a mandate of their local energy commission.

A change came recently, when the Edison Electric Institute (EEI),[340] which is the association of US shareholders owned electric

---

[339]DeBlasio D (June 28, 2012) *Interconnection Standards for the Smart Grid*, IEEE 2030 Working Group, *PV Magazine*.

[340]www.eei.org.

companies, issued a report in January 2013 entitled *Disruptive Challenges: Financial Implications and Strategic Responses to a Changing Retail Electric Business*."[341] This extremely interesting publication realizes what the title of this chapter indicates, except that it describes it more definitely. It indicates that PV—a disruptive technology—is emerging, which could directly threaten the centralized utility model.

The utilities' aversion for the distribution-level PV systems may be also the result of the utilities' "corporate culture." The destructive effect of the "corporate culture" to prohibit companies moving into a new but related field can be shown in many examples.

The EEI paper mentioned that Kodak's film and photographic business was replaced by digital technology, including movie theaters which are discontinuing to project films and the projection machinery is now digital. Kodak because of their "corporate culture" did not invest in digital technology and at the end filed for bankruptcy in 2012.

The EEI paper also mentions the case of AT&T. There is a similarity between the US AT&T, the nickname of which was "Ma Bell" and the electric utilities, namely they were monopolistic and regulated businesses.

Ma Bell's corporate culture was to rent telephone sets and collect a monthly fee for them and collect money for each call. When other organizations started to make and sell telephone sets, Bell was very seriously fighting the connection of those telephones to their wires in order to prevent their use. Ma Bell's corporate culture did not allow making any changes, for example to slowly discontinue the rental business and make and sell telephones. Today corporations exist, which make and sell telephone sets, but none of them is a descendent of Bell and the telephone set rental business practically does not exist anymore.

The electric utilities corporate structure is the central power station to sell the produced electricity to customers. The new idea of the distributed electric power generation, such as solar, did not agree with the utilities' corporate culture and

[341]http://www.eei.org/ourissues/finance/Documents/disruptivechallenges. pdf.

also resisted the connection of PV-generated electricity to their grid. In order to fathom the revolutionary changes, RE electric sources, and especially PV are causing to the electric utilities it is best to look at how the utilities were started and how they wound up where they are now.

Starting with the twentieth century, in every major locality a factory was established to produce electricity and wire up the town to sell the product to retail customers. As the electricity demand grew, more power plants were built. Utilities in various locations also connected to each other to buy or sell electricity if for some or other reason they had over-capacity or shortages. The transfer of electricity between generating facilities was accomplished by high-voltage wires. These great numbers of low and/or high voltage wires are collectively called the "grid." The central electric factory and the wiring required large investments. To have the privilege to be able to wire up the locality to provide it with electricity and to give the company a monopoly, the localities either owned the utility or they were regulated by local authorities over pricing of the electricity, making sure that the electric factory should also be profitable to attract investors and also to make sure that it will be able to enlarge the facility as the customer base is going to expand.

Bringing all this down to an understandable level, utilities make money by receiving a guaranteed rate of return on the investments they have made in the infrastructure, generating plants, and the electric grid they built. If they make new investment by building more power plants, more wiring that is excellent for the company's rate of return translated to an understandable word "profit." All this is paid by the rate payers. PV built on roofs and feeding electricity into the grid reduces the need to build more power plants which would boost the income and profit of the utilities. And that makes understandable the utilities aversion to PV.

Similarly, to AT&T which insisted that non-AT&T equipment could not be connected to their monopolistic telephone lines, the utilities such as "Ma Bell" were fighting the connection of distributed systems such as PV to be connected to the electric grid. As it was with AT&T, new technology started to change their system, this happened to the utilities also that the above outlined simple and idyllic situation was disturbed by regulations that electricity generated by non-utility entities, for example PV had to be accepted by the utilities to be connected at any point to their grid. This was the "Public Utility Regulatory Policies Act" (PURPA) passed in November 1978 by the United States Congress. This law mandated that utilities were required to purchase electricity generated by independent power producers (IPP). Utilities still resisted. Later (in 2003), as mentioned before, a standard was accepted (IEEE 1547) for the interconnection of distributed renewable technologies to the electric grid. The utilities should have no more excuse to prevent RE to be connected to the grid. But the situation did not become totally resolved. The US Federal Energy Regulatory Commission (FERC) is continuously improving the rules how smaller (up to 5 MW) PV systems which are the majority, could be easily connected to the grid.

The resistance of the utilities to let electricity generated by PV or other RE sources be connected to the grid was not a unique USA phenomenon. The situation was also the same in other countries. For example, in Germany in 1991 a law was needed as in the USA to force the utilities, in spite of their resistance, to accept the electricity produced by RE.

The similarity between AT&T and the utilities ends with the rules that they have to accept non-AT&T made telephones and the utilities have to allow to connect RE sources of electricity to their grid. The similarity ended because AT&T was a national monopoly and was broken into pieces, while both the US and the German utilities are like quilts assembled from thousands of small and several large producers. In the USA at present (August 2013) there are 3262 electricity providers of which 193 are investor-owned and have (MWh) 54.5% of the customers, while the 2006 publicly owned is servicing only 15.4%. All the others 30.1%. In Germany there are four large electric utilities providing electricity to 80% of the entire German electricity market, and there are more than 1000 other electricity providers.

The recent EEI report indicates that the utilities starting to realize that their business model is threatened not only their rate base and as a consequence their profit may decrease but because the new challenges caused by a major increase of the utilization of PV and other RE generating systems the entire existing utility concept may have to go through changes.

The utilities probably believed that the extremely large capital needed to enter the over 100-year-old centralized utility business is protecting it like the walls of a major fortress but they have not realized that the decentralized nature of PV is making cracks in the ramparts and ultimately the utility business will be different from what it was before.

Cracks were coming from many different directions:

(a) In most cases, strong public support for RE, such as in Germany, and in many other countries, including the USA forced the utilities to accept RE especially PV to generate some of the electricity they provide. In the USA, various states are mandating the utilization of certain percentage of RE electricity to be installed or purchased by the utilities selling electricity in that state. In Europe, the EU was setting the targets for the utilization of RE.

(b) To enter the RE electricity generating business as an IPP requires relatively small capital. For this reason and especially because of the decentralized nature of PV, where everybody's roof could be turned into a mini-utility, many new entries are possible. Another possible problem for the utilities is that an independent entity could become a large-scale IPP by buying up many PV fields and/or contracting many of those fields and establishing itself as for example an electricity source, thereby reducing the utilities' revenues.

(c) Net metering in the USA is currently in place in 43 states and the District of Columbia. This allows customers having their PV system to supply its own electricity but if it utilizes less than the PV system is producing, the electric meter turns backward reducing the previous usage charges. As more and more PV systems are being connected to the grid, as a result, utilities are actually selling the IPP's electricity thereby reducing their income. Utilities are trying to achieve a limitation of net metering to only up to 5% of their own production.

(d) Utilities also started to lose business because the electricity production of a PV system coincides with the peak power usage. The large number of PV mini-utilities feed electricity in the grid and larger customers are also installing systems and utilities cannot use their own "peak power stations" to provide electricity, depriving them from collecting the expensive "peak power rate." If a customer is using its own electricity utilities cannot do anything about it, but it certainly results in declining income. In many instances investors are making a contract with large stores that the store is only providing its roof but invest nothing, while the investors acting like IPPs are selling the produced electricity at a fixed price to the store on a long range (e.g., 20 year) contract.

More and more of the thousands of large and small electric utilities realized their problem in what these penetrations into their business can mean. This is well summarized in SEPA Hamm' presentation in 2011,[342] which was 2 years before the EEI report was published.

The total capacity of the installed PV systems in Germany by the end of 2011 was 24,700 MW and remarkably the four large German *utilities did not buy any* (Table 41.1).

Table 41.1 Germany (2011): Installed capacity and investment

Ownership	MWp installed MW (2011)	Invested money (estimate)
Private	24,700	US $80 billion
German utility owned[343]	0	0

In Germany, more and more property owners and investors choose to generate power with their own solar facilities, and large utilities are seeing this as a loss of income.

As proof of that the cracks in the ramparts of the utilities started seriously affecting their business, let's quote Dr. Johannes Teyssen Chief Executive Officer and Member of Management

---

[342]Hamm J (2011) *Speech at the Solar Power Industry (SPI) Conference*, Dallas, TX.
[343]Wolfe P. *Private Communication*, philip@wiki-solar.org.

Board of E.ON, the second largest German utility. E.ON is little involved in wind energy projects but not in PV. Dr. Teyssen already noticed in the company's 2011 annual report the existence of PV, what he calls "distributed generation": "Distributed generation also has the potential to play an important role in the energy world of the future. A team of experts drawn from across our company is currently designing a strategy for propelling E.ON's growth in this area."

Only 9 months later at the end of the third quarter of 2012, in E.ON's third quarter report, Dr. Teyssen realized that the future when the distributed generation will play an important role in the energy world must have been started a few years before, because he sounded the alarm bell as he realized that "in most European markets, the gross margin for gas fired units is approaching zero or is indeed already negative. One factor is that the demand for electricity remains very low. But another *key factor is that renewable-source electricity is being fed into the grid during peak load periods.*"

The situation also provides some interesting observations about the major German electrical utilities and their future. They expressed some interest in wind energy as that was a little bit closer to their business model. On the other hand, PV was viewed as an expensive technical curiosity to be researched to find a breakthrough to become inexpensive. The utilities cannot be blamed they adopted this idea, because as described in this book that was the prevailing opinion of "PV experts." On the other hand people in the PV industry knew that big demand was needed to trigger mass-production to make PV inexpensive. The PV industry people also believed that for the mass production no new breakthrough or new materials were needed only some development of the already known process because everything was ready, technology, automatic machinery even was known how to get inexpensive Si raw material. When the demand would become high, it could be implemented fast.

The FiT created the needed demand and the world-wide production of PV modules which was in 1997 only 114 MW, 8 years later in 2004 at the beginning of the German FiT was 1000 MW and again 8 years later in 2012 as the result of the improved FiT law reached an incredible 31,500 MW. *This 1-year production of PV is equivalent to about the 24 hour a day electric output of 5 nuclear power plants.* Obviously neither the "PV experts" nor the German utilities expected this.

German utilities did not consider another important matter: the true decentralized nature of PV that could be installed as a small or large electricity generator. This meant that people do not have to have a huge amount of money to become a utility, since one can become a utility by installing less than 1 kW but with more money it could be bigger 5, 10, 100 kW, 1 MW even 500 MW or more. The most unbelievable result was that by 2012 in Germany a very large number of individuals became mini-IPPs and cumulatively 20,000 MW of these small utilities were installed, the electric output of these mini-PV systems is equivalent to the capacity of about three full time operating nuclear power plants. Also, in addition even the installation of very large above 10 MW PV utilities started to grow. The problem became for German utilities that PV generates the largest amount of electricity during daylight hours, replacing the utility-generated electricity during peak power usage as noticed by Dr. Teyssen because this hit the utilities in their soft underbelly.

Interestingly the big utilities started only in 2011 to consider the situation, as it is stated in E.ON's 2011 annual report: "designing a strategy for propelling E.ON's growth in this area." By that time, the total installed PV electric systems output in Germany would be an amount to make it the fifth largest utility in the country. It seems that even the second largest German utility E.ON realized only in 2011 that PV is not anymore a nuisance which takes away some business powering garden lights, but became a substantial source of electricity which is damaging E.ON's peak load business and they may have realized that this is only the beginning of lost business.

These major utilities could maybe resort to the old approach and try to somehow restrict PV by political means, but this time this may not work. There was a strong public support for PV, which became even bigger considering the extremely large number of homeowners and farmers with a total of 20,000 MW PV on their roofs. Furthermore, the German Government will consider everything very carefully as at the beginning of 2011 Germany had 133,000 jobs associated with production and deployment of PV[344] which is maybe twice as many workers as are employed by E.ON in Germany.

The situation that the large German utilities are facing is very similar to what was happening in the present telephone business that also started with a central telephone station connected to customers by wires. It was a monopolistic situation which was, as mentioned, regulated by localities and governments similar to the electric utilities. Same way as the utilities, it was a regulated stable and profitable business. The emerging cell (mobile) phones which were not connected to the local central station by wire, but could be used anywhere in the world, became an unregulated business, where the profit is limited only by competitors. Telephone companies realized this during the nascent stage of the cell phones and they had the vision and overcame their "corporate culture" of the wired phone business, established independent separate corporate structure for their unregulated cell phone business and put their money into that business and in some instances they were even trying to unload the wired telephone business which was not anymore a growth area. As a matter of fact, Verizon lost 45% of the wired customers over the past 5 years. In May 2009, Verizon sold its wired telephone business in 12 mostly western states to Frontier Telecommunication Corporation.

Facing this doomsday forecast described above, the four major German utilities opened their eyes and realized that PV, the technical curiosity they ignored for so many years, became one of the major electricity producers in Germany, now even threatening their business. Maybe they took notice of the changes the telephone companies made and suddenly decided with or without the studies made by "team of experts," that they should do what the old saying teaches: "If you cannot beat them, join them." Utilities have suddenly realized that instead of fighting PV, they could sell decentralized PV systems and also develop a maintenance business of the profitable decentralized PV systems. And with this they were able to develop a truly "win–win solution."

[344]Palz W. *PV Policies and Markets*, Springer Verlag, New York.

For utilities it would be a win situation to design and install the PV system on the roofs of homes and barns. The utilities would also offer electricity storage system, which recently is also promoted and subsidized by the German Government, for houses. The utilities could offer regular maintenance of these systems and offer insurance. All of these would constitute a new business for them. They may lose the homes and farms as electricity customers, which was a regulated business but they will stay as customers needing maintenance of their PV systems and if they have surplus electricity they can sell to and if they need electricity they can buy from the utility.

The PV industry and homeowners will also win. The utilities will become one of the big customers of the PV industry and in their own interest they will not cause problems in interconnections. The homeowners will be happy to wind up with an uninterrupted electric source and an assurance that they will pay the same amount for electricity for 20 years. They will not be affected by the fuel cost fluctuations and not going to have blackouts because of grid overloading or a tree falling somewhere on a power line or a lightning strike into a transformer.

If I would have published this book 2 years ago and would have written what you are going to read; I would have called it utopia. Nobody would have believed it. But this unbelievable story actually happened. The four large German utilities turned suddenly "green."

Recent advertisements of these four large German utilities' decision about becoming "green" can be seen if one goes to a German-oriented search engine, for example "google.de" and see that all of the four major German utilities appear "solarized": RWE Solar, E.ON Solar, EnBW Solar, and Vattenfall Solar.

RWE Solar Web site declares: "Switch now to smart energy. Make your roof a profit center" E.ON Solar: "You should be your own electricity producer." EnBW Solar: "Innovative all-round-carefree-package for Photovoltaic customers." Vattenfall Solar: "Without moving parts, noise and emission-free, thousands of plants in this country provide electric power."[345] It is hard to believe, but they are all talking about "PV."

---

[345]The quotations are translations from the German websites of these companies.

All of these utilities offer their customers the possibility that they would install a PV system on their property and connect them to the grid. They would also offer a 20-year guarantee and, if needed, insurance. On the other hand, it seems that none of these German utilities made any changes in their business model; they still did not plan to purchase any PV systems for themselves.

It is interesting to compare the utilization of terrestrial PV systems in the USA and in Germany. The total deployed PV in the USA by the end of the first half of 2013 was over 10,000 MWp.[346] In Germany by the end of 2012 a total of 32,300 MW was installed. The interesting similarity is that in both countries large amount of PV was installed for residential purposes.

The German FiT system for PV is very simple, very understandable, and is guaranteeing a 20-year return of the investment with proper interest. In the USA, however, there are many complicated programs. Incentives, subsidies, tax advantages, FiT, and net metering are being used with various expiration dates in the 50 states each having their own rules and there are also many independent city governments involved in regulating the distribution and pricing of electricity. Many states have also mandatory RE portfolio standards. In addition, the Federal Government regulations and policies are also involved. There are also expenses and time consuming rules for the interconnection to the grid, which as mentioned, the FERC is trying to speed up for smaller (up to 5 MW) installations. There are states where the regulations were designed to encourage and help the utilization of solar power. These states, California, New Jersey, Arizona, Colorado, and New Mexico had the most PV deployed in the USA (2011). Obviously to introduce a nation-wide FiT system like in Germany is most likely not feasible.

Comparing the German to the US utilities, it is interesting, that the German large utilities until just recently had no interest in PV, but in the USA for one or other reasons some of the utilities, large and even smaller ones, have participation in PV. In many cases, the utilization of certain percentage of PV in the

---

[346]NPD Solarbuzz North America PV market quarterly report.

electricity sold by the utilities was mandated by the state or locality. Despite that a lot of details can be obtained from the SEPA reports,[347] it is fairly complicated to establish how the various utilities are obtaining the PV electricity sold by them in their electricity mix. It is, however, strange, that according to SEPA,[348] out of the 3269 US utilities, only 6 owned in 2012 a "measurable portion" of solar projects. Reviewing SEPA's report listing the 10 utilities with the largest cumulative solar megawatts, they are all investor owned except in 2012 one publicly owned, SMUD managed to become the eight on the list. It is an interesting decision by the utilities, that instead of setting up their own PV systems they elected to obtain electricity from IPPs by Power Purchase Agreement (PPA), but the great majority of the purchased electricity came from "Customer-sited PV" systems very similarly as in Germany. Some of the utilities, including the largest (PG&E), encourage private homes to install PV systems and help them to obtain the permit to connect their system to the grid and buy from them on net-metering basis. According to the cited SEPA report, "each utility has taken very different paths to solar development, influenced by differences of geography, state policy, and utility decision-making."

The utilities probably selected the method, not to own many PV systems to produce electricity and buying what they need because of mandating or other regulations, because in their corporate setup they would have had to establish a new organization to administer the PV systems, buying, installing, and maintaining them. For the utilities buying the electricity only by a PPA or net metering, all of this is avoided, since they are only buying the produced electricity the IPP's problem is to produce it. However, this decision of utilities is probably going to have consequences. Like it was shown in Germany, PV's decentralized nature and that one can become an IPP with

[347]SEPA is providing very useful information about the progress of utilities utilizing PV and helping to connect to their system. (www. solarelectric power. org).

[348]2012 SEPA Utility Solar Rankings (June 2013) http://www.solarelectric power.org/sepa-utility-solar-rankings.aspx (page 15).

relatively little capital could lead to the loss of business for utilities and consequently will cause their major transformation.

As the EEI report indicates, loss of business for the utilities is already happening in the USA as private IPPs and also large corporations are setting up their own PV systems and cutting out the utility. The IPP can sell the electricity directly to a customer and companies can set up PV systems to produce electricity for themselves. They are doing it as it is more cost-effective than to buy electricity from local utilities, furthermore, it gives them price stability for at least 20 years compared to the price volatility of utilities because of their fluctuating fuel prices. SEIA and Vote Solar prepared a report[349] in which they listed major companies deploying PV on the roof of their facility totaling 321 MW which is estimated to be about a $50 million loss to the utilities. The companies installed the PV systems as the result of serious financial considerations. The fast decreasing cost of PV energy has made solar an increasingly desirable investment for American businesses. SEIA stated, "During the first half of 2012, over 3,600 non-residential PV systems came online, an average of one every 72 minutes."

In summary, based on the above, what could the future of the electrical utilities look like?

The answer for Germany is easy. The decentralized nature of the electricity generation of PV systems seems to transform the present status of the German electrical utility business. The 1000 small local electricity providers will gain, new IPPs will emerge based on the multitude of IPPs, and the four major utilities market share is believed to shrink from the present 80% to under 50%. The four large German utilities had the opportunity to become major players in the "PV rush" but until recently (see Chapter 42) they ignored it. All of the people and companies which decided to install PV were their customers. These people were all wired to them.

Now the four large German utilities suddenly realized their opportunity to extend their business to offer PV system installation for homes and businesses. This may open up for them

---

[349]http://www.seia.org/research-resources/solar-means-business-top-commercial-solar-customers-us.

a major unregulated business opportunity. They were obviously late as over a million of PV roofs were already built, several companies participated in this boom gaining a lot of experience and publicity, but on the other hand with the low PV prices the business just accelerated and the utilities as newcomers had an opportunity. The major German utilities were now starting to advertise to install PV for private houses. Utilities advantage is that they had the list of prospective customers as they were presently customers for electricity and they had money to finance this new business opportunity. The utilities had also an important trump card to get into the PV business. They were offering a 20-year electricity delivery program and they could say that the utility will be surely in business 20 years from now to fulfill their obligations and guarantees, but whether the other installers will be still around is not that sure based on experience.

Interestingly RWE, the largest of the four utilities realized there are already millions of private homes equipped with PV which provide a large and a new business opportunity for RWE which has been described in Chapter 42.

For the US utilities, the crystal ball is much cloudier to make a prediction. In the USA, at present 10 GW PV is deployed. Comparing to Germany, because of the size of the country and its population it can be expected that 10–20 times this could be deployed in 5–10 years.

And this is the point the utilities, like the telephone companies, have a choice:

- They may not be able to overcome their "corporate culture" and are going to try to preserve their present business model, relying on a new tariff structure, new service charges, etc., or
- following the telephone example, continuing the present only wired and regulated "corporate structure" but also branch into the unregulated PV electricity generating business like the German utilities did.

There is a small window for the utilities to work out a strategy of how to participate in this. Obviously, compared to Germany, the USA is located over an extremely large natural gas reservoir which during the coming 5–10 years will be developed by fracking. The utilities also have a large lobby available to tilt the table, but that will be limited by the existence of the large work force employed by the PV business and a population which wants to limit global warming.

As the German example showed utilities, the first problem may come from the changes caused by PV in the peak power business. Changes may also come, because as shown, companies realize that they can be their own electric power producer and also new IPPs may emerge assembling large number of decentralized PV power systems. Because of the complicated regulatory system existing in the USA and the number of utilities, it is hard to predict, but the decentralized nature of PV will also cause the transformation of the US utility business. Some of the utilities are anticipating this and are rearranging their business accordingly.

Chapter 42

Energy Independence and the Wireless Environment

Reading the title of this chapter, the reader would think that after getting to the end of writing this book, in which everything was based on facts, the writer wanted to find out if he is capable of writing science fiction. This chapter is not an escapade into science fiction, but the realization that we are already living in the beginning of the "wireless age."

Being in the center of major cities in the world, especially in the USA and in Europe, one sees no wires, they are all underground. Getting out of the center to the suburbs, the wires appear overhead. They mostly carry electricity and some are for telephone or cables for TV/Internet, except the new fiber optic cables, which, being more fragile, are perhaps underground. How one can even think that we are already living in the beginning of the "wireless age" when in some places one sees the sky through a large web of wires?

We can begin the story of the wireless environment soon after we started to manufacture PV modules at Solarex, when I received a telephone call from the West Coast and the person on the phone told me that he had a small electric store and would

Sun above the Horizon: Meteoric Rise of the Solar Industry
Peter F. Varadi
Copyright © 2014 Peter F. Varadi
ISBN 978-981-4463-80-5 (Hardback), 978-981-4613-29-3 (Paperback), 978-981-4463-81-2 (eBook)
www.panstanford.com

like to buy a few of our largest (at that time it was a 30 Wp) PV modules. I quoted him a price, which at that time was quite high. He said it was OK. I also told him that we needed the payment in advance as we did not know him. He said that was no problem and that he was going to transfer the money. The money arrived, we shipped the few modules. About a week later, he called again and ordered about twice as many as he ordered before. Same routine: Money arrived and we shipped. When that happened two more times with increasing number of modules being ordered, I asked him what was he doing with those modules?

"Oh!" he said, "We have many big mountains in this area and some of the people got tired of the hectic city life and built themselves a house in the wilderness in the mountains far from everything. Also, there are many people who build a house to spend weekends away from the dirty city where they work. They have as much wood as they want from the incredible amount of trees, they can use that for heating and cooking, but there is no electricity for light and to listen to the radio. They bought large batteries for that, but they have to bring them to my store practically weekly to charge them. Those batteries are very heavy and to drive from wherever their house is to me is a long trip. I read about PV but the company making PV for space applications here in California makes only very small units and they are very expensive. I found out about your company, and you make larger modules which are much cheaper. When I bought your first shipment, we tried those PV modules mounted them on one of the houses, hooked up to the large battery and the system worked beautifully. The batteries are recharged by the PV modules. They have to be careful not to totally discharge the battery but they do not need to bring them to me anymore. They love it, so all of those people who have their houses in the wilderness far from electric wiring are buying them."

We called it our "wilderness home" business, and I think this was the beginning of the wireless world of electricity.

When we opened our business in Switzerland with all of those mountains, we thought that it will be a good market to sell PV to the "wilderness homes," which there is called "chalet." It turned out that we were right. People were very happy to buy the PV modules.

Suddenly our UK office received substantial orders from Norway from a gas bottling company. It turned out that Norway had a substantial number of "wilderness" weekend houses. The gas bottling company sold the bottled gas, for cooking and heating. These houses used batteries for lighting and radios and TVs. The problem for the owners was the same as in the USA and Switzerland they had to get the batteries periodically recharged, which was a "heavy" duty project for the owners. The gas bottling company was afraid that they may pay to bring in a power line and the area houses may get connected to the grid and then they may not need the bottled gas and as a result lose their business. The bottled gas company decided to offer the "wilderness" people solar modules to charge their batteries. They did. The result was very successful the "wilderness" people did not pay to be connected to the grid. They stayed "wireless."

Suddenly we started to get large orders from our Spanish representative—Ecasolar SA–in Madrid and we found out that they were selling it to farmers for their houses. When I visited our representative, they arranged for me a trip to Alicante to see the solarized "wireless village." The small village was on top of a hill outside of Alicante full of small olive tree farms. The farmers who owned the small olive tree farms lived only there when they harvested the olives. Other times they lived in another farm somewhere south west where they had orange tree orchards. They lived there when they harvested the oranges. The olive tree farms—despite that they were close to Alicante—had no electricity and probably, because they would have used only very little of it, the utility did not want to waste money to bring the power line to the village.

We had our Spanish representative's customer also with us. He had an electric shop near the city. His story in some respect was the same as the shop owner in California told me on the telephone. The farmers living there during the harvest were reasonably affluent people and they had TV as well as a radio but to operate them they acquired a large battery used for trucks. They were able to operate their radio and TV but only a very small light, because they had to take the battery for recharging even with the radio and TV and this little light once a week. When the electric shop owner noticed our Spanish representative's advertisement, he immediately bought a solar module—by that time we made larger modules, 60 Wp—tried it and making some calculation he realized that he could install at least three fluorescent lights besides of the radio and TV in a house. By that time, we had already made charge controllers, which were small boxes containing electronics, which could be set to stop charging the battery when it was fully charged and turn off the electricity for the lights or TV not to let the battery totally discharge.

He mentioned this system to the first farmer who came to recharge his battery. The farmer bought it immediately and the electric shop owner took a steel pipe, hammered it into the ground next to the farmer's house mounted the solar module on top of it facing south, installed the battery and the charge controller and installed one fluorescent light on the porch and two in the house. The battery was obviously fully charged when the porch light was turned on that evening. The entire village came from their dimly lit houses to admire the spectacle of the bright porch. They came every evening and could also see that while the farmer and his wife were sitting on the brightly lit porch, their two kids were watching TV in the house.

So when the farmers took their battery next time to recharge and carried it from their trucks to the electric shop, swearing all the way how heavy the damned thing was, they immediately ordered the solarization of their houses.

The shop owner ordered the solar modules and charge controllers from our representative in Madrid and as soon as he received them he and his helper worked for 2 weeks and completed the first electrified village with wireless environment.

In Europe and in North America people, live in a wired environment. There are only few spots where wireless electricity is needed for lights, radio, and TV (Internet did not exist in those days). But there are vast areas on the Earth where one quarter of the Earth's population is living, which is not reached by electric or telephone wiring and there would probably not be enough copper on Earth, to wire them up.

In 1992, Anil Cabraal of the World Bank ASTAE division came up with the idea which was named "Solar Home System"

(SHS). The SHS' function is basically the same, what we called "wilderness home" system, with the difference that it was not purchased in individual components but was provided with a proven design and all of the components—PV solar panel, storage battery, battery charging controller, and various end-use equipment such as fluorescent lamps, switches, and wiring—were matched to optimize performance. The components were in a box so that people who needed an SHS could easily install it. The idea was realized as a World Bank Group program and the result of the program was described by Martinot and Cabraal,[350] "Twelve World Bank Group projects provide basic 'energy services' such as lighting, radio, television, and operation of small appliances to rural households without access to electricity grids through the use of SHS."

The utilization of SHS in developing countries was so successful that by now it is estimated that at least 4 million of SHSs are in use in Asia, Africa, and South America. An example is in Bangladesh, where Grameen-Shakti[351] has enabled as many as over 1 million to light up homes, shops, fishing boats, etc. The program started in 1996 and with the decreasing price of PV more and more SHS were installed. Presently in Bangladesh they are installing more than 1000 SHS per day. The Grameen Shakti story is very well described in Nancy Wimmer's book.[352]

The beauty of PV is that it is not only that electricity is generated where there is a need for it, but also that it is modular, which means that when more electricity is needed one can add to an existing system as many modules as are needed to provide enough electricity. People who started with a simple SHS system can double or triple its electric output by adding more PV modules and storage batteries.

But there exists a large population on Earth for whom the starting point of electrification is only some light in the evening

---

[350]Martinot A, Cabraal A (2000) *World Bank Solar Home Systems Projects: Experiences and Lessons Learned 1993–2000*, World Renewable Energy Congress VI, Brighton, UK; Elsevier Science, Oxford UK.

[351]Shakti G (Grameen Energy), http://www.enfsolar.com/news/Grameen-Shakti-Reaches-One-Million-Solar-Home-System-(SHS)-Installations-in-Bangladesh.

[352]Nancy Wimmer: *Green Energy for a Billion Poor*, MCRE Verlag UG (www.mcreverlag.de).

to replace the extremely dangerous kerosene lamps. For these people, the World Bank Group developed the program "Lighting Africa," which was also initiated by Cabraal. This program provided a reliable PV-powered and affordable lantern which was also tested according to the World Bank specification.[353] In Africa, about 7 million (January 2013) of these solar lanterns are being used.[354] Worldwide, including Asia, there are probably another 7 million in operation.

Interestingly, people in the developing countries, not connected to the electric grid, leapfrogged over our wired twentieth century and started with the wireless twenty-first century, living in a wireless environment. The effect of these SHS and solar lanterns went far beyond just to provide light to replace kerosene. It can also be used to charge cellular phones, run televisions, radios and cassette, and CD players and now also to have access to Internet. The possibility to receive TV broadcasts and the Internet was feasible because those can now be obtained everywhere wireless from satellites.

The wireless world described above was never wired. The legitimate question is, if our world is already wired, can it ever become wireless?

Based on the experiences with telephones, the answer is *Yes*. Telephones—and this is may be a surprise for a lot of young people—until recently needed wires. The telephone wires were and are still mounted on poles, crisscrossing cities, countryside, going from the streets to every building and house and inside of the building/house wires were strung inside or outside of the walls. Telephone companies charged and still are charging "wire maintenance" fee, which customers religiously pay thinking what they could do if the wire goes wrong. But by now fewer and fewer people are using "corded" phones. The demise of wired phones can be observed in that sometime ago many "public phones" existed on streets or in buildings, and today

[353]Cabraal A (May 21, 2009) *Solar Lantern Testing to PVGAP PVRS11A* (ESMAP technical paper #078), Lighting Africa QA Workshop, Nairobi. Recent version: http://www.lightingafrica.org/, Lighting Global QA Protocol, January 22, 2013.

[354]Lighting Africa, Newsletter, Issue 28/January 2013, www.lightingafrica.org.

they are as rare as a white raven; "public phones" practically disappeared and many houses or apartments do not even have one phone which is attached to the wall outlet with a wire. However, 5 years ago, at least a hundred of various telephone sets were displayed to be used in connection with wired outlets. Today one can perhaps see a dozen; on the other hand, hundreds of different types of wireless cell (mobile) phones are displayed. Children under the age of 10 probably never used a wired phone. And the mobile phone age started only about 25 years ago.

First it was used mostly in automobiles, because it required a heavy big box which was installed in the trunk of the car. I bought my first cell phone for my car about 20 years ago. At that time it was a novelty. I bought it because when I was driving from Philadelphia, PA, to Washington, DC, I reached the stretch of Interstate 95 in Delaware, where there are 6 lanes in both directions and I got a flat tire. I pulled off the road and waited in my air-conditioned car that a police car will drive by and help me by calling a service station to come and put on my spare tire. When one drives 5 miles over the speed limit the police appears immediately, in this case no police car came for more than half an hour. I realized that I cannot spend the entire day sitting and waiting for help. After I finished changing the tire (still no police on the horizon) and sat in my air-conditioned car to recover, I was surprised that I did not got a stroke under the blazing sun at 100°F (38°C) making all that work which I never did in my life. I had heard already that wireless cell phones existed and there and then I decided that the next day I was immediately going to get a cell phone installed in my car and I can call for help myself. I figured that paying the fee for my safety was well worth. They installed a fairly big box in the trunk of the car and a telephone receiver next to the driver's seat. At that time I did not know anybody who had a cell phone. But very soon I got used to it. It was a convenience calling home, office, or friends especially when I was in a traffic jam.

In not even 20 years after I acquired the first cell phone mounted in my car, the cell (mobile) phones got much cheaper, smaller, and the recent ones called "smart phones" incorporate Internet, e-mail, messaging, GPS, photographic and movie cameras, and many other features. According to the estimate of the International Telecommunication Union,[355] the astounding statistics of how many cell phones and how few "fixed" (wired) phones were used in 2012 is shown in Table 42.1.

Table 42.1 The International Telecommunication Union statistics of how many cell phones and how many "fixed" (wired) phones were used in 2012

	Global[1]	Developed nations	Developing nations[2]	Africa
Mobile cellular subscription (millions)	6835 m	1600 m	5235	1048
Per 100 people	96.2%	128.2%	89.4%	63.5
Fixed telephone lines (millions)	1171	520 m	652 m	12 m
Per 100 people	16.5%	41.6%	11.1%	1.4%

[1]World' population 7.1 billion according to ITU.
[2]Including China and India.

Considering all of the "developing countries," 89.4% (in 2011 was only 87.2%) of the telephones were wireless cell phones. An interesting addition to this story is that as described above the World Bank/IFC joint program "Lighting Africa" was to replace kerosene lights with solar lanterns. In Africa in 2012, 63.5% of the population had mobile-cellular phones, but only 1.4% had fixed telephone lines. People living there acquired cell phones to talk to each other and to friends, relatives in Africa or all over the world. Cell phones, especially the older types, are relatively cheap gadgets but they had to be periodically recharged. In Africa, as was described, there is no electricity at many places, but the area had cell phone connections. So people acquired solar lanterns, took out the light bulb and used the solar electricity to recharge their cell phones. They could use kerosene for light in evenings, but to charge their cell phones they needed electricity. Manufacturers of solar lanterns, who realized

---

[355]http://www.itu.int/ITU-D/ict/statistics/also: http://mobithinking.com/ mobile-marketing-tools/latest-mobile-stats/a#subscribers.

this, added an outlet where cell phones could be connected to be charged. It is evident that people in those parts of the world simply bypassed the age of wired phones and also wired electricity. They lived in a wireless environment.

As it was described, telephone company management probably realized that cell phones are not regulated and unlikely could ever be. For this reason, only the sky and competition will be the limit to their profits. Telephone companies are getting out of the regulated and old-fashioned wired phones and the wired digital subscriber line (DSL) Internet connection and are jumping into the not regulated cell phones which are wireless and also into the wireless Internet connection business.

The Internet also started to be connected via telephone wires. Some of us maybe remember that it required a "Modem," an electronic device that made it possible to use a telephone line to connect it to the Internet. The data transmission was very slow. The telephone lines were still used, when the DSL system was introduced, which was digital and much faster. At that point, the telephone lines could be used for voice and also for digital data transmission.

The wireless Wi-Fi system was started about in 1999. In about a little over 10 years, Wi-Fi connection practically eliminated the need of telephone wires for Internet. Several satellite operators are offering high speed Internet connections too. Wi-Fi connections are now available in developed as well as in developing countries in many areas where there is no electricity. Today practically all of the cell phones are able to be connected to the Internet. The Internet also became wireless.

The "Developing World" entered the twenty-first century by becoming wireless. The population living on big areas lacking electricity is acquiring the advances made in the world during the twentieth century, radio, telephone, TV, and Internet are for them wireless and the electricity, which is needed for everything, is also wireless.

The question is, will the developed part of the world, which was wired during the twentieth century become wireless?

As we have seen, cell (mobile) telephone subscriptions in developing countries by now have outnumbered fixed lines by a ratio of 6:1; in developed countries, it is 3:1. In the areas where cables are used for TV and the Internet, those cables will disappear, because they will be not of copper, but fiber optics and that has to be underground as the glass fibers are too fragile. And the rest of the places will receive TV from satellites. The Internet can be connected via Wi-Fi without wires. Wi-Fi is now covering bigger and bigger areas including entire cities.

This left us only with the large amount of wiring to provide electricity for other usage.

It is interesting that in many developed countries the biggest numbers of PV installations are for private homes and farms. In Germany it is estimated that by now 1 to 2 million houses and farms have PV systems installed, which could provide enough electricity for the needs of the entire house, but as houses have no electricity storage, those PV systems are connected to the grid and the produced electricity is sold to the utilities. We have seen that the SHS used in developing countries always had a battery for the storage of electricity, but at present there is no electricity storage for the PV systems of homes and farms, because for a house in a developed country would require quite large electric storage capacity which was expensive. The disconnection from the utility and avoiding the need of grid expansion may happen only when proper and reasonably priced electric storage systems will become available. Surprisingly we are approaching that time very fast.

Electricity storage (Annex. 12) is needed for the utilities as well as for establishing the independence of private or business users of electricity. Users' independence, which means, independence from the electricity supplied by the utilities is an extremely interesting subject. Millions of people in the developing countries are already using solar power, which makes them independent from utilities. In the developed world, many millions and also companies with PV electricity available to them could be and are going to be able to become independent from utilities. The difference is that a home in Europe or in the USA requires much more electricity than a SHS. Sometime ago in an e-mail correspondence, Wolfgang Palz predicted that the next major advance of PV would be the electric independence of homes and stores from the utility.

This may surprise you, but it can be stated that his prediction became reality in the spring of 2013 because the beginning of the "disconnection from the utility" for homes and farms in Germany was actually started with the introduction of reasonably priced storage systems. How did that happen?

Hermann Scheer, the father of the German FiT system, foresaw this problem already in 2006 when he organized the first Eurosolar International Renewable Energy Storage Conference in Germany. Since then the need for the development and utilization of a proper and reasonably priced electric storage system came in the forefront. Several types of electric storage systems are available. They are described in Annex. 12. From several possible electric power storage systems for PV, there are two electric batteries which presently are mostly used: Sealed Lead acid and lithium-ion (Li-ion) batteries. To encourage the use of electric storage for homes, the German Federal Ministry of Environment has announced recently the implementation as of May 1, 2013, incentives for electric power storage systems for PV installations with a capacity lower than 30 kW.

In order to make a home independent from the electric utilities, they need a sufficient-size PV system on the roof and a sufficient-size battery and electronic power management as described in Chapter 12 and as described in connection with small home systems.

One indication that this is really happening is that on Intersolar's European exhibition 3 years ago, only 10 battery manufacturers were present. In 2013, more than 200 energy storage companies exhibited.[356] The Energy Advisory Service (IHS predicts[357] that residential PV storage will hit 2.5 GW by 2017. Also, predicts that the business for electricity storage systems for PV will grow to $19 billion in 2017 from 200 million in 2012.[358] IHS predictions are supported that recently even the German large utilities are supporting the idea.

---

[356]Runyon J (July 18, 2013) RenewableEnergyWorld.com.

[357]http://www.pv-tech.org/news/residential_pv_storage_to_hit_2.5gw_by_2017_ihs.

[358]http://www.pv-tech.org/news/ihs_germany_to_lead_explosion_in_global_pv_storage_market.

To achieve the independence for homes and farms from the electric utilities, the most unbelievable thing happened in the spring of 2013 when the largest German utility, Rheinisch-Westfälisches Elektrizitätswerk AG (RWE), realized that there is a huge market for storage systems in Germany for the already existing over a million homes and farms equipped with PV systems on their roof feeding their electricity into the grid. It would be an excellent market for selling to those people an electricity storage system and making them independent from the grid and as an advantage for the utility this would eliminate that the PV on their roof would be fed in at peak power time into the grid. These mini-utility owners may want to become independent, especially by utilizing the German Government's subsidy for the storage of electricity. The utilities would lose customers, but offering them to install the PV system—as described in the previous chapter; furthermore, service and insurance would convert them into a new but different customer. Most surprisingly, RWE, the largest electricity provider utility in Germany, carry on their Web site an advertisement (see Fig. 42.1) explaining to their electricity customers the advantages of becoming independent from the electric utility.

At present, RWE offers (Fig. 42.2) three different sizes of storage systems utilizing Li-ion batteries for homes that have 3–8, 5–12, or 8–20 kWp PV installations. RWE also predicts a 20-year life for the system.[359] The seriousness of this is also indicated by the German company Bosch, one of the World's largest electric appliance manufacturers, offering since 2012 an electric energy storage system (BPT-S 5 Hybrid storage) utilizing Li-ion battery[360] manufactured by Saft, a very large battery manufacturer in France. The electronic "brain" is a "Smart" PV system controller, which, when the sun is shining, directs the electricity produced by the PV system to the appliances in the house which need power; if not needed, it directs it to recharge the Li-ion battery. If the battery is fully charged, it will "sell" excess electricity to the grid and "buy" if electricity is needed

---

[359]http://www.energiewelt.de/web/cms/de/1641350/energieberatung/solarenergie-photovoltaik/solarstrom-speichern/rwe-homepower-solar/.

[360]http://solarstrom-tag-und-nacht.de/.

from the grid. The system can be remotely controlled by the owner's i-Phone. Bosch is giving a 20-year guarantee for the system.

RWE HomePower solar	RWE HomePower solar
Solarstrom speichern und dann nutzen, wenn Sie ihn brauchen. Wenn Sie eine <u>Photovoltaikanlage</u> betreiben, kennen Sie das Problem: Der meiste Strom wird dann erzeugt, wenn der Verbrauch am geringsten ist—nämlich tagsüber, wenn die Familie beruflich, in der Schule oder sonst wie unterwegs ist. So nutzen Sie nur einen kleinen Teil Ihres Solarstroms zum Eigenverbrauch. Den überwiegenden Teil müssen Sie ins öffentliche Netz einspeisen, was durch die sinkende <u>Einspevergütung</u> langfristig immer unattraktiver wird. Nachts und in sonnenarmen Stunden, wenn Ihre PV-Anlage wenig bis keinen Strom produziert, müssen Sie sogar noch zusätzlichen Strom einkaufen.Was bisher fehlte, war ein <u>Solarstromspeicher</u>, der es möglich macht, den Strom genau dann zu nutzen, wenn man ihn braucht. Diese Unabhängigkeit bietet Ihnen ab sofort **RWE HomePower solar**.	**Solar electricity storing and using it when it is needed.** When you are using a *Photovoltaic* system you know the problem: The majority of the electricity is being produced, when the utilization is the minimum—during the day, when the family is working, in school or is on the way to go somewhere. Accordingly you are using yourself only a small amount of the solar electricity. The predominant part of the electricity will be fed into the grid, which because of the sinking *Feed-in Tariff* which will be in the long term is increasingly unattractive. During the night and during cloudy hours when the PV system is producing no or insufficient electricity you have to purchase additional power. What you should have is a ***solar electricity storage*** that makes it possible to utilize the solar electricity exactly when you need it. This Independence offer you as of now the **RWE HomePower solar.**

Figure 42.1 The RWE advertisement reproduced and translated.

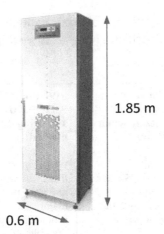

Figure 42.2 RWE HomePower solar system—control cabinet including batteries.

Figure 42.3 shows a schematic of a "smart" PV system controller.

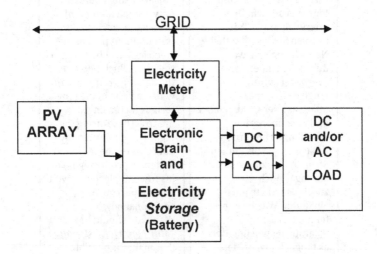

Figure 42.3 Schematic of the "smart PV system".[361]

For example, a home in the Washington, DC, area—where houses are equipped with air conditioning system—requires less than 8,000 to 10,000 kWh/year. This means an average 20

---

[361]Varadi P (23 September 2013) www.energypost.eu.

kWh/day. Considering an 8–10 kWp PV system on the roof of the house, one of the RWE or Bosch storage systems would be sufficient. Eric Daniels[362] has applied a similar concept in his patent (2013) covering solar synchronized loads (SSL).

Electric washing machines were first used in about 1910. Home dishwashers are available since 1937. The year 1938 was the beginning of the use of electric clothes dryers. Window air conditioners were first used in 1931 and central air conditioners in houses about 1970. Today no new housing development would be imagined not using all of those electric appliances.

Electric energy independent homes in the "wilderness" and in developing countries started to exist since 1974, but the electric energy independent homes in developed countries started to be commercially available in 2012. It is highly possible that within 5–10 years, every new housing development in Europe and in America will offer electric energy independent homes.

The result will be the reduction or even elimination of the large number and size of power lines and transformers from the landscape. Homes may be connected to power lines; the purpose will be selling some electricity when, for example, the inhabitants are on vacation or buying some electricity if for some reason they need it, but mostly this wire will not be used. The large cables to developments will not be needed, as the wires will be used only occasionally as described before.

The utilization of solar energy was the last step in freeing ourselves from most of the ugly wires which were disfiguring our beautiful sky. Furthermore, the energy-independent home will provide the owner not only with an uninterruptible electric power source but also with partial financial stability because the price of electricity for the entire home will be unchanged at least for 20 years in spite of inflation and in spite of the price fluctuation of the fuel prices.

---

[362]Eric Daniels—"Solar synchronization loads for photovoltaic systems" (US patent No. 8,373,303).

Epilogue

The year 2013 is the 60th anniversary of the discovery by Daryl Chapman in 1953 at Bell Laboratories of the first useful way to convert light into electricity utilizing pure silicon. It is also the 55th anniversary of the occasion when the first solar photovoltaic cell powered spacecraft, Vanguard 1, was put into the orbit around the Earth. Also, it is the 40th anniversary of the oil embargo and also when the terrestrial PV industry started to manufacture PV cells to convert light into electricity to be used on Earth. The year 2013 is the year when 100 GW of solar electric systems have already been deployed, producing about the same amount of electricity as 20 nuclear power plants.

Controlled use of fire was mankind's biggest invention which separated us from the animal world. The controlled use of fire opened up for mankind the entire Earth, independently from climate, to be able to live on.

Since the discovery of the controlled use of fire, a very large number of important inventions brought us to this date. If one would search for another invention which changed or can change most of the human life like the controlled use of fire did, it would probably be the invention of the utilization of photovoltaic, which opened up for mankind our entire solar system.

Mankind invented ways to send objects to the universe, but without PV cells it would be like the Sputnik operating only for 22 days or as one of the transmitters of Vanguard 1 operating 20 days until their batteries were exhausted. But Vanguard 1's other transmitter powered by PV operated for 6 years when its transmitter with no PV power supply failed. PV made the difference enabling mankind to explore and also to operate machinery in our solar system.

The possible utilization of PV could be beyond our imagination. On Earth it provides us electricity converting the energy from the Sun. We were also able to put the PV-powered objects into orbit around our planet such as artificial moons. Our big Earth, of which we learned to know only from explorers who brought back stories and perhaps pictures from faraway places, seems to shrink as one can see and even talk to anybody at any locations. With the help of these PV-powered satellites, it is now possible to even navigate safely from any point to any other point on Earth.

PV also makes it possible to explore our entire solar system and the Universe. It makes possible to operate machinery on another planet such as Mars or on the Moon. It is commensurable in importance to the controlled use of fire, which made it possible for mankind to live at any place on this Earth, because PV made it possible for us to leave our lovely blue planet, the Earth.

Annex 1

Metallurgical and Polycrystalline Silicon[363]

The production of thin silicon wafers for semiconductor devices or for solar cells requires several steps. Their description is according to the numbers given in Fig. A1.1.

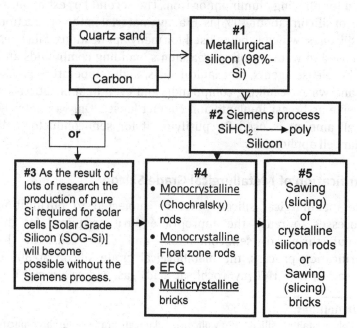

Figure A1.1 Production of metallurgical grade silicon.

---

[363]Other detailed descriptions are: Ryningen B. *Silicon Production and Purification*, http://www.suite101.com/content/silicon-production-and-purification-a232890. Kopecek R. *What is Behind c-Si Material?* http://isc-konstanz.de/fileadmin/doc/What%20is%20behind%20c-Si%20material_R.%20Kopecek%20v3.pdf.

Silicon[364] occurs in nature in the form of quartz or quartz sand. From these, the metal silicon is produced by mixing the quartz or pure quartz sand with carbon and heating the mixture in an electric furnace between carbon electrodes to a very high temperature (1900°C). At this temperature, the resulting silicon is in the liquid state, because its melting point is 1410°C.

$$SiO_2(s) + C(s) \rightarrow Si(l) + CO_2(g),$$

where (s) = solid; (l) = liquid; (g) = gas.

When it cools down, the dark silver gray silicon hardens. This material is called metallurgical silicon, which is only 98% pure. The price of which is very low, $2–4/kg.

The majority of the metallurgical silicon material (55%) is used for alloying aluminum or iron. The second largest application of silicon (about 40%) is as a raw material in the production of silicones, which is being used for many applications. Silicones are used in waterproofing treatments, molding compounds and mold-release agents, mechanical seals, high temperature greases and waxes, caulking compounds, and even in applications as diverse as breast implants and contact lenses. Only a relatively small amount is used for purifying it for semiconductor and solar cell applications.

Purification of Metallurgical Grade Silicon

The metallurgical silicon must be subjected to purification processes to reach the appropriate purity required for the manufacturing of semiconductors and solar cells. In the purification process, the metallurgical grade silicon at 300°C is reacted with HCl (hydrochloric acid) gas, and as a result it is

---

[364]**Definitions:**

- Polycrystalline silicon (Polysilicon), solar cell grade or greater purity pieces of silicon formed from multiplicity of crystals obtained by the Siemens or by other process.
- Monocrystalline silicon, a silicon block (rod), which consists of a single crystal. This may be produced by the so-called Chochralsky method.
- Multicrystalline silicon "bricks" crystal consisting of several large silicon crystals produced when hot liquid silicon is cast into a container the cooling of which is regulated.

transformed into trichlorosilane ($HSiCl_3$), which has a boiling point of 31.8°C, which means that above that temperature it is a gas which can be subjected to more cleaning steps.

$$Si + 3HCl \rightarrow HSiCl_3 + H_2$$

Manufacturing Polycrystalline Silicon (Siemens Method)

Mostly the so-called "Siemens" method is used for the manufacture of high-purity (99.9999999%) polycrystalline silicon needed for the production of "semiconductors," transistors, and integrated circuits. Only 99.9999% or somewhat lower purity is needed for solar use.

In the Siemens method, the purified $HSiCl_3$ (trichlorosilane) gas is introduced into a reactor chamber in which a heated (1150°C) pure silicon rod is positioned. The trichlorosilane gas decomposes and pure silicon is deposited on the hot Si rod and the HCl gas is exhausted. This process is called chemical vapor deposition (CVD). The resulting 99.9999999% (or 99.9999%) purity Si material is called "polycrystalline silicon" (Fig. A1.2).

Figure A1.2 Polycrystalline silicon produced by the Siemens process.

SoG-Si—Solar Grade Silicon

The "solar grade silicon" (SoG-Si—Solar Grade Silicon—99.9999%) is produced the same way as described above. The only difference is that the trichlorosilane is not subjected to so much cleaning process than the one used for the production of semiconductors. Therefore, the price of this solar-grade polycrystalline silicon material is lower than that used to produce semiconductors.

To produce SoG-Si besides the polycrystalline silicon made by the Siemens process, a material called "tops and bottoms" of the silicon rods of ultra-pure monocrystalline silicon obtained by the Czochralski (Cz) crystal pulling process or by the float zone (FZ) was also used. These "tops and bottoms" contain small amounts of impurities and dislocations and are therefore cut off. This cut-off and polycrystalline silicon, which is not pure enough for the electronics industry, was used in the solar industry, and up to about the year 2002, was enough to cover the material need for the solar cell industry. After 2002, however, the demand for feedstock surpassed the supply. This led to silicon shortage and substantial price increases to $500/kg. Seeing the demand, manufacturers added large additional capacity to produce polysilicon, but the demand for it decreased. This led to substantial price erosion, and in May 2012, the polysilicon prices dropped as low as $23/kg.

All of the companies (Table A1.1) are using the Siemens process, except MEMC, which uses silane in a fluidized bed reactor (developed by Ethyl Corporation in the late 1980s).[365]

REC uses the "monosilane" process (developed in the late 1970s by Union Carbide Corporation).[366]

Just to consider the incredible increase of the size of the PV industry, one should be reminded that in the 1980s, it used only the left-overs of the semiconductor polysilicon requirements. However, in 2010, to produce crystalline solar cells, the PV industry needed four and a half times as much polysilicon as the entire semiconductor industry needed.[367] A considerable amount of capacity was added and recently an oversupply of solar-grade silicon resulted in substantially decreased prices.

---

[365]http://www.memc.com/assets/file/technology/articles/Ad_11_2000. pdf.

[366]http://www.scribd.com/doc/25870757/Polysilicon.

[367]Bernreuter J (May 2011) *Sun and Wind Energy*.

Table A1.1 polycrystalline silicon manufacturers

Manufacturer	Country	Web site
ELKEM	Norway	http://www.elkem.no
GCL-PolyEnergy Holdings	China	http://www.gcl-poly.com.hk
Hemlock Semiconductor	USA	http://www.hspoly.com
LDK Solar	China	http://www.ldksolar.com
MEMC	USA	http://www.memc.com
Mitsubishi	Japan	http://www.mpsac.com
OCI Company	South Korea	http://www.oci.co.kr/eng/
REC Silicon	Norway/USA	http://www.recgroup.com
ReneSola	China	http://www.renesola.com
Sumitomo	Japan	http://www.sumcousa.com
Tokuyama	Japan	http://www.tokuyama.co.jp
Wacker	Germany	http://www.wacker.com

When Solarex started, in 1973, the Jet Propulsion Laboratory funded the research at several large companies to develop a less expensive SoG-Si procedure than the Siemens process. These studies revealed that it would be possible to develop such a manufacturing process, but in the years 1970–1980, the manufacture of silicon solar cells for terrestrial photovoltaic cells was not on such a scale as it was predicted in 1973, and the semiconductors did not use up all the polycrystalline silicon manufactured for them, and as mentioned before, the surplus capacity of the silicon manufacturing plants was sold cheaply to the solar manufacturing companies.

At that time, because of the availability of cheap silicon, there was not enough need to invest more money in a non-Siemens SoG-Si research and development. The SoG-Si development was shut down until the late 1990s, but in early-2000, it became apparent that a great need for cheap polycrystalline silicon would become a reality. Therefore, the research started again. This research was supported by the European Union,[368] and also

---

[368]European Union's "Altener" program: "A Task Force for the Creation of a Consortium to set up a SoG-Si Production Plant" 2002.

by European private firms[369] engaged in research. Therefore, the possibility of inexpensive SOG-Si, the so-called "upgraded metallurgical silicon," (UMG) which has purity between 99.999 and 99.9999, is now very likely. ELKEM is already producing some amount, and the US Calisolar and the Chinese Propower Renewable Energy are also supplying some of this material.

---

[369]Elkem Solar AS, Norway; International Solar Energy Research Center Konstance, Germany; 6N Silicon Inc., Canada; Umoe Solar AS, Norway. http://isc-konstanz.de/fileadmin/doc/valencia08-peter-paper.pdf.

Annex 2

Float Zone and Edge-Defined Film-Fed Growth Processes

A.2.1 Float Zone Process

The "float zone" (FZ) process is another method for growing single-crystal silicon.

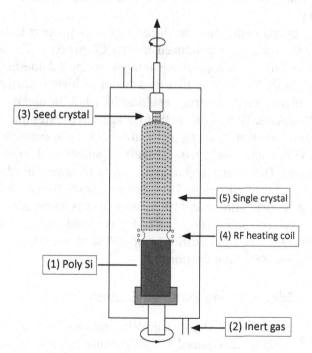

Figure A2.1 The Float Zone (FZ) monocrystal growing process.

The FZ process consists of the following steps (see Fig. A2.1).

- A polysilicon rod (1) is mounted vertically inside a chamber, which may be under vacuum or filled with an inert gas (2).
- A "seed crystal" (3) is touching the upper part of the polysilicon rod.
- A needle-eye coil that can run through the rod is activated to provide RF power (4) to the rod, melting a 2 cm-long zone in the rod. This molten zone can be maintained in stable liquid form by the coil. The coil is then moved through the rod, and the molten zone moves along with it. FZ growing equipment can also use a stationary coil, coupled with a mechanism that can move the silicon rod through it (as shown in Fig. A2.1).

The movement of the molten zone through the entire length of the rod purifies the rod and forms the near-perfect single crystal (5).

The purity of an ingot produced by the FZ process is higher than that of an ingot produced by the CZ process. There are two principal technological advantages of the FZ method for PV Si growth. The first is that as a result of higher purity and better micro defect control, resulting in 10–20% higher solar cell efficiencies. The second is that faster growth rates and heat-up/cool-down times, along with absences of a crucible and consumable hot-zone parts, provide a substantial economic advantage. The main technological disadvantage of the FZ method is the requirement for a uniform, crack-free cylindrical feed rod. A premium cost (100% or more) is associated with such poly rods. Currently, FZ Si is used for premium high-efficiency cell applications and CZ Si is used for higher-volume, lower-cost applications.

A.2.2 Edge-Defined Film-Fed Growth

The disadvantages of the Czochralski and the FZ systems are that the equipment required to manufacture single-crystal rods are very expensive, the process is very slow and the silicon rods had to be sliced to produce wafers resulting in considerable material losses. For this reason, the idea arose that the molten silicon could be pulled not to form single crystal rod but to form a very thin ribbon. Mobil-Tyco Solar Energy Company made the experiments funded by the Jet Propulsion Laboratory[370] (JPL) of California. Mobil-Tyco Solar Energy developed the edge-defined film-fed growth (EFG) process and solar cell in 1973.

SCHOTT Solar[371] was the last company using the EFG process to manufacture solar cells,[372] in which an octagonal tube is pulled directly from the silicon melt. Thus, an about 6 m-long silicon tube has a wall thickness of 330 μm in conformity with the thickness of the later cells. A laser cuts the tube into 100 mm × 100 mm wafers, which is processed to become a solar cell.

Since 2012, no solar cells have been made with the EFG process.

---

[370]Large area silicon sheet by EFG, DOE/JPL 954355-79, July 15, 1979.

[371]Schott discontinued its Silicon solar cell manufacturing production in 2012.

[372]http://www.actec.dk/Schott-Solar/pdf/efg-100x100.pdf.

Annex 3

Qualification Testing of PV Modules and PV Standards[373]

History of Qualification Testing of PV Modules[374]

- JPL Blocks I–V (1975–1981)—all crystalline Si.
- European Union Specifications 501–503 (1981–1991).
- SERI IQT (1990)—modifications for thin films (a-Si).
- IEEE 1262 (1995–2000)—all technologies.
- IEC 61215 (Ed 1—1993, Ed 2—2005)—crystalline Si.
- IEC 61646 (Ed 1—1996, Ed 2—2007)—thin films.

Photovoltaic Standards[375]

A. International Electrotechnical Commission[376] (IEC) Photovoltaic Standards

IEC 60891 Photovoltaic devices—Procedures for temperature and irradiance corrections to measure I–V characteristics.

IEC 60904-1 Photovoltaic devices—Part 1: Measurement of photovoltaic current-voltage characteristics.

IEC 60904-2 Photovoltaic devices—Part 2: Requirements for reference solar devices.

---

[373]Updated by John Wohlgemuth (NREL), July 2013.
[374]Wohlgemuth J (19 June 2011) *Reliability, PV Cells, Modules and Systems*, Tutorial, 37th IEEE PVSC, Seattle, WA.
[375]Provided by Wohlgemuth J (2013) NREL.
[376]www.iec.ch.

IEC 60904-3 Photovoltaic devices—Part 3: Measurement principles for terrestrial photovoltaic (PV) solar devices with reference spectral irradiance data.

IEC 60904-4 Photovoltaic devices—Part 4: Reference solar devices—Procedures for establishing calibration traceability.

IEC 60904-5 Photovoltaic devices—Part 5: Determination of the equivalent cell temperature (ECT) of photovoltaic (PV) devices by the open-circuit voltage method.

IEC 60904-7 Photovoltaic devices—Part 7: Computation of the spectral mismatch correction for measurements of photovoltaic devices.

IEC 60904-8 Photovoltaic devices—Part 8: Measurement of spectral response of a photovoltaic (PV) device.

IEC 60904-9 Photovoltaic devices—Part 9: Solar simulator performance requirements.

IEC 60904-10 Photovoltaic devices—Part 10: Methods of linearity measurement.

IEC 61194 Characteristic parameters of stand-alone photovoltaic (PV) systems.

IEC 61215 Crystalline silicon terrestrial photovoltaic (PV) modules—Design qualification and type approval.

IEC 61646 Thin-film terrestrial photovoltaic (PV) modules—Design qualification and type approval.

IEC 61683 Photovoltaic systems—Power conditioners—Procedure for measuring efficiency.

IEC 61701 Salt mist corrosion testing of photovoltaic (PV) modules.

IEC 61702 Rating of direct coupled photovoltaic (PV) pumping systems.

IEC 61724 Photovoltaic system performance monitoring—Guidelines for measurement, data exchange, and analysis.

IEC 61725 Analytical expression for daily solar profiles.

IEC 61727 Photovoltaic (PV) systems—Characteristics of the utility interface.

IEC 61730-1 Photovoltaic (PV) module safety qualification—Part 1: Requirements for construction.

IEC 61730-2 Photovoltaic (PV) module safety qualification—Part 2: Requirements for testing.

IEC 61829 Crystalline silicon photovoltaic (PV) array—On-site measurement of I–V characteristics.

IEC 61836 Solar photovoltaic energy systems—Terms, definitions, and symbols.

IEC 61853-1 Photovoltaic (PV) module performance testing and energy rating—Part 1: Irradiance and temperature performance measurements and power rating.

IEC 62093 Balance-of-system components for photovoltaic systems—Design qualification natural environments.

IEC 62108 Concentrator photovoltaic (CPV) modules and assemblies—Design qualification and type approval.

IEC 62109-1 Safety of power converters for use in photovoltaic power systems—Part 1: General requirements.

IEC 62109-2 Safety of power converters for use in photovoltaic power systems—Part 2: Particular requirements for inverters.

IEC/PAS 62111 Specifications for the use of renewable energies in rural decentralized electrification.

IEC 62116 Test procedure of islanding prevention measures for utility-interconnected photovoltaic inverters.

IEC 62124 Photovoltaic (PV) stand-alone systems—Design verification.

IEC 62253 Photovoltaic Pumping Systems—Design qualification and performance measurements.

IEC 62257-1 Recommendations for small renewable energy and hybrid systems for rural electrification—Part 1: General introduction to rural electrification.

IEC 62257-12-1 Recommendations for small renewable energy and hybrid systems for rural electrification—Part 12-1: Selection of self-ballasted lamps (CFL) for rural electrification systems and recommendations for household lighting equipment.

IEC 62257-2 Recommendations for small renewable energy and hybrid systems for rural electrification—Part 2: From requirements to a range of electrification systems.

IEC 62257-3 Recommendations for small renewable energy and hybrid systems for rural electrification—Part 3: Project development and management.

IEC 62257-4 Recommendations for small renewable energy and hybrid systems for rural electrification—Part 4: System selection and design.

IEC 62257-5 Recommendations for small renewable energy and hybrid systems for rural electrification—Part 5: Protection against electrical hazards.

IEC 62257-6 Recommendations for small renewable energy and hybrid systems for rural electrification—Part 6: Acceptance, operation, maintenance, and replacement.

IEC 62257-7 Recommendations for small renewable energy and hybrid systems for rural electrification—Part 7: Generators.

IEC 62257-7-1 Recommendations for small renewable energy and hybrid systems for rural electrification—Part 7-1: Generators—Photovoltaic generators.

IEC 62257-7-3 Recommendations for small renewable energy and hybrid systems for rural electrification—Part 7-3: Generator set—Selection of generator sets for rural electrification systems.

IEC 62257-8-1 Recommendations for small renewable energy and hybrid systems for rural electrification—Part 8-1: Selection of batteries and battery management systems for stand-alone electrification systems—Specific case of automotive flooded lead-acid batteries available in developing countries.

IEC 62257-9-1 Recommendations for small renewable energy and hybrid systems for rural electrification—Part 9-1: Micro power systems.

IEC 62257-9-2 Recommendations for small renewable energy and hybrid systems for rural electrification—Part 9-2: Micro grids.

IEC 62257-9-3 Recommendations for small renewable energy and hybrid systems for rural electrification—Part 9-3: Integrated system—User interface.

IEC 62257-9-4 Recommendations for small renewable energy and hybrid systems for rural electrification—Part 9-4: Integrated system—User installation.

IEC 62257-9-5 Recommendations for small renewable energy and hybrid systems for rural electrification—Part 9-5: Integrated system—Selection of portable PV lanterns for rural electrification systems.

IEC 62257-9-6 Recommendations for small renewable energy and hybrid systems for rural electrification—Part 9-6: Integrated system—Selection of Photovoltaic Individual Electrification Systems (PV-IES).

IEC 62257-12-1 Recommendations for small renewable energy and hybrid systems for rural electrification—Part 12-1: Selection of self-ballasted lamps (CFL) for rural electrification systems and recommendations for household lighting equipment.

IEC 62446 Grid connected photovoltaic systems—Minimum requirements for system documentation, commissioning tests, and inspection.

IEC 62509 Battery charge controllers for photovoltaic systems—Performance and functioning.

IEC 62716 Photovoltaic (PV) modules—Ammonia corrosion testing.

IEC/TS 62727 Photovoltaic systems—Specification for solar trackers.

B. PV Gap Recommended Specification (PVRS)

PV GAP developed the below PVRS documents to specify technical requirements for conformity assessment purpose for products not yet covered by IEC standards (Table A3.1). This PVRS have been transferred to the IECEE at the end of 2009 and made available to the IECEE members to determine whether they would like to use such specifications whilst awaiting the future IEC standards.

Table A3.1   PV recommended specifications

References	Description
PVRS 5A	Lead-acid batteries for PV energy systems (modified automotive batteries). General requirements and methods of test.
PVRS 6A	Charge controllers for PV stand-alone systems with a nominal system voltage below 50 V. Specification and testing procedure.
PVRS 7A	Lighting systems with fluorescent lamps for PV stand-alone systems with a nominal system voltage below 24 V. Specification and testing procedure.
PVRS 8A	Inverters for PV stand-alone systems. Blank detail specification. Specification and testing procedure
PVRS 10	Code of practice for installation of PV systems.
PVRS 11A	Portable PV solar lanterns. Design qualification and type approval.

Annex 4

Early Photovoltaic Pilot Applications in the European Union

Dr. Wolfgang Palz

Former EU Commission Official, Brussels

The European Commission has carried out R&D programs on Photovoltaic Solar Energy Conversion (PV) since the mid-1970s. The first one was decided in 1975 by the European Council of Ministers as part of a larger New Energy Program. The author of this paper was in charge of their preparation and implementation through multinational contracts with the industry and academia until 1996. Several hundred million dollars were awarded from the Commission during this time for cost-sharing activities on PV. The Commission was always involving the Governments of the EU Member Countries. To some extent, it was a co-coordinator for all activities going on at that time in Europe.

Since its commencement, the European PV Program put a great emphasis on the development of innovative Pilot Plants. It was felt that it was not good enough to develop only the solar cells. Their application for practical electricity generation made it necessary to demonstrate "turnkey" solar systems that were tailor-made for any application of practical use. The Commission took the initiative for a whole range of such systems. Since the early 1980s, they were developed, built, and operated jointly by the leading European PV experts under the common roof of the so-called "European PV Pilot Program." An overview is given in this article.

- *PV integrated in solar buildings (those were some of the first ones in Europe; now in 2011 more than 1 million of them are in operation).*

Grid-connected PV House in Munich, Germany

Three kilowatt of PV were integrated into a modern building, mostly built from glass. The building served as a private residence. It was built in 1982 and put in operation in 1983. The building's architect was Thomas Herzog. ISE Fraunhofer, Siemens, AEG, and others built the project. It had no storage batteries. The inverter efficiency was 94% and the overall efficiency of the system was 7%.

PV House in Bramming, Denmark

The project had 5 kW of PV on a "Danish Standard House." It was put into operation in August 1984. Prime contractor was Jutland Telephone. The house was built from wood. The PV modules were architecturally integrated into the South slope. The system was grid-integrated but also had a battery: As this type of system was also designed for export, the stand-alone mode was also to be demonstrated.

PV House in Sulmona, Italy

The project was supposed to develop PV application on a modern "prefabricated house." Architect was the German star architect, Thomas Herzog. The project had no follow-up as the responsible company went broke after completion of the project.

Villa Guidini near Venice, Italy

It is aneighteenth century villa converted into a public garden and museum. The 2.5 kW PV array from the Italian company Helios was set up in the garden. The modules were lightweight to simulate the possibility of roof mounting. The angle of inclination could be adjusted. The system was grid-connected; it also had a large battery.

Rappenecker Hof, a Hiker's Inn, Black Forest Near Freiburg, Germany

The 3.8 kW PV system is integrated in the building's roof. It is a farmhouse from the seventeenth century. ISE Fraunhofer, the project leader claims that this system from 1987 was Europe's first inn supplied with PV. It is stand-alone and includes a battery. In later years, ISE converted the project to a showcase for new energies by adding to the diesel generator a small wind turbine and a fuel cell system fed from gas bottles. Excess energy is employed for water heating.

Berlin PV in a Residential District

A 10 kW system was installed in 1989 by BEWAG and the Institute of Prof. Hanitsch, Berlin's solar pioneer. The PV array from Siemens and AEG was not building integrated. It was grid connected but also had a storage battery. It had also an MPPT controller. The consumers included a heat pump, house appliances, lighting appliances, and a charging station for an electric car.

- *Stand-alone PV for power supply to communities and islands from 1982/83.*

Agia Roumeli in Crete, Greece

A 50 kW PV system associated with a battery and a diesel generator supplied electricity to the remote village of Agia Roumeli on the South coast of Crete. The inverter had a rating of 40 kVA. The system had been built by the state utility PPC. After some years it was dismantled.

Vulcano Island, Italy

ENEL built this ground-mounted 80 kW PV system to supply electricity to the small village grid. The system was associated with a large storage battery and an inverter. Thanks to its high quality, the system was operating without major problems for almost 30 years to date.

Kaw Village, French Guyana, Near the Equator at 4° N, South America

The 35 kW PV system was associated with a 40 kVA diesel generator. It was built by SERI Renault and operated by the French utility EdF. The consumer loads were lights, refrigerators, fans, etc., for private homes, schools, and the church.

Pellworm Island, North Sea, Germany.

The PV system had a capacity of 300 kW and was at that time the largest in Europe and was situated at the highest latitude (54°N). It was built by Telefunken and operated by the local utility Schleswag. The plant supplied power to the recreation center of the local community and visitors. Excess power was charged to the local grid. It had also a large storage battery, an inverter, a controller, and a data acquisition unit. The system gave total satisfaction. Hence, in the following years the PV array was enlarged and a few large wind turbines added.

Rondulinu, Corsica, France

Rondulinu is situated 25 km North of Ajaccio. The PV plant had a capacity of 44 kW. It was designed by the University of Corsica and built by Leroy-Somer. It was not connected to any grid and just supplied electricity to the few families living in the hamlet. Power served also for a milking system, public lighting, and a water pump. The system was associated with a battery, an inverter, and a 25 kVA Propane-powered generator.

- *Other tailor-made PV systems for different applications from 1982 to 1983, all with batteries and inverters.*

Zambelli, Italian Alps, 70 kW, Water pumping.

Tremiti, Island in the Puglia region, Italy, 65 kW, Water desalination.

Giglio Island, Italy, 30 kW, Water treatment, refrigeration.

Chevetogne, Belgium, 40 kW, Swimming pool pumps.

Fota Island, Ireland, 50 kW, Dairy Farm pumps.

Nice Airport, France, 50 kW, Air traffic control equipment.

Mont Bouquet, France, 50 kW, TV/FM transmitter.

Terschelling Island, the Netherlands, 50 kW, Merchant marine school.

Marchwood, Coast near Southampton, UK, 30 kW, Feeding to the grid.

Annex 5

US and Canadian Coast Guards Solar Modules for Navigational Aids

Design Criteria and Environmental Tests

US Coast Guard Specification No. 368 (November 30, 1981).

Canadian Coast Guard Specification MA 2055 Issue E (November 15, 1982).

Annex 6

Space Solar Cell Conversion Efficiency of Light to Electricity

The increase in the efficiency of solar cells used for space applications 1960–2013 is shown in Table A6.1.

Table A6.1 Efficiency measured at air mass zero (AMO)—28°C

Year	Efficiency (%)	Remarks	Reference
1960	10	Silicon	377,378
1970	10–12	Silicon	377,378
1972	14	Comsat Labs, Violet Cell, Silicon **Figure A6.1**	379
1978	12–14	Silicon	377
1981	12–14	Silicon	377
1988	14	Silicon	377
1990	20	GaAs	378
1999	23	EMCOR (Dual junction)	380
2001	27.5	EMCOR (Triple Junction—ATJ) **Figure A6.2**	381
2002	27.1	Sharp (Triple Junction-501)	382
2003	28.7	Sharp (Triple Junction-502)	383
2006	28.5	EMCOR (Triple Junction—BTJ)	392
2009	29.6	EMCOR (Triple Junction—ZTJ)	392

(*Continued*)

Table A6.1 (*Continued*)

Year	Efficiency (%)	Remarks	Reference
2010	29.5	Spectrolab (Triple Junction—XTJ)	384
2011	33	Estimate: EMCOR (3J IMM)	391
2012	35	Estimate: EMCOR (4J IMM)	391
2013	37	Estimate: EMCOR (5J IMM)	391

[377]Bailey SG, Viterna LA (2011) Role of NASA in photovoltaic and wind energy, *Energy and Power Generation Handbook* (Rao KR, ed), ASME Press, available from Barnes and Noble books.

[378]Billerbeck W (2006) *Systems Engineering Seminar Spacecraft Primary Power*, Presented at NOAA.

[379]Lindmayer J, Allison J (1972) *An Improved Silicon Solar Cell-The Violet Cell*, Conference Record of the Ninth IEEE Photovoltaic Specialists Conference, IEEE, pp. 83–84.

[380]Fatemi NS, *Satellite Market Trends and the Enabling Role of Multi-Junction Space Solar Cells*. navid_Fatemi@emcore.com.

[381]EMCORE Corporation, Albuquerque, NM. http://www.emcore.com/space-photovoltaics/space-solar-cells/

[382]JAXA Specification (October 2002) NASDA-QTS-2130/ 501.

[383]JAXA Specification (June 2003) NASDA-QTS-2130/502.

[384]Spectrolab, a Boeing Company, Sylmar, CA, http://www.spectrolab.com/DataSheets/cells/PV%20XTJ%20Cell% 205-20-10.

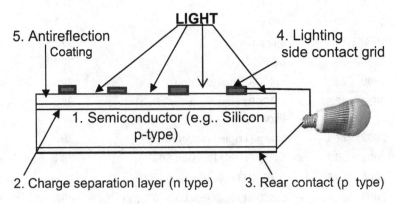

Figure A6.1 Space Solar cell made of silicon (used until 1980).

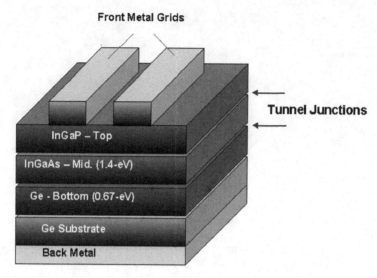

Figure A6.2 Emcore's three junction solar cell made for space use (2010).

Annex 7

Satellite Navigation Systems in Use or in Development in Various Countries

- Global Positioning System (GPS) of the USA Government. Fully operational and available to civilians since 1994.

- GLONASS—Russia's global navigation system. Fully operational worldwide. It was made fully available to civilians in 2007.

- Galileo—a global system being developed by the European Union and other partner countries, planned to be operational by 2014.

- Beidou—People's Republic of China's regional system, currently limited to Asia and the West Pacific. COMPASS—People's Republic of China's global system, planned to be operational by 2020.

- IRNSS—IRNSS-1A, the first of seven satellites of the Indian Regional Navigation Satellite System was launched on 1 July 2013, covering India and Northern Indian Ocean.

- Quasi-Zenith Satellite System (QZSS)—Japanese regional Global Positioning System covering Japan and Oceania. Three satellites is slated for launch before the end of 2017.

Detailed information is available online.[385]

---

[385]http://en.wikipedia.org/wiki/Global_positioning_system#Other_systems.

Annex 8

Regional PV Manufacturing Shares[386]

Paula Mints

Table A8.1 PV industry shipments, 1997–2011

Year	Total MWp						Total shipments (MWp)	Annual growth (%)
	US	Europe	Japan	ROW	China	Taiwan		
1997	47.9	20.5	27.4	14.8	1.1	2.3	114.1	38
1998	50.5	28.0	36.4	16.0	1.3	2.7	134.8	18
1999	56.2	29.8	68.4	17.6	1.8	1.8	175.5	30
2000	75.6	58.0	95.8	17.6	2.5	2.5	252.0	44
2001	95.3	84.7	144.7	21.2	3.5	3.5	352.9	40
2002	106.0	177.0	233.1	27.7	—	11.0	554.9	57
2003	94.5	175.6	344.4	47.3	—	13.5	675.3	22
2004	136.5	272.9	545.8	52.5	10.5	31.5	1,049.7	55
2005	126.7	422.0	717.8	70.4	28.1	42.2	1,407.4	34
2006	158.8	654.9	873.2	99.2	158.8	39.7	1,984.6	41
2007	245.8	983.4	921.9	153.7	491.7	276.6	3,073.0	55
2008	384.4	1,670.0	1,228.0	439.3	1,166.0	604.1	5,491.8	79

(Continued)

---

[386]Mints P, *Principal*, Solar PV Market Research. https://twitter.com/PaulaMints1, www.paulamspv.com.
Solar 2012: A Tale of Two Points of View, Renewable Energy World, 22 November 2011.
The Solar PV Ecosystem, A Brief History and a Look Ahead, Renewable Energy World, 20 November 2012.
©2012 Copyright Solar Power Market Research—The Tables and chart were reprinted with Paula Mints' permission.

Table A8.1 (*Continued*)

Year	Total MWp						Total shipments (MWp)	Annual growth (%)
	US	Europe	Japan	ROW	China	Taiwan		
2009	456.0	1,430.0	1,300.0	1,100.0	2,519.4	1,107.9	7,913.3	44
2010	1,040.0	2,610.0	2.080.0	2,415.0	6,438.2	2,819.2	17,402.3	120
2011	707.4	1,650.6	2,829.5	3,536.9	10,846.5	4,008.5	23,579.3	35

Table A8.2 PV industry shipments (in %), 1997–2011

Year	US	Europe	Japan	ROW	China	Taiwan	Total shipments (MWp)	Annual growth (%)
1997	42	18	25	13	1	2	114.1	38
1998	38	21	27	12	1	2	134.8	18
1999	32	17	39	10	1	1	175.5	30
2000	30	23	38	7	1	1	252.0	44
2001	27	24	41	6	1	1	352.9	40
2002	19	31	42	5	0	2	554.9	57
2003	14	26	52	7	0	2	675.3	22
2004	13	26	52	5	1	3	1049.7	55
2005	9	29	51	5	2	3	1407.7	34
2006	7	31	44	5	8	5	1984.6	41
2007	8	32	29	5	16	9	3073.0	55
2008	7	31	22	8	20	11	5491.8	79
2009	5	18	16	14	32	14	7913.3	44
2010	6	15	12	14	37	16	17402.3	120
2011	3	7	12	15	46	17	23579.3	35
14-Year CAGR							**46%**	

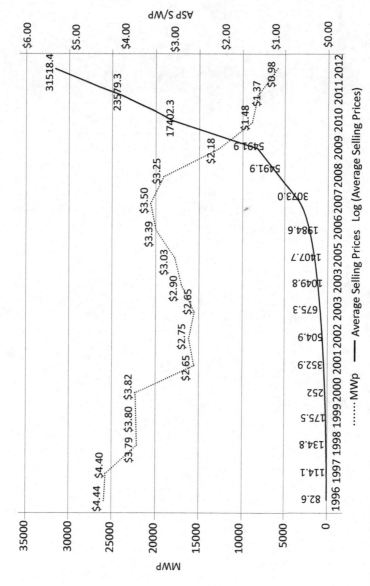

Figure A8.1 PV module prices and shipments, 1996–2012.

Annex 9

Twenty-Five-Year Life Prediction of PV Modules

In Chapter 40 (bankability), it was suggested that the module to be acceptable must fulfill the following requirements:

(1) Products (e.g., modules) are to be tested by an accredited PV testing laboratory to the latest Qualification Test Standards (IEC 61215 and 61646). These standards were not designed or covered for reliability of PV modules. But today it is the best known method to forecast that the module will provide at least 80% of its rated electric output up to a 20–25-year life.

What can be expected in the future?

More accurate test procedures are being worked on and at present a draft of this exists which contains the procedures that do not differ greatly from those performed under IEC 61215. The difference is that the modules are subject to the test conditions for much longer.[387] One of the most important elements is the "accelerated stress testing," which if the module is able to survive, will give an indication that it could survive 20–25 years.[388] An important factual support for this claim is that the BP

---

[387] *Photon Magazine* (2012) (2), p. 83.

[388] Wohlgemuth JH, Cunningham DW, Nguyen AM, Miller J (6–10 June 2005) *Long-term Reliability of PV Modules*, 20th European Photovoltaic Solar Energy Conference, Barcelona.

Solar/Solarex's database indicates that the return of their multicrystalline silicon modules in the time frame between 1994 and 2003 was 0.13%, which represents approximately one module failure every 4200 module years of operation.[389]

(2) The manufacturing facility has to have ISO 9001:2000 certification for the production facility certified by an accredited ISO Registrar. This step verifies that a real manufacturing facility exits having Quality Management system and the inspectors may take samples for testing which eliminates cheating of the test data

What can be expected in the future?

Recently, an NREL "Proposal for Guide for Quality Management Systems for PV Manufacturing: Supplemental Requirements to ISO 9001:2008" was published.[390] "While ISO 9001 is used as an industry standard for documenting quality programs, it addresses generic elements of a quality management system and does not cover specific details of interest to the PV industry. The purpose of this proposal is to be a guide to manufacture PV modules in maintaining module consistency."

This proposal will be submitted to IEC/ISO for formal adoption as a guideline that lays out best practices for design, manufacturing and selection, and control of materials to be utilized in the manufacturing of photovoltaic modules that have met the requirements of IEC 61215 or 61646. In the meantime, it provides the PV community with a way to begin benefiting from the more robust quality programs, even before the standard process is completed.

(3) Random unscheduled inspection may also be able to take samples for testing.

---

[389]Wohlgemuth J (March 2003) *Long-term Photovoltaic Module Reliability*, DOE Solar Program Review Meeting, Denver, CO.

[390]Technical Report NREL/TP-5200-58940 (June 2013) http://www.ftc.gov/bcp/about.shtm.

(4) Displaying a PV Quality Mark (PV GAP or Golden Sun). This would also assure that the manufacturer indicates that the product, for example the PV module was produced fulfilling all of the above requirements.

Annex 10

PV GAP Training Manuals

光伏质量管理

生产厂家质量控制培训手册

为世界银行准备

由光伏环球认证机构（ＰＶ ＧＡＰ）编写

Geneva, Switzerland

Peter F. Varadi 博士, PV GAP 主席
Ramon Dominguex (Dominguez   Associates; Rockville MD)
Deborah McGlauflin (Insight in Action, Inc.; Annapolis, MD, USA)

® 1999.09   **PV GAP QM 1.0**

国家经贸委世界银行 / GEF 中国可再生能源项目办公室
上海交通大学太阳能研究所
北京市计科能源新技术开发公司
翻译、审核
一九九九年十 月十八日

Figure A10.1 Quality Management Training Manual for the 1999
training course in China.

Figure A10.2 Quality Management Training Manual for the 2005 training course in China—English and Chinese cover pages.

Annex 11

China General Certification Center Photovoltaic Product Certification

11.1 Photovoltaic (PV) Product Certification Overview[391,392]

China General Certification Center (CGC) is the National Certification Body (NCB) of China, and it is authorized by Certification and Accreditation Administration of China (CNCA) to conduct PV product certification.

Figure A11.1 The Chinese "Golden Sun" quality mark.

---

[391]The text in this Annex was copied from CGC's web site: http://www.cgc. org.cn/eng/news_show.asp?id=3.

[392]©Peter F. Varadi, 2012, USA.

Table A11.1 Standard/specification and implementation rules for Golden Sun certification

No.	Scope of certification	Standard/ specification	Implementation rules for certification
1	Crystalline silicon terrestrial photovoltaic (PV) modules	IEC61215-2005—Crystalline silicon terrestrial photovoltaic (PV) modules—Design qualification and type approval; IEC61730-2:2004—Photovoltaic (PV) module safety qualification—Part 2: Requirements for testing	CGC-R47005:2012A—Implementation rules for photovaltaic products certification-crystalline silicon terrestrial photovoltaic (PV) Modules
2	Thin film silicon terrestrial photovoltaic (PV) modules	IEC61646-2008—Thin-film terrestrial photovoltaic (PV) modules—Design qualification and type approval; IEC61730-2:2004—Photovoltaic (PV) module safety qualification—Part 2: Requirements for testing	CGC-R47013:2012/A Implementation rules for photovoltaic products certification—thin film silicon terrestrial photovoltaic (PV) modules
3	Concentrator photovoltaic (PV) modules and assemblies	IEC62108-2008—Concentrator photovoltaic (PV) modules and assemblies—Design qualification and type approval	CGC-R47008:2012/A—Implementation rules for photovoltaic products certification-concentrated photovoltaic (PV) modules
4	Grid-connected photovoltaic (PV) inverter	CGC/GF004:2011 (CNCA/CTS0004-2009A)—Technical specification of grid-connected PV inverter	CGC-R46016:2012—Implementation rules for photovoltaic products certification-grid-conncected photovoltaic (PV) inverter

5	Photovoltaic (PV) combiner box	CGC/GF002:2010 (CNCA/CTS0001-2011)—Technical specification of PV combiner box	CGC-R46009:2012—Implementation rules for photovoltaic products certification-photovoltaic (PV) combiner box
6	Charge-discharge controller and DC/AC inverter of stand-alone System	GB/T 19064-2003—Solar home system specifications and test procedure	CGC-R46008:2012—Implementation rules for photovoltaic products certification-controller and inverter of stand-alone photovoltaic (PV) system
7	Junction box for terrestrial solar cell modules	CGC/GF002.1:2009 (CNCA/CTS0003-2010)—Technical specification of major components of terrestrial solar cell modules—Part 1: Junction box	CGC-R46005: 2012—Implementation rules for photovoltaic products certification-junction box for terrestrial solar cell modules
8	Connector for terrestrial solar cell modules	CGC/GF002:2009 (CNCA/CTS0002-2012)—Technical specification of major components of terrestrial solar cell modules—Part 2: Connector	CGC-R46006: 2012—Implementation rules for photovoltaic products certification-connector for terrestrial solar cell modules
9	Lead-acid storage battery used for energy storage	GB/T 22473-2008—Lead-acid storage battery used for energy storage	CGC-R46006:2012—Implementation rules for product certification-lead acid battery for energy storage
10	Stand-alone photovoltaic (PV) system	IEC 62124—Photovoltaic (PV) stand-alone systems—design verification	CGC-R46007:2012—Implementation rules for photovoltaic products certification—Stand-alone photovoltaic (PV) system

(Continued)

Table A11.1 *(Continued)*

No.	Scope of Certification	Standard/ Specification	Implementation Rules for Certification
11	Grid-connected photovoltaic (PV) system	CGC/GF003.1: 2009(CNCA/ CTS0004-2010)— Acceptance inspection of grid-connected photovoltaic (PV) system—Basic requirements	CGC/R46020:2012— Implementation rules for grid-connected photovoltaic (PV) system certification

Remarks

(1) CNCA is the acronym for Certification and Accreditation of the People's Republic of China; CTS is the acronym for Certification Technical Specification.

(2) In China, it requests that if there are no national standards or industrial standards, certification bodies can make their own certification specifications which should be checked and approved by CNCA.

Annex 12

Electricity Storage Systems

Utilities and users of electricity have several reasons to utilize storage for electricity.

Users of Electricity

- to establish uninterruptable electricity in case of power outage.
- to have storage system for solar PV generators

Utilities

- to reduce the need for fossil fuel plants used for peak demand times
- to provide reserve capacity when electricity supply becomes suddenly unavailable
- to provide electricity to maintain voltage and frequency following large disturbance
- to provide energy because of system failure until it is restored

Several storage systems are already utilized, and others are in the development stage.

Batteries (rechargeable)

Batteries are presently the most used systems especially by the users of electricity. Recently more and more battery storage systems are offered for residential and industrial PV systems.

These storage systems are also offered with multi-purpose electronics to buy or sell electricity when attached to the grid and also to direct and regulate the electricity supply for equipment attached to them.

Sealed Lead-Acid Type Batteries

Sealed lead_acid type batteries are widely used to store electricity in small or large PV systems. They are available in capacity as small as 10 Wh and as large as 3 kWh. These types of batteries are used for many applications and, therefore they are manufactured in many countries. The smaller batteries are presently used in very many small applications, for example in Uninterruptible Power Supplies (UPS), weather stations, road side telephones, traffic signals, etc. The larger ones (1 kWh or larger) are used for automobiles and trucks and, therefore they are also available everywhere as they are manufactured on every continent and in many countries. The larger ones, being used in cars are also used in the millions of Solar Home Systems (SHS), mini grids, and now to store electricity for residential as well as large industrial PV systems.

They are charged during daytime and provide electricity all day long. They are sealed and need no maintenance. Lead acid batteries because they are manufactured in large quantities are relatively inexpensive. The disadvantage of lead acid batteries for storing large amount of electricity is they are heavy and require more space.

Lithium-Ion Batteries

Lithium-ion batteries are now being used in the majority of electronic devices, for example computers, telephones, cameras, and very importantly they are also used in the all-electric automobiles. They are more expensive than the lead acid types; however, because their mass production for automobiles is increasing, it is expected that their price will be much lower than it is now. They are ideal for electric storage for large PV systems. They require less space than the lead acid batteries, are much lighter, and are assembled for the all-electric cars in clusters delivering more voltage.

They are going to be ideal for houses, stores, or offices. Several companies (utilities as well as PV module manufacturers) are now offering electricity storage systems utilizing lithium-ion batteries.

Compressed Air

Compressed air energy storage (CAES) is an electric energy storage, in which the electricity to be stored is used to compress air. When the electricity is needed, the compressed air is released, heated, and is driving gas turbines to generate electricity. This electricity storage is used for utilities and at this time about 400 MW of CAES capacities is installed worldwide.

Pumped Hydro

Pumped hydro is an electrical energy storage system where the electricity to be stored is used to pump water to an elevated container and used when needed to flow down to drive a turbine to produce electricity. This system is mainly used for power stations.

Flywheel

Flywheel electricity storage is actually a very simple system in which the electric energy to be stored is fed into an electric motor which spins up a flywheel thereby converting the electrical energy into the mechanical rotation of the flywheel. In order to retrieve the stored energy, the process is reversed with the motor that accelerated the flywheel acting as a break extracting energy from the rotating flywheel. The flywheel must be contained in a strong container to protect the environment in case the flywheel fails.

The flywheel was first used in 1950 when it was installed in an autobus in the city of Yverdon-les-Bains in Switzerland. The flywheel in the bus was accelerated at the end station of the bus and was providing the electricity for the bus during its route across the city. Flywheels have limited installation in the electric grid. There is one for example in Stephentown, New York, a 20 MW installation where it is operated continuously storing and returning energy to the grid.

As discussed an average house in the USA requires about 20 kWh/day electrical energy. The utilization of flywheels to store the electrical energy generated by a PV system in a house is feasible and those flywheels can be purchased from several manufacturers.[393]

Hydrogen Storage

Hydrogen storage is being considered for many applications, including PV-powered homes and also for automobiles. The production of hydrogen using electricity is accomplished by electrolysis, in which water is being used to produce hydrogen and oxygen. The oxygen is released and the hydrogen is stored by various ways. To produce electricity the hydrogen is introduced into a fuel cell, where it combines with oxygen, producing electricity and water. Recently a new system was proposed to simplify the machinery in which the electrolysis and the fuel cell chambers are combined.[394]

---

[393]Hoffman A (January 2011) *Using Flywheels to Supply Residential Electricity Demand*, Distributed Energy Newsletter.
[394]Liptak BG (2009) *Post-oil Energy Technology: After the Age of Fossil Fuels*, CRC Press, Boca Raton, FL.

Index